FUZZY EXPERT SYSTEMS

Editor

Abraham Kandel, Ph.D.

Department of Computer Science
and Engineering
University of South Florida
Tampa, Florida

CRC Press
Boca Raton   Ann Arbor   London   Tokyo

Acquiring Editor: Russ Hall
Production Director: Richard Sales
Coordinating Editor: Andrea Demby
Cover Design: Chris Pearl

Library of Congress Cataloging-in-Publication Data

Fuzzy expert systems/edited by Abraham Kandel.
    p.   cm.
  Includes bibliographical references and index.
  ISBN 0-8493-4297-X
  1. Expert systems (Computer science)   2. Fuzzy systems.
 I. Kandel, Abraham.
 QA76.76.E95F88   1991
 006.3'3—dc20

91-17459
CIP

Direct all inquiries to CRC Press, Inc., 2000 Corporate Blvd., N.W., Boca Raton, Florida 33431.

© 1992 by CRC Press, Inc.

International Standard Book Number 0-8493-4297-X

Library of Congress Card Number 91-17459

Printed in the United States 2 3 4 5 6 7 8 9 0

PREFACE

This edited volume addresses the specific area of fuzzy expert systems and the choices knowledge engineers and expert systems designers must make to succeed in the field of artificial intelligence. To a large extent the field of uncertainty managment, in general, and fuzzy set theory, in particular, is not driven just by an interest in exploring fuzziness and managing uncertainty in what appears to be a vague and imprecise environment, but also by a practical interest in producing better expert systems and knowledge-based systems. In this book we examine the choices within the context of specific and general purpose fuzzy expert systems.

In the first chapter Hall and Kandel identify the basic features of the evolution from expert systems to fuzzy expert systems, investigating the different choices provided by fuzziness. The choices here involve deciding how to handle fuzziness by both machines and humans and how to exploit fuzziness at each level of imprecision. Some of these features may not be visible to the user, but they largely determine the performance of a given expert system implementation. Their selection is based on expected performance benefits vs. the complexity associated with their inclusion. The method of handling imprecision in the system leads us through that evolution to fuzzy expert systems.

In Chapter 2, Schneider and Kandel illustrate the evolutionary process described earlier by discussing the design principles of a general purpose fuzzy expert system using fuzzy expected values and fuzzy expected intervals. It is also shown how to use them as an integral part of the fuzzy expert system.

Chapter 3, by Bouchon-Meunier, discusses inferences with inaccuracy and uncertainty in expert systems. This is the first step in the previously discussed evolution. The knowledge base of an expert system may contain uncertainties and inaccuracies. They are either associated with the inference rules given by the experts or deduced from the observation of facts which do not fit exactly the conditions expressed in the premises of the rules. A good tool to cope with these problems is one of the fuzzy implications introduced in fuzzy logics. In this chapter Bouchon-Meunier indicates some reasons supporting the choice of these inferences and the corresponding combination operator.

The representation within the framework of approximate reasoning of relational type rules is illustrated by Yager in Chapter 4. A relational production rule consists of a rule in which one of the antecedent requirements involves the satisfaction of a relationship between two variables. This concept is used by Yager to represent his view of expert systems and the issues involved in some of their trade-offs.

Chapter 5, by Di Nola, Pedrycz, and Sessa, is devoted to the problem of reduction of the knowledge base in rule-based expert systems. The authors first point out that this problem is placed in the main stream of procedures of expert's knowledge (or its implementation) validation. Subsequently, some algorithms leading to its resolution are given in detail and their complementary character is also underlined. The entire analysis is performed treating the knowledge base in the form of a fuzzy relation, while an inference mechanism is given by means of fuzzy relation equations. This chapter represents, like the previous chapters, important stages in the evolution of expert systems to fuzzy expert systems.

In Chapter 6, Rocha, Giorno, Leäo, and Theoto use a fuzzy automata model to describe neural networks and perform experiments in the acquisition and use of knowledge by experts. This is an interesting attempt to shed light on some of the problems of knowledge engineering and presents a set of ideas concerning theoretic vs. expert knowledge.

Chapter 7 continues the evolutionary exposition via a survey by Sage, putting things in perspective. I felt the need of such a contribution in order to bridge some gaps in the presentation so far. The purpose of this chapter is to provide an overview of recent research into the use of structured information processing frameworks under conditions of imperfect information. By imperfect information, Sage means information that is incomplete, uncertain, imprecise, inconsistent, incoherent, or some combination of all. The central goal of this presentation is the development of conceptual approaches that have ultimate potential for operational implementation as aids to enhance human decision making in a variety of realistic areas through intelligence based support to humans. First, Sage discusses the principal approaches taken. A discussion of research accomplishments is followed by a brief discussion of potential applications of the results.

In Chapter 8, Whalen and Schott deal with linguistic inference and the difficult problems involved in linguistic approximation to determine the linguistic form of the final and intermediate deductions. Basically, this chapter describes FLING, a system which facilitates knowledge acquisition by translating a diagrammatic influence network into an executable knowledge representation for a rule-based, linguistic reasoning system. The cognitively convenient diagrammatic representation can be understood by the domain specialist and improves the communication process for knowledge acquisition. Initially designed for failure diagnosis in a fuzzy network, FLING is extended to represent a generalized transitive digraph.

The first major input to FLING is a definition of the influence network; this consists of a list of nodes, their connecting arcs, and the combining operator for the incoming arcs of each node. The other major input is a list of indicator nodes. The output of FLING is a rule-based system in LISP, compiled from the two major inputs. The inputs to this new system are the user's linguistic assessments of the specified indicator nodes; its outputs are inferred linguistic assessments of the remaining nodes in the digraph. Depending on where, in the casual chain of the digraph, the indicator nodes fall, the compiled system will operate in diagnosis mode, simulation mode, or a combination of these.

In Chapter 9, Dubois, Lang, and Prade take a fresh look at advances in automated reasoning using possibilistic logic. In the framework of mechanical theorem proving under uncertainty, an extension of the resolution principle to possibilistic logic is proposed; for this purpose they use clauses weighted by a degree which is a lower bound of a necessity or a possibility measure. Two resolution rules are presented. The refutation method is generalized and resolution strategies dealt with. In case where only lower bounds of necessity measures are involved, a semantics is proposed in which the completeness of the extended resolution principle has been proved. Moreover, deduction from partially inconsistent knowledge bases can be managed and displays some forms of nonmonotonicity. The resolution patterns and the refutation strategy presented in this chapter are implemented in a system known as POSLOG (for POSsibilistic LOGic) on a microcomputer.

Kosko's Chapter 10 concludes the first section of this book — the theory part. In this chapter, Kosko investigates encoding and decoding properties of several fuzzy associative memories (FAMs). A metrical FAM maps arbitrary input patterns into the nearest stored patterns. The author reviews all fuzzy theory concepts used in the analysis. These include new definitions, theorems, and geometrical interpretations of fuzzy entropy, fuzzy subsethood, and the demarcation of fuzzy theory from probability theory in the context of unit hypercubes and Boolean cubes. Zadeh's *compositional rule of inference* is interpreted by Kosko as the operation of a subset FAM where a subset FAM F is sought that stores patterns as fuzzy eigensets. Fuzzy cognitive maps (FCMs), signed fuzzy digraphs with feedback, are also examined as FAMs by operating on their connection matrix representation. FCMs offer a fuzzy neural network alternative to traditional expert systems. An expert represents his knowledge of some uncertain causal domain by indicating the degree to which variable phenomena, causal concepts, causally increase or decrease each other. The underlying connection matrix is interpreted by Kosko as an FAM network with excitatory and inhibitory causal (synaptic) connections. An FCM inference is a *resonant* state of causal activity. The resonant state is a *hidden pattern* among the FCM edges. The hidden patterns presumably represent how the expert would answer different what-if questions (input states). An important task is to decode these limit-cycle hidden patterns. Another task is to combine, arbitrarily, many FCMs, with different causal concepts, given by different experts with varied levels of expertise. This section of the book is concluded with a discussion of the FCM adaptive learning laws and their relations to imprecise expert systems.

The second part of the book, dealing with applications of fuzzy experts systems, begins with Chapter 11, by Hudson and Cohen, who describe the role of approximate reasoning in a medical expert system. The authors developed EMERGE, a rule-based expert system for the analysis of chest pain in an emergency room environment. It was designed to be machine-independent, to provide rapid decision making, and to permit easy replacement of the rule base. EMERGE utilizes certainty factors to indicate the seriousness of the patient's illness. In this chapter, new techniques are discussed which have been incorporated into the EMERGE system to permit weighting of rule antecedents, indication of the degree of presence of symptoms, and generalization of logical inference techniques.

In Chapter 12, Hall and Kandel develop a methodology for building fuzzy expert systems. The developed methodological theory is applied and the results are embodied in a reusable fuzzy expert system called Fess. Fess is a multi-knowledge source expert system. It may be used to emulate several cooperating experts or any hybrid environment. The knowledge sources are connected by a blackboard architecture which enables them to be loosely coupled. Further, the system has been designed to facilitate the distribution of knowledge sources across several processors. The system may interact with the user in a questioning mode via a simple natural language interface.

In Chapter 13, Mancini and Bandler take a look at environmental design as a very much top-down activity, and consider the mental processes at work in the inception and germination of the *grand idea* or, more democratically, of a *generative kernel* of a building task. As both an arrival and a departure, this is, in the authors' opinion, the *sine qua non* of creative designing. If one accepts that the built form — the aim of the design task — is to be a framework, both physical and psychological, for meaningful experiences, then that task requires a high degree of empathy with the eventual users of the building. The creative designer is one who has lived that built form in advance and has provided this condensed account of it.

The roles of the two modes of psychical activity — the primary and secondary processes — are investigated with the intent of bringing the contribution of the computer to the design development as near as possible to the functioning of the primary process. This contribution is considered in connection with two fundamental and difficult states in the design process: the development and matching of the spatial organization, on the one hand, and the formal visual one, on the other, with the two facets of the kinetic structure of the activities to be housed in the built form.

Chapter 14, by Friedman and Kandel, is devoted to the application of a fuzzy expert system to numerical analysis. In this presentation a two-dimensional fuzzy intelligent differential equation solver is discussed. It can handle 2-D elliptic partial differential equations defined over general domains. The system replaces the domain by a finite element net, which is then used by an intelligent finite element package to solve the problem. The system has been implemented on a network of computers, having the supercomputer CYBER 205 and, later, CRAY YMP as its main numerical tool.

In Chapter 15, López de Mántaras, Agusti, Plaza, and Sierra introduce a unique approach to the propagation and combination of uncertainty performed by simple accessing of pre-computed matrices.

The main objective of this chapter is to describe the MILORD system. MILORD is an expert systems building tool consisting of two inference engines, an explanation module, and a knowledge elicitation module. The system allows the performance of three different calculi of uncertainty on an expert-defined term set of linguistic statements about certainty.

Each calculus corresponds to specific conjunction and disjunction operators. The internal representation of each linguistic statement is a fuzzy interval on the interval [0,1]. The different calculi of uncertainty applied to the elements in the term set give another fuzzy interval as a result. A term from the term set is assigned to the resultant fuzzy interval by means of a linguistic approximation process thereby keeping closed the calculus of uncertainty. One of the main advantages of this approach is that, once the linguistic statements have been defined by the expert, the system computes and stores the matrices corresponding to the different conjunction and disjunction operators for all the pairs of linguistic statements. Therefore, when MILORD is applied, *the propagation and combination of uncertainty is performed by simply accessing the precomputed matrices.* The knowledge elicitation is based on the Personal Construct Theory of Kelly referenced in the chapter. This methodology elicits the personally relevant conceptual constructs of an expert, over his domain of expertise, through a man-machine dialogue. The constructs are freely chosen by the expert and are the way in which pairs of elements can be described as being either alike to different. From an initial domain and problem-solving specification, the expert's discriminating constructs and their contrastive set (the linguistic values assigned by the expert when applying the constructs to domain elements) are elicited. This leads to a repertory grid of cross-references between constructs and domain elements. An interactive analysis of the repertory grid is performed in order to:

1. Detect poorly characterized elements
2. Detect elements that the expert has not yet taken into account.

These problems are solved introducing new constructs or elements, respectively. The final result is a personal fuzzy repertory grid that links domain elements with their relevant constructs by means of the linguistic labels of the contrastive set.

In Chapter 16, Cohen and Hudson address the issue of medical decision making via the use of unmultidimensional polynomials. Automated decision making aids have found numerous applications in medical fields. A number of techniques have been utilized in these systems. In this chapter a method is described which utilizes a class of orthogonal functions in pattern classification. The objective of the method is twofold: to determine correct classification of collections of data and to determine the relative importance of parameters which contribute to the decision. The classification may involve two or more categories. The method is illustrated in a medical application for diagnosis of coronary artery disease using exercise testing. This chapter includes explanation of the use of this technique, to extract information directly from a data base and then combine this information with heuristic rules obtained through expert consultation, to design a prospective decision-making model in the form of an expert system.

In Chapter 17, Hawkes, Derry, and Kandel address the business of producing Intelligent Tutoring Systems (ITS) or Intelligent Computer-Assisted Instruction (ICAI) and the techniques that educators and computer implementors must make to succeed in the automated tutoring marketplace.

Chapter 18, by Togai and Watanabe, originated at Duke University, where Togai received his Ph.D. However, Bell Laboratories made possible the implementation of an expert system on a chip. An important factor in this implementation is the work of Togai with a chip person — Watanabe. Between the initial formulation of this *fuzzy expert system chip* and the final solution are several levels of interesting procedures discussed in detail by the authors.

As shown so far, one of the most difficult research issues in expert systems is the method of coping with uncertainty in the knowledge representation and reasoning process. In Chapter 19, Tzvieli claims that one reason for the unsatisfactory state of affairs is the narrow scope of application of the existing methods. Each system of modeling uncertainty is based on some assumptions (axioms), which do not hold in all circumstances. The contribution of this chapter is in suggesting a *formal logical system,* in which the assumptions should be made explicit. It is fairly easy, once the logical system is understood, to make the substitutions necessary to tailor a logical system to the circumstances of the application.

Tzvieli's chapter elaborates on an example of a probability-based logical system PL, in which one assumes that the probabilities, that different formulas are satisfied, are independent of each other. The author proposes a generalization of the logical interpretations of the connectives and quantifiers to this case. The notions of a formula, a structure, assignment, satisfaction, deduction, etc., are generalizations of the corresponding first order ones. A method of assigning semantics to PL is proposed, and consistency, soundness, and completeness of PL are studied. In particular PL is shown to be compatible with the first order predicate calculus.

The last chapter, Chapter 20, by Schneider, Perl, and Kandel, concludes this volume, which has only scratched the surface of the techniques and models associated with delivering a marketable fuzzy expert system. This article demonstrates a fuzzy intelligent autonomous control system that manages the operation of a communication station in a multi-link, multidrop communication network. It accepts the communication request from the operator, assesses the communication conditions on the assigned frequency bands, estimates the predicted performance of the available equipment under the prevailing conditions, operates external sensors to collect and update the various channel conditions, selects the optimal communication mode and frequencies, sets the operating modes of the selected equipment and, finally, performs the communication automatically or via the operator.

The system accepts certain preliminary information from the human operator, from utility programs running on the same processor and from external sensors. These sensors are operated periodically, or as necessary, to update the information concerning channel conditions, network connectivity, and equipment status. The system is an expert system that operates on three distinct knowledge bases in order to reach an optimal solution for the communication problem. It is written in Turbo Pascal and runs on an IBM compatible PC, which functions as the communication station controller.

In summary, this and all other nineteen chapters in this volume will take you through the details pertaining to fuzzy expert systems design and applications. I trust you will find the presentations interesting and useful.

Abe Kandel
Tampa, Florida
November, 1991

THE EDITOR

Abraham Kandel, Ph.D. is Professor and Chairman of the Department of Computer Science and Engineering at the University of South Florida in Tampa. He also holds the title of the CSE Endowed Chair at USF. Previously, Dr. Kandel held the position of Professor and Chairman of the Computer Science Department at the Florida State University in Tallahassee. He was also the Director of the Institute of Expert Systems and Robotics and the Director of the Center for Artificial Intelligence at FSU. He received his Ph.D. in EECS from the University of New Mexico, his M.Sc.E.E. from the University of California, and his B.Sc.E.E. from the Technion-Israel Institute of Technology. He is a senior member of IEEE - the Institute of Electrical and Electronics Engineering, and a member of the ACM - the Association for Computing Machinery, AAAI, NAFIPS, IFSA, and ASEE.

Dr. Kandel is an advisory editor to the international journals *Fuzzy Sets and Systems, Information Sciences, Engineering Applications of Artificial Intelligence,* and *Expert Systems,* as well as the editorial advisor to the *Reston Computer Science Series* (Simon and Schuster). He has authored or co-authored over 200 papers and authored, co-authored, edited, or co-edited 17 books on a wide variety of aspects of computer science and engineering.

CONTRIBUTORS

Jaume Agusti, Ph.D.
Centre D'Estudis Avancats
Consejo Superior De Investigacions
 Cientificas
Blanes, Girona, Spain

Wyllis Bandler, Ph.D.
Department of Computer Science
Florida State University
Tallahassee, FL

Bernadette Bouchon-Meunier, Ph.D.
National Center for Scientific Research
 (CNRS)
Claude-Francois Picard Laboratory
University of Paris
Paris, France

Moses E. Cohen, Ph.D.
Department of Mathematics
California State University
Fresno, CA

Sharon Derry, Ph.D.
Psychology Department
Florida State University
Tallahassee, FL

Antonio Di Nola, Ph.D.
Mathematics Institute of Architecture
 Faculty
University of Naples
Naples, Italy

Didier Dubois, Ph.D.
Institute of Information Research of
 Toulouse
Paul Sabatier University
Toulouse, France

Menahem Friedman, Ph.D.
Nuclear Research Center-Negev
Department of Mathematics and
 Computer Science
Ben-Gurion University
Beer-Sheva, Israel

Fernando Giorno, Ph.D.
Scientific Center
IBM Brazil
Rio de Janeiro, Brazil

Lawrence O. Hall, Ph.D.
Department of Computer Science and
 Engineering
University of South Florida
Tampa, FL

Lois Wright Hawkes, Ph.D.
Computer Science Department
Florida State University
Tallahassee, FL

Donna L. Hudson, Ph.D.
Section on Medical Information Science
University of California
San Francisco, CA

Abraham Kandel, Ph.D.
Department of Computer Science and
 Engineering
University of South Florida
Tampa, FL

Bart Kosko, Ph.D.
Department of Electrical Engineering
 Systems
School of Engineering
University of Southern California
Los Angeles, CA

Jérôme Lang, Ph.D.
Institute of Information Research of
 Toulouse
Paul Sabatier University
Toulouse, France

Beatriz F. Leäo, M.D., Ph.D.
Medical Informatics Division, Research
 Unit
Cardiology Institute
Rio Grande do Sul, Brazil

Ramon López de Mántaras, Ph.D.
Spanish Council of Scientific Research
  (CSIC)
Group of Research on Artificial
  Intelligence and Logic (GRAIL) of the
  Centre of Advanced Studies of Blanes
  (CEAB)
Blanes, Girona, Spain

Vasco Mancini, Ph.D.
Architecture and Environments
Colchester, U.K. and
Department of Computer Science
Florida State University
Tallahassee, FL

Witold Pedrycz, Ph.D., D.Sc.
Department of Electrical Engineering
Computer Engineering Program
University of Manitoba
Winnipeg, Canada

Joseph M. Perl, Ph.D.
Digital Communication Group/DSP
Hertzliya, Israel

Enric Plaza, Ph.D.
Centre of Advanced Studies of Blanes of
  the Spanish Council of Scientific
  Research
Blanes, Girona, Spain

Henri Prade, Ph.D.
Institute of Information Research of
  Toulouse
Paul Sabatier University
Toulouse, France

Armando F. Rocha, M.D., Ph.D.
Center of Informatics on Health
EPM
Sao Paulo, Brazil

Andrew P. Sage, Ph.D.
School of Information Technology and
  Engineering
George Mason University
Fairfax, VA

Mordechay Schneider, Ph.D.
Computer Science Department
Florida Institute of Technology
Melbourne, FL

Brian Schott, Ph.D.
Decision Sciences Department
Georgia State University
Atlanta, GA

Salvatore Sessa, Ph.D.
Architecture Faculty
Mathematics Institute
University of Naples
Naples, Italy

Carles Sierra, Ph.D.
Centre of Advanced Studies of Blanses
Spanish Council of Scientific Research
Blanes, Girona, Spain

Marly Theoto, Ph.D.
Laboratory of Information, School of
  Nursing
EEUSP
Sao Paulo, Brazil

Masaki Togai, Ph.D.
Togai InfraLogic, Inc.
Irvine, CA

Arie Tzvieli, D.Sc.
Bell Communication Research
Piscataway, NJ

Hiroyuki Watanabe, Ph.D.
Department of Computer Science
University of North Carolina
Chapel Hill, NC

Thomas Whalen, Ph.D.
Decision Sciences Department
Georgia State University
Atlanta, GA

Ronald R. Yager, Ph.D.
Machine Intelligence Institute
Iona College
New Rochelle, NY

To Nati, Shmulik, Uri, David, and Daniel

Adhuc neminem cognovi poetam
qui sibi non optimus videretur;
sic se res habet: te tua,
me delectant mea

Cicero, Tusculanae disputationes, V, 63

TABLE OF CONTENTS

FUZZY EXPERT SYSTEMS THEORY

APPLICATIONS OF FUZZY EXPERT SYSTEMS

Fuzzy Expert Systems Theory

Chapter 1

THE EVOLUTION FROM EXPERT SYSTEMS TO FUZZY EXPERT SYSTEMS

Lawrence O. Hall and Abraham Kandel

TABLE OF CONTENTS

I. CLASSICAL EXPERT SYSTEMS

Expert systems are computer programs that emulate the reasoning process of a human expert or perform in an expert manner in a domain for which no human expert exists. They typically reason with uncertain and imprecise information. There are many sources of imprecision and uncertainty. The knowledge that they embody is often not exact in the same way that a human's knowledge is imperfect. The facts or user-supplied information is also uncertain.

An expert system is typically made up of at least three parts: an *inference engine,* a *knowledge base,* and a *global* or *working memory.* The knowledge base contains the expert domain knowledge for use in problem solving. The working memory is used as a scratch pad and to store information gained from the user of the system. The inference engine uses the domain knowledge together with acquired information about a problem to provide an expert solution.

Expert systems have modeled uncertainty and imprecision in various ways. MYCIN[32] uses certainty factors, while CASNET[36] uses the most significant results of tests. Most of the methods of dealing with uncertainty and imprecision in expert systems have been ad hoc, in the sense that there is no underlying theory to support them. They have been validated only via empirical testing.

They typically use some form of high level rules. Blind search of the solution space is avoided and high performance, approaching or surpassing an expert's, is obtained. Reasoning can be done by symbol manipulation. They show some *intelligence.* Expert systems embody fundamental domain principles and weak reasoning methods. They have difficulty or complexity associated with them. They can reformulate a problem and some reason about themselves. They can be described as computer programs that use domain knowledge and reasoning techniques to solve problems normally requiring a human expert for their solution. Expert systems may perform a task that humans do not normally perform, such as missile guidance or planning for a robot. An expert system may be able to perform expertly in an area in which there are no human experts.

In addition to the three basic components, most expert systems also include an explanation facility. Some expert systems have separate natural language generation and/or interpretation facilities. Some expert systems have interfaces to mechanical devices, such as an expert system to monitor the performance of a nuclear reactor. Others do very basic learning about their domain. They may have several knowledge sources. A knowledge source is made up of at least the three basic parts mentioned before. Knowledge sources communicate with each other and with the overall system controller via a device called the blackboard. Messages are posted and received via this blackboard. The information on a blackboard may cause processes to be activated or be used in some other manner. A possible configuration of an expert system is shown in Figure 1.

A blackboard may be looked at as a communications coupling device. It provides for loose coupling between several knowledge sources in an expert system. Each knowledge source may be viewed as an individual expert. Therefore, an expert system of several knowledge sources is made up of a set of cooperating experts. The experts communicate by writing messages on the blackboard and reading messages from it. A blackboard has a loose resemblance to a mailbox in classical message-passing systems.

One important feature that many expert systems incorporate is that of the explanation facility. This facility enables the expert system to explain its reasoning to the user. The user of the system will be able to trace the knowledge used by the system and, in some cases, determine the motivation for a question that the system has asked. An explanation facility enables a user to determine why information is being asked for and how both intermediate

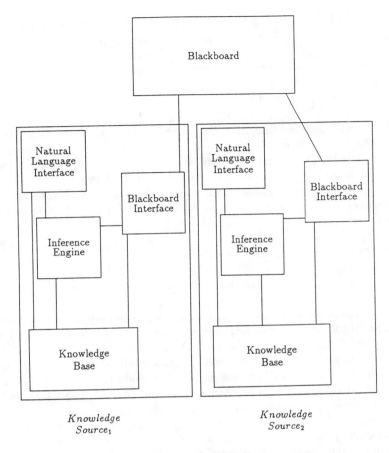

FIGURE 1.   A two knowledge source expert system.

and final conclusions are obtained. It is a very important facility to have when the knowledge base is being debugged. It enables inconsistencies, errors, and omissions to be easily identified for correction.

A consultation expert system that does not have an explanation facility has little chance of being accepted by experts. They will need to know how a conclusion is arrived at, as well as what the reasoning is behind the intermediate steps. An expert system which gives advice to a human must be able to satisfy any misconceptions or skepticisms. There are expert systems which do not need an explanation facility such as Macsyma, which solves calculus and algebra problems. However, most interactive expert systems will need some explanation ability to gain user acceptance.

In order to understand the functioning of an expert system one must see how knowledge is represented in the knowledge base. It is also necessary to understand how the inference engine makes use of the knowledge to come to expert conclusions. Knowledge representation and the inference process will be the subjects of the following sections.

A. KNOWLEDGE REPRESENTATION

There are many methods of *knowledge representation* and a thorough description of them is given in *The Handbook of Artificial Intelligence I.*[5] We intend to discuss those that have had a major initial impact upon expert system development. The three most popular knowledge representation schemes are rules, semantic networks, and frames.

Under the heading of rule-based systems comes MYCIN which uses production rules as its knowledge representation scheme. The general form of a rule is shown in Figure 2.

IF PREMISE **THEN** CONCLUSION

FIGURE 2.   General form of a rule.

Both the premise and the conclusion are normally given some truth value, often called a certainty. The premise is usually some restricted sequence of clauses connected by the connectives **and** or **or.** The connectives often serve as min and max operators, respectively. Often, rules are set up in some LISP functional format, or a Prolog Horn clause, or the premise may look like a fully parenthesized logical expression. The conclusion may be an action to be taken or a clause, to be added to working memory, which is in some other premise. An actual rule is shown in Figure 3.

IF class is gymnosperm and leaf shape is needlelike **THEN** family is cypress

FIGURE 3.   Example rule.

The term *semantic networks* encompasses a class of knowledge representation formalisms. They are made up of nodes and arcs between them. The nodes usually represent objects, concepts, or situations in the domain. The arcs represent the relations between the particular type of node. A semantic network may be viewed as an acyclic weighted graph. The relational arcs often have weights, indicating the strength of the relation associated with them. The relations are typically not two-valued, but multivalued. CASNET[36] is an important example of an expert system which uses the semantic network knowledge representation formalism. Figure 4 shows the structure of a semantic net. In the figure the N_i's stand for nodes, which could be disease states, for example. The weighted links between them determine how they relate under some appropriately defined relation. As an example, relation between disease states could be a causal one.

A frame is a data structure used for representing a stereotyped situation. It is organized much like a semantic net in many cases. It can provide built-in inheritance properties. A frame is made up of a set of slots. Slots may contain procedures, data, or be pointers to other frames. We, therefore, may have nested frames. A frame system may be implemented in the context of an object-oriented knowledge representation scheme.[22] These systems, such as the common LISP object system, allow inheritance through the concept of super- and subclasses. Methods, which act as procedures, will define what actions may be taken when a slot relations are familiar to most. An n-ary relation is an n-tuple of objects. The relation holds with some truth value. A rule may be described as an implication relation, for example. Almost any piece of knowledge may be described as a relation, although it may require some sophistication to accomplish. Relations provide an extremely flexible scheme for knowledge representation. Prospector[14] has used relations to represent knowledge about prospecting for minerals. An example of a five-ary basketball team lineup relation is given in Figure 6. The first configuration given is the normal one for a team, the second is only occasionally used.

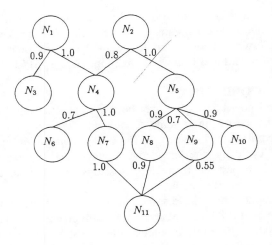

FIGURE 4. Semantic net structure.

slot₁	value₁
$slot_1$	$value_1$
$slot_2$	$value_2$
...	...
...	...
$slot_n$	$value_n$

FIGURE 5. Generic frame.

R(guard, guard, forward, forward, center) = 1
R(guard, guard, forward, guard, center) = 0.7

FIGURE 6.   Relation between types of player in a lineup.

B. THE INFERENCE ENGINE

The *inference engine* uses the knowledge in a particular representation to come to some expert conclusion or offer expert advice. It contains the system's general problem-solving knowledge. It is responsible for determining what piece of knowledge to use next and scheduling other necessary actions. It will take all actions indicated, as necessary, by a piece of knowledge which is found to be true, due to the current facts presented to the expert system.

The inference engine is responsible for determining when to ask the user a question and when to search the knowledge base for the information. It must ensure that questioning is done in a concise logical manner. In most expert systems it must be responsible for dealing with imprecise and uncertain information.

An inference engine makes use of a special type of expert system knowledge called *meta-knowledge*. Meta-knowledge is knowledge about the system knowledge. This may include such things as how best to utilize various knowledge chunks, which pieces of knowledge to use first, whether a piece of knowledge should be inferred or asked of the user, and when to stop processing. This knowledge may be in the knowledge base in the same representation as other knowledge; it is just used in a different manner.

Inference engines operate primarily in one of two ways. They may be data driven, known as *forward chaining,* or they may work backward from conclusions, known as *backward chaining.* A forward chaining system begins with some data and moves down the inference chain until it reaches a final node, frame, or conclusion. A backward chaining system begins with some final node, frame, or conclusion and works backward until it finds a complete path of evidence for one of the stopping states. MYCIN is a backward chaining expert system. XCON[25] is a forward chaining expert system. There are some expert systems which use a combination of forward and backward chaining, such as SPERILL II.[27]

C. KNOWLEDGE ACQUISITION

We have discussed how knowledge may be represented in an expert system and how that knowledge may be used to infer some conclusions. The question that arises concerns the methodology of acquiring the knowledge. *Knowledge acquisition* is a bottleneck in expert system development. Typically, the knowledge engineer and the expert system developer must sit down with an expert, in many long sessions, and extract the expert's domain knowledge for use in the expert system.

The knowledge engineer must have some understanding of the expert's area so that he can converse in the expert's lingo. The expert will often not be able to provide general problem-solving heuristics. Example problems often must be presented and the heuristics explicated in the process of the expert solving them. The initial heuristics gathered tend to be incomplete and often not quite correct. Therefore, the process of gathering knowledge is incremental and coincides with the development and testing of the expert system.

In order to develop an expert system, a willing and patient expert must be found for the knowledge acquisition process. The initial sessions with the expert will be an intense process of finding the relevant information and getting at it. Further sessions will be concerned with filling gaps in the system knowledge and correcting any errors that may have crept in. Incorrect knowledge is acquired due to the fact that the knowledge engineer must do a lot of interpretation and weeding out in acquiring the heuristics from the expert. Also, if several experts contribute knowledge, it is occasionally the case that they contradict one another to some degree. The knowledge engineer is responsible for weeding out the inconsistencies in the knowledge base.

Some tools have been developed to automate the knowledge acquisition process, such as one based on Personal Construct Theory.[8] These tools are of limited utility, but show some future promise. More useful are the tools that aid the knowledge acquisition process, such as TEIRESIAS[11] which aids knowledge acquisition and refinement for the expert system building tool EMYCIN.[34] It checks phrasing of clauses in rules and for completeness in the rules, as well as for contradictions. SEEK[28] gives advice about rule refinement during the development of a diagnostic-type expert system. It will help refine rules represented in the EXPERT[37] language. EXPERT is an expert system building tool. Some of the commercial expert system shells have interfaces which aid knowledge acquisition, in a manner much like TEIRESIAS.

II. EXPERT SYSTEM CLASSIFICATION

Expert systems are commonly described in terms of their application domain, knowledge representation mechanism, inference methods, and special features. We examine them from a different angle. We classify them based on their usability. That is whether they are in common use solving everyday problems. An expert system may belong to class one, class two, or class three. We shall not comment on expert system building or developing tools, which are a separate kind of entity.

A. CLASS ONE SYSTEMS

Definition: A class one expert system is in common use. It is commercially viable. It is a program that has earned acceptance by the user community.

Let us examine some examples of class one expert systems. We look at three systems that were near the first to enter this class. Dendral, Macsyma, and R1 are the examples of this class. R1 has evolved into XCON.[25] Note that there are excellent examples of expert systems which have not made it to this class.

The three class one examples perform mass spectrographic analysis, symbolic mathematical manipulation, and configuration of VAX 11/780 computer systems (XCON encompasses the whole VAX family of systems), respectively. Their domains are diverse. Note that none of these is a consultation system. Their interaction with the user is minimal. Each gives a concrete solution. It is rather straightforward and noncontroversial to determine the correctness of their solutions. They all have manageable and well-defined problem domains. A limited, focused domain currently is a characteristic of the more successful expert systems. Less knowledge is needed for a limited problem, and it is more easily verified than the knowledge base of a large domain. Experts were available for consultation on each system that has made it into class one.

Dendral has been heavily used by scientists and has proven itself a viable tool. Since it was difficult to get expert chemists to explicate all their knowledge in Dendral domain of expertise, Meta-Dendral was developed. It is an expert system designed to form rules from empirical observation. Meta-Dendral has discovered several good heuristics for mass spectrographic analyses and itself qualifies as useful expert system.

We will describe R1/XCON, in some detail, as it is one of the best examples of a class one system. This was the first major commercial expert system. XCON is a rule-based or production system. Match is used rather than generate-and-test as a problem-solving method. This means that rather than exploring several hypotheses until an acceptable one is found, it uses its knowledge of the task domain to generate a single acceptable solution.

Two important domain independent lessons came out of the implementation of XCON. An expert system can perform a task by simply recognizing and taking the proper action at each step. Of course this implies that at each step one can determine whether an action is consistent with the acceptable performance of the task. They also found that it was quite easy to refine and extend system knowledge when the system was implemented as a production rule system.

The problem-solving method of Match can be considered a weak search method. Its search space consists of all instantiations of the variables in the form. The form is the body of domain specific knowledge that defines acceptable sets of instantiations. The form will hold constraints for task satisfaction. A partial ordering on decisions must exist so that the application of an operator will affect only aspects of the solution that have not yet been determined. The subtask of placing modules on the unibus does require search which is done via generate-and-test with match used as a heuristic guide.

The rules in the original working version of the R1 expert system were broken up into domain-specific and general rules. These two categories were further subdivided. The smallest set of rules covered computation (47 rules), while the largest set covered component association (156 rules). These are both examples of sets of domain-specific rules. Today XCON contains many more rules and much more sophistication.

Now let us summarize the characteristics of problems that have class one expert system applications. These problems have a limited domain which is reasonably narrow. Experts are available and able to explicate their knowledge about solving problems in the domain. They have minimally imprecise input and output. It is straightforward to determine the correctness of their output. They have minimal interaction with the user. There are a large number of these systems now coming into use.

B. CLASS TWO SYSTEMS

Definition: Class two expert systems are those with good, even expert, performance, which have not gained wide acceptance by the average user.

Class two systems include consultation or diagnostic expert systems which do not explain themselves well enough to satisfy the user. These systems may not gain expert acceptance because they do not ask his diagnosis, but simply give their own. It is conjectured that if they would comment on the expert's diagnosis they would gain wider acceptance. Some class two systems have not acquired quite enough expertise to be in class one. Human prejudice against machines may be holding back some.

There are many medical consultation expert systems in the class two category. Internist/Caduceus,[29] MYCIN, CASNET, and PUFF/Centaur are all examples. Their domains are internal medicine, bacterial infections, glaucoma, and pulmonary physiology, respectively. We will examine Internist/Caduceus and PUFF/Centaur in some detail as follows.

Internist/Caduceus covers at least 85% of internal medicine which is a large domain. It takes into account and tries to distinguish multiple diseases. It has been known to correctly diagnose complex test cases that have stymied human experts! It reasons under uncertainty, although in a somewhat ad hoc fashion. It is slow to respond, can require large chunks of input from the user, and does not have an impressive user interface.

The user interface has probably been its greatest drawback. Any awkwardness simply magnifies the fact that one is dealing with a machine. If we are working with a new machine we want to know exactly how and why it comes up with its answers. Of course, speed is a factor that will also impede acceptance. Doctors will need time to become familiar and comfortable with the system, especially since a life may be at stake.

PUFF was originally developed with the use of the expert system building tool EMYCIN. EMYCIN allows the development of expert systems with MYCIN-like control and knowledge structures. Centaur is a newer version of PUFF with control knowledge represented within prototypes. Prototypes are basically frames of data. They allow context-specific control since control knowledge is represented in each prototype. This enables the separation of control knowledge from other knowledge in the system. Centaur asks the user fewer questions than PUFF. It is able to determine more things for itself via more powerful inference.

Prototypes are a very general construct. A typical consultation is represented as a prototype with the different stages of the consultation represented in control slots. Prototypes make it natural for Centaur to have a review task. This task will allow the user to specify a prototype which then has its typical features revealed as well as any control knowledge associated with it.

Prototypes were developed to solve some of the problems that the production rule-based system PUFF faced. The problems which prototypes help to solve are representing prototypical patterns, adding new rules or modifying old ones, altering the order in which information is requested during the consultation, and explaining system performance. The explanation of system performance is crucial to the acceptance of medical diagnostic systems, which points out the importance of prototype improvement. Prototypes, it should be noted, are a fancy name for a frame-based knowledge representation scheme.

The control structure or formalism used in Centaur is that of hypothesize and match. An attempt is made to match the representation of classes of hypotheses against the actual data in the case. Prototypes represent classes of hypotheses. An attempt to instantiate a prototype to determine whether there is a match to the data may require new information, which can result in questions being asked of the user. The system may be able to infer the answer to a question from other known pieces of data. It always tries to avoid asking the user a question.

The performance of the system appears to be equivalent or just below that of an expert in several performance studies. The explanation facilities of the system are good. It reasons under uncertainty using a graded method, which is somewhat ad hoc. Their certainty factors are added to upon the occurrence of supporting data and reduced by data that conflicts with the hypotheses. The expert system does not appear to ask the user an unreasonable number of questions. It is used at a California medical center.

Possibly the major factors impeding its widespread acceptance are user familiarity and confidence. We believe another significant problem exists. A doctor will learn from experience as new cases come before him, while PUFF/Centaur will not. It must be taught. This involves acquisition of the new knowledge from an expert. If knowledge acquisition is difficult, then it may not occur or it may be done incorrectly. Yet the system must always perform as well as possible (at least when it is working on us!). If the system becomes invalid, since it lacks some new information, people will be very wary of using it as anything more than a showpiece. We must have a method of acquiring new knowledge in a timely, efficient, and correct manner in order for consultation systems in dynamic fields to be ultimately accepted.

Sacon operates in a very focused domain. There is a large structural analysis program called Marc which is very useful for structural engineers. However, it is difficult to use and people often need an expert's help to use it or learn its use. Sacon performs the task of the Marc program expert. It enables the user of Marc to use its features more fully, efficiently, and correctly. It interacts a fair amount with the user on a very limited domain, one computer program. If it helps one to use Marc in a manner that increases productivity, then it is working correctly. Hence determining its correctness is not difficult. Its performance matches that of a human consultant and it has been validated to an extent by field use. It was written with the use of an expert system building tool called EMYCIN, with is LISP based.

The knowledge of structural analysis keeps growing, and a structural analysis program is not immune to change. Other analysis programs that are easier to use limit the usability of this expert system. These problems have kept Sacon from becoming a widely available tool.

In many class two expert systems it is difficult to determine whether a solution is correct. This is especially true of the medical expert systems in which even experts may disagree over treatment plans. Even an acceptable solution may not have the desired effect upon the patient, thus leaving a machine open to blame. Imprecision and uncertainty play an important part in these systems.

In the future it is expected that many class two systems will become widely usable. The future is near for some of the smaller domain systems that do not deal heavily with the user. The acceptance may be some time in coming for medical expert systems. Consultation aid systems will move up faster.

C. CLASS THREE SYSTEMS

Definition: A class three system is one that has not gained even limited acceptance by the appropriate user community. It is incapable of reaching true expert performance.

While it might seem that expert systems in this class are dregs of technology, quite the opposite is true. Many have made great contributions to the advancement of expert system technology. This type of system should not be attempted by a commercial enterprise since it is doomed to a type of failure, usually because the problem tackled is too ambitious. Most systems in this class have been pushing at the frontiers of knowledge and have pushed them back.

The Hearsay I, II[16,30] expert systems are the best examples of class three systems. They address the domain of speech understanding. Hearsay II is the second version of Hearsay I. An example of a very high level class three system is AM[11] which operates in the domain of elementary mathematical research.

Hearsay I, II pioneered the concept of multiple cooperating knowledge sources. It uses a blackboard concept to provide for message passing among the knowledge sources. Hypotheses are posted on the blackboard. Each knowledge source (KS) may put a hypothesis on the blackboard. The hypotheses are placed on the blackboard so that all knowledge sources may have the opportunity to use them in their processing or modify them if necessary. In general, all knowledge sources will not make use of each hypothesis. The KS has knowledge about which hypotheses are useful for it and which are to be left alone. A KS has two major components: precondition and response frames. A precondition will provide the scheduler with a response frame, which is an estimate of the action that the KS will take. Preconditions are fired by the blackboard handler when certain primitive changes, that the precondition has specified, occur. The knowledge sources allow for knowledge separation into some reasonable set. They can be grouped together into modules in order to share code or data.

At any time during a run of Hearsay there may be many pending tasks to execute. The tasks can be both invoked KSs and triggered preconditions. The tasks are prioritized by a scheduler. The scheduler selects the highest priority task for execution. The priority of any task is an estimate of the usefulness of that task in recognizing the utterance that is being processed. Some KSs called policy KSs can be used to impose global search strategies on the basic priority scheduling mechanism.

Hearsay addresses a very large problem domain. It deals with unreliable data or knowledge. The problem is very complex. This is probably the largest and most difficult task yet undertaken by an expert system. The expert system can understand connected discourse from a 1000 word vocabulary. Its skill rivals that of a 10 year old. While this is impressive, it does not come very close to expert performance. In very limited domains, such as about the game of poker, the system can interpret over 90% of the utterances. This is still short of expert performance.

The builders of Hearsay I, II still have difficulty with focus of control. Initially the knowledge sources were independent, but it was found that the cost was too high. Parallelism of KSs can be exploited with some modifications, and this might help the expert performance level. There were problems in adequately evaluating partial solutions. They had difficulty finding the combination of knowledge sources that would optimize system performance. Hearsay must reason under uncertainty, but its methods seem weak. They also had other difficulties to overcome. These difficulties probably led to some of the innovative problem-solving techniques introduced by the generations of Hearsay expert systems.

AM is designed to carry on simple mathematics research. It starts with elementary ideas in set theory. It uses heuristics to generate new concepts by combination of ideas. It uses a worth function with values from 0 to 1000 for ideas. It has done an excellent job at rediscovering known concepts starting from its initial knowledge base. It has also discovered two concepts that were not previously known. Given these facts, why is AM a class three expert system?

It has come across two concepts which, when observed and worked with by a mathematician, became sharp new ideas. However, AM is unable to prove the validity of a new claim the way an expert would. It cannot explain its discoveries in a satisfactory way. In general, it will not concentrate on a new discovery and build around it the way a true researcher would. This means it will not lay out the explicit groundwork from the old to the new and the reasons why a new conjecture is important.

AM searches a space of possible conjectures that can be generated from elementary ideas. It chooses the most interesting conjectures and pursues that line of reasoning. Its control hierarchy is that of plan, generate, and test. Its main limitation is its inability to synthesize powerful new heuristics for the new concepts it defined. It is unable to build upon its discoveries the way a true researcher would. The methodology behind good research is unclear, which makes knowledge acquisition for this type of expert system very difficult. Also, the question of building intuition into an expert system is currently overwhelming. These problems make this type of expert system a difficult undertaking.

Class three systems all deal with broad domains. Speech is routinely done by people after many years of learning. Research is practiced by skilled people after years of experience through toil. Class three systems attack difficult problems for a nonthinking and learning machine to solve. They will often require multiple knowledge sources. They have difficulty with focus of control. They tend to generate large search spaces. They tend to use a hypothesize, generate, and test form of control. Choosing the correct knowledge sources and then gathering them can be a major chore for these systems. The experts for these systems are not as well defined as in other classes. Uncertainty and information error generally play important roles in class three systems. Currently, uncertainty is not handled in a uniform or optimal manner.

Class three systems will be the ones that force major innovations to occur in the field of expert systems. They are currently just beyond our reach in terms of solution. They have already brought some important contributions to the field, most notably Hearsay.

D. CLASS DIFFERENCES AND OTHER COMPARISON METHODS

We will now point out some differences between the classes of expert systems that we have discussed. The domains of the systems become larger as you move along the class scale. Expert systems in classes two and three tend to deal with more imprecision than those in class one. The more imprecision inherent in a domain, the worse the class an expert system in that domain to which it will likely belong. The more knowledge necessary for expert system development, the worse the class to which it will generally belong.

In class three the experts are not as easy to define as in other classes. There is often a lot of interaction with the user, which needs to be smoothed out, in class two. Class one systems tend to be small, focused and not consultation systems. The classes that have been defined provide a snapshot of expert system development. Their characteristics will have meaning in the future, but it is likely to be different from the current.

A good way to compare expert systems individually is to use the activity structures developed by Kohout.[3] These structures provide a uniform set of windows into the operation of diverse expert systems. As you examine them in terms of the structure provided and the one we propose, their differences and similarities will become apparent. The information gleaned provides a valid method for a comparison and evaluation of diverse systems. There are six activity structures. They are broken up into two sets of three. One set is known as information handling structures and the other as constraint structures.

The information handling structures deal with the control structure, knowledge structure, and inference structure of the expert system. The constraint structures deal with the protection structures, human-computer interaction structure, and technology-dependent constraints upon the expert system. By applying these structures to expert systems, their strengths and differences become apparent.

In order to completely describe expert systems one other structure should be included. That is an uncertainty handling structure. This structure is contained in all of the other structures, yet is important as a separate entity. Many expert systems deal heavily with imprecision, and the explication of their methods can tell a lot about the usefulness and robustness of the system.

III. SUITABLE PROBLEMS FOR EXPERT SYSTEMS

We will summarize the types of problems for which expert systems may currently be the solution. They are certainly the solution for class one problems, which include part of the broad class of control problems. They can solve many large problems which are appropriately restricted. The restriction must rule out areas which involve common sense and analogical reasoning, as well as those that require extensive learning. An expert must be available, and the complexity of the problem must not be extreme in the eyes of an experienced knowledge engineer.

Expert systems can be useful in limited domains in which there is not a human expert. These include such things as robot activities. An expert system is best applied to problems that have limited domains and well-defined expertise. They are most likely to be accepted in areas that do not require extensive user interaction. As with humans, they work best in environments that do not have a large amount of erroneous and uncertain information in them.

Expert systems are the solution for consultation systems of very limited domains. They are most accepted if they critique the expert's or practitioner's conclusions rather than coming to independent conclusions. Explanation facilities, the ease of knowledge acquisition, and making the users feel comfortable with the systems are the major obstacles for consultation systems. It is currently difficult to develop a viable expert system for a dynamic problem area.

In a problem area that appears to be suitable for development, the use of an expert system building tool to provide a prototype for one or two problems will show whether the expert system is appropriate. In the near future, expert systems will be able to handle larger domains and those with a large amount of uncertainty. However, they currently are not useful in the extremes of the two environments just mentioned.

A brief discussion of the problems that currently are not amenable to expert system application will further clarify those that are. Problems that require extensive learning or reasoning by analogy are currently out of reach. Currently, the knowledge about a problem must be easily explicatable by the human expert. We do not yet know how to acquire and adequately represent common sense knowledge. Expert systems are not the solution for problems in the social sciences due to their great complexity and the fact that they are not well modeled. They are often not solvable by human experts and tend to involve a great amount of uncertainty. Problems in meteorology, in which no expert is very precise, do not lend themselves well to expert system development. This indicates that an expert system is not appropriate as we get above a certain amount of uncertainty in our problem area.

Hayes-Roth[18] provides some insight into the problems which are currently beyond the scope of an expert system. We do not know how to reason from first principles, the core set of ideas which generally enable a skilled person to re-create many special cases. We are unable to capture the nonverbal understanding of physical operations that even very young children demonstrate. For instance, we know what will happen when we knock that expensive, fragile vase to the hard floor and we cringe in anticipation as it begins to fall. The effort to put this naive physics knowledge into knowledge-based systems appears extremely large. The representation of meta-knowledge is still weak. Most of our meta-knowledge today is involved in influencing our architectural design. Meta-representation is also lacking. That is, we lack a high level method of uniformly representing our knowledge before it is brought into the formalism appropriate for our application. This means that the wrong representation structure may be chosen, when a structure is not clearly indicated, and has no easy method of recovery.

IV. IMPRECISION IN EXPERT SYSTEMS

Imprecision is not a big problem in some expert systems domain such as XCON. In a domain with little imprecision a class one expert system can likely be developed. In the domains in which a medium or large amount of imprecision is inherent, the development of an expert system is quite difficult. Therefore, imprecision is an important aspect in the evolution of expert systems.

There are several sources of imprecision and uncertainty in an expert systems domain. The solution to problems in the domain may be imprecise. For instance, this is true of most medical problems. A solution to a problem may actually be made up of two seemingly separate parts. An expert system must be able to recognize this as well as ways in which solutions may overlap. When a solution is imprecise, it must be presented to the system user in a manner which indicates the uncertainty in it. A composite solution must be skillfully put together and the certainty associated with it made plain when it is presented.

The questions that are asked to determine the solution are often imprecise. An imprecise question will usually lead to an answer which is also imprecise. An expert system must be able to phrase questions in a manner which captures their somewhat vague meanings when they arise. The expert system must also be capable of accepting and providing an interpretation for an answer which contains some uncertainty, and needs to be able to adequately explain its reasoning in the face of the uncertain information which it has gathered.

The process of acquiring knowledge is quite imprecise. It is likely that the knowledge acquired does not exactly capture the expert's, especially since the expert is often not aware of all the tools he uses in his reasoning process. The knowledge that he reasons with may itself contain uncertainty. This must be effectively captured if we are to emulate the expert's reasoning process. Typically, the reasoning process of an expert is an imprecise one. We mean that the way in which the knowledge is used to make inferences and decide upon a problem solution is often not a precise process. A reliable methodology is necessary to capture it. As the knowledge engineer captures the reasoning process some variability may also creep in. What the knowledge engineer sees as the reasoning process may not exactly match the true process of the expert. Since the expert is not always aware of the actual reasoning process he uses, the differences may not be discovered. Glaring errors will be found, but the expert system must be capable of successfully recovering from the small inconsistencies which remain.

The knowledge representation language will also introduce some uncertainty into our expert system. This type of imprecision is inherent in our current technology in the following sense. If knowledge is not expressed in some formal language, the meaning cannot be interpreted exactly. Since knowledge engineers have not been able to develop or make universal use of a suitable formal representation language, a knowledge representation scheme must be chosen that appears to adequately capture the knowledge about a problem domain. Since the scheme that is chosen may not provide an exact match of the expert's knowledge representation, the expert system must be able to deal with this imprecision in a consistent manner. Fuzzy reasoning techniques can provide the basis for representing the imprecision inherent in an expert's knowledge. It can be used in conjunction with probability theory and Dempster-Shafer evidential theory. If the situation is probabilistic one, a nonfuzzy method would be used.

Some types of problems do not lend themselves to modeling by probability. For example, the case of a statement *the age of Joe is about 30,* could be modeled as a probability. However, in the fuzzy world it could be modeled by the membership function for *about 30.* This could give $.2/27 + .5/28 + .8/29 + 1/30 + .8/31 + .5/32 + .2/33$ with all other ages being outside the set. This could be done with the use of a fuzzy number. It would appear a more natural method of describing such a term.

Incomplete information also adds uncertainty to our system. We cannot always extract all the information that we could use from the user. In this case, we still wish to come to a conclusion if possible. This is true even if the conclusion is less certain. Common sense reasoning about physical life includes many inferences from incomplete information. This is a problem which has only begun to be attacked but appears to fit in with fuzzy reasoning. In any event, an expert system needs to be robust enough to overcome cases in which it has incomplete information. This may be done either by reasoning over, or going around, the gaps in its knowledge. The incorporation of fuzzy techniques in an expert system should allow either of these methods to be applied.

The data that the expert system acquires as it attempts to solve a problem often comes from error-prone humans. Therefore, it cannot be expected to be 100% precise. Even if the answer is related with precision, it may not be a binary response. Since people do not think in black and white, the answer will usually lie somewhere between the extremes. As stated earlier, it is extremely important that a general purpose expert system be able to accept and properly interpret imprecise information provided by the system user. In cases where the expert system received data from sensing devices or some type of equipment, the well-known problem of noise comes into play and puts imprecision into the equation. In addition the equipment may be out of adjustment. The expert system interface to external equipment must be able to accept that readings are imprecise or fuzzy and effectively deal with this fact.

Uncertainty also occurs in the knowledge base where we may have weak implications, in the form of implication relations with middle range certainties, given us by the expert. In translating the expert's statement to the implication, we may lose the meaning of it somewhat. The expert system must be able to accept and use the weak implications together with other sources of information to come to a conclusion about the problem under investigation.

Another type of uncertainty arises in the collection of knowledge from different sources, experts, literature, etc. There may be conflicting, redundant, subsuming, or missing chunks of knowledge. In rule-based systems this type of uncertainty has led to the compilation of the rules into a network where they may be examined for these problems. Missing chunks of knowledge become evident as the system is tested. Pieces of knowledge which subsume others may be found by the knowledge engineer or through traces of the system operation. Conflicts are also normally found by system testing, but it is conceivable that a fuzzy function, which operates on the knowledge base, could be used to indicate conflicts.

A. EXAMPLES OF IMPRECISION

In the following some examples of the different types of imprecision presented previously will be given. The likely source(s) of the imprecision or uncertainty in the examples will also be discussed.

It can be the case that a person is unsure of a particular piece of causal information. An example:

> IF Z is tall, THEN Z played on a basketball team (0.8).
> Z is tall (0.9).

The numeric values are certainty values associated with the rule and fact, respectively. In this statement, it may be that enough information has been gathered to reliably assign a probabilistic uncertainty to the statement. However, this is often not the case, and the value may be generated by a well-designed fuzzy membership function[17] for people who have played basketball. To determine whether a person is tall, a fuzzy membership function on heights may provide us this information. It may not be as exact as doing the work necessary to derive a reliable probability, but since fuzzy inferencing will still work reliably with small errors, the correct answers can still be determined.

Another type of imprecision comes from a linguistic statement in which the boundary is not clearly defined. For example:

> Mary is middle-aged.
> IF the inventory is high, THEN production should be slowed.

The terms middle-aged, high, and slowed are imprecise terms. In the context of fuzzy set theory, they are called fuzzy terms. This type of information is modeled well by linguistic fuzzy sets. It is true that they could be modeled with the Dempster-Shafer theory of evidence, but it is our contention that fuzzy sets provide a more intuitive method and will provide accurate answers.

Both uncertainty and imprecision may exist in some statements. For instance:

IF the wind is high THEN the sailing should be good (0.8).

In this case, the fuzzy information is contained in the terms high and good. The overall statement only holds most (0.8) of the time.

The uncertainty could also be fuzzy. In the statement,

Ed is a masterful sailor, (0.8 to 0.9),

a fuzzy range with which it may be believed is given. The range could be interpreted as about 0.85.

It is not our intention to cover anything other than how Fuzzy Set Theory may be used in expert systems. There are places where the uncertainty may be probabilistic in nature, and probability-based theories will provide the best models. Most often it will be the case that fuzzy sets may be used, possibly in conjunction with probability, to model the uncertainty and imprecision inherent in real world problems.

There is imprecision in most areas of an expert system. Most of the interesting domains to which an expert system may be applied have some significant amount of imprecision in them. It should be noted that some of the applications with the least amount of imprecision in them have been most successfully attacked. Those with large amounts of imprecision have not been as successfully solved and some are beyond the current reach of expert systems. This indicates that uncertainty and imprecision are very important aspects of an expert system. They must be handled in a sound manner both theoretically and practically.

V. THE EVOLUTION TO FUZZY EXPERT SYSTEMS

The method of handling imprecision must be excellent for an expert system to succeed in becoming a useful tool. It also needs to be natural so that the knowledge flows freely from the expert. There are currently quite a few different ways that imprecision may be handled in an expert system. The ones that are complete tend to allow imprecision and uncertainty to be managed in only a small degree or in an awkward fashion. Most methods of handling imprecision are probability based. It is interesting that experts often do not think in probability values, but in terms such as much, usually, always, sometimes, etc. While it is possible to give these terms probabilistic values, this book is concerned with the use of fuzzy values which capture their meaning in a natural fashion.

We will briefly examine the way the expert systems MYCIN[32] and CASNET[36] deal with uncertainty. MYCIN introduced the concept of certainty factors. This concept has been widely applied, in several different forms, to the handling of imprecision in expert systems. Each MYCIN rule has a strength, called a certainty factor, lying in the interval [0,1] associated with it. When a rule is fired its premise is evaluated and a numeric value between -1 and 1 is associated with the premise. If the premise value is outside the previously specified threshold interval (e.g., 0.2 and -0.2, respectively), then the action part of the rule is evaluated and the conclusion is made with a certainty which is the premise value times the certainty factor of the rule. Evidence for hypotheses is combined into measures of belief and disbelief. The belief measure is a value in the range [0,1], and the disbelief measure is in the range $[-1,0]$. A hypothesis is believed or disbelieved if the hypothesis is above or below the respective threshold. It is informative to note that in other expert systems with MYCIN base reasoning, the method of combining has been modified. Centaur[1] is an example.

CASNET is a semantic net-based expert system. Each node, or state, as it is known in this context, in the net has a forward and reverse weight associated with it; these weights are the strengths of causation between the nodes. This enables the net to be traversed from bottom to top or top to bottom. The weights correspond to the following interpretations of causal strength: sometimes, often, usually, almost always, and always. Since time is an element of the glaucoma disease model of CASNET, some states when confirmed will cause the rest of the path, of which they are a part, to be considered for treatment. Rules are used to associate observations with states. They have a confidence value, between −1 and 1, assigned to them. Prescribed methods for combining confidence values have been defined. Certainty factors are associated with each state. They denote the belief that the state has occurred. States or nodes are considered confirmed when their certainty factor is above a threshold. Observations are usually tests. The observation with the strongest positive or negative result is used to confirm or deny a state. The combination of observations and partially confirmed pathways allow inference under uncertainty to be done and conclusions with treatment plans obtained.

These methods and others are reasonably effective in specific cases. In fact, many variations of the MYCIN method of dealing with uncertainty are currently in use. We argue that fuzzy methods can, in some areas, replace the classical methods and provide better performance. In others they may augment and work in concert with probability-based methods. The need for a theory-based method of dealing with uncertainty brings us to fuzzy set theory. Before we delve into that, the Dempster-Shafer theory of evidence[12,31] must be mentioned. This theory is probability based and it has been speculated to be useful in building expert systems. A problem with this theory is that rule of evidence combination may create too large a certainty measure of the combined evidence about a fact, if a normalization is used to eliminate or hide a contradiction.[43] This method is certainly valuable in a domain in which most of the uncertainty may be effectively described by probabilities. However, where the domain is not amenable to the estimation or measurement of probabilities, fuzzy set theory offers an alternative.[44] Where fuzzy sets or linguistic terms best describe items in a domain, the theories of possibility and necessity sets may be applied more naturally than probabilities.

We shall first describe what is meant by a fuzzy expert system. We will then indicate why they are more useful than others where applicable and give some examples of fuzzy expert systems.

By fuzzy expert system we mean an expert system which incorporates fuzzy sets and/or fuzzy logic into its reasoning process and/or knowledge representation scheme. Several expert systems have been developed which incorporate fuzzy techniques.

The theory of fuzzy sets and fuzzy logic is well-founded and strong. The theories have been in existence for over 25 years and have been shown useful in several control applications among others. Fuzzy logic is employed in the control of trains in Japan.[26] It has been very successful, and an expert fuzzy controller is on-line in the city of Sendai, Japan. It provides smoother changes in velocity than nonfuzzy systems have yet achieved. Fuzzy set theory is used in the cement kiln controller of Mamdani.[24] In both of these cases, linguistic fuzzy variables are used to do internal calculation and provide the control strategy.

The theoretical basis behind fuzzy techniques will allow us to deal with uncertainty in a manner that is well supported. The theory properly used will allow fuzzy reasoning schemes to be developed and applied to a wide range of problems without constant minor changes. A unified, theoretically sound set of methods can be developed for reasoning under certainty in expert systems.

Whalen and Schott[39] use a fuzzy linguistic logic system with production rules to suggest appropriate forecasting techniques for sales predictions in their expert system. They use possibilities and work in a backward chaining manner. They start with the possibilities of all conclusions equal to one and attempt to reduce the possibilities until they find an irreducible set. SPERILL II,[27] an expert system for damage assessment of existing structures, uses fuzzy sets to represent imprecise data. FLOPS[10] is a fuzzy rule-based production expert system. Input to the system is a vector all of whose components are fuzzy sets. As output it produces a fuzzy set of conclusions. Incoming information is pattern matched against rules to provide a fuzzy set of fireable rules, which are then executed. They continue this process until they can no longer find rules to fire and then produce a set of fuzzy conclusions. FLOPS has been sold commercially for several years now. It has been successfully used to model several different expert domains.[8]

The Z-II system[23] is a fuzzy expert system shell, which effectively deals with both uncertainty and imprecision. It has been used to construct several expert systems. The domains have been medical diagnosis, psychoanalysis, and risk analysis. It has been found that the experts felt it allowed them to express their information in a natural manner. It allows knowledge to be expressed in fuzzy linguistic terms. An example rule is the following:

> IF your interest in analyzing the human body is high,
> THEN your overall interest in medicine should be high (0.95).

As long as the linguistic terms are defined by some reasonable membership function, this is a very effective knowledge representation format for imprecise and uncertain information.

Another important type of system for which fuzzy sets provide a powerful basis is common sense reasoning. The *theory of usuality* put forth by Zadeh[44] provides a tool to incorporate some common sense information into expert systems. The concept of usuality relates to events which are usually true or have a high probability of occurrence. A usuality quality proposition may have an implicit usually. It is then called a *disposition*. Some common dispositions are *snow is white, Florida is a warm state,* and *windy days are good for sailing.* Now usually may be interpreted as a fuzzy quantifier, which is basically a fuzzy proportion based on an implicit or explicit σ count. The theory of usuality may be applied to ordinary situations and provide a method to effectively represent knowledge about events or items which are often true. This includes many common sense concepts.

Possibly the biggest weakness of nonfuzzy methods of dealing with imprecision and uncertainty is their handling of linguistic terms. Fuzzy set theory provides a natural method for dealing with the linguistic terms by which an expert will describe a domain. An imprecise numeric term can be effectively described by a fuzzy number.[20] Other terms are simply mapped to and from fuzzy sets. Hence, the use of fuzzy set theory in expert systems has caused an evolution of systems.

The failings of probability in situations where little or no *a priori* information is known provide an arena for the use of fuzzy expert systems. The rest of this book will show that their use can provide an enhancement for systems which must do reasoning with much uncertainty and imprecision.

REFERENCES

1. **Aikens, J. S.,** Prototypical knowledge for expert systems, *Artif. Intell.,* 20, 163, 1983.
2. **Aikens, J. S., Kunz, J. C., and Shortliffe, E. H.,** PUFF: an expert system for interpretation of pulmonary function data, *Comput. Biomed. Res.,* 16, 199, 1983.
3. **Bandler, W. and Kohout, L. J.,** Activity Structures and Their Protection, Proc. 1979 Int. Meet. Soc. Gen. Syst. *Res.,* Louisville, KY, 1979, 239.
4. **Bandler, W.,** Representation and Manipulation of Knowledge in Fuzzy Expert Systems, presented at workshop on fuzzy sets and knowledge-based systems, Queen Mary College, University of London, England, 1983.

5. **Barr, A. and Feigenbaum, E. A.,** *The Handbook of Artificial Intelligence I,* William Kaufmann, Los Altos, CA, 1981.
6. **Bennett, J. S. and Engelmore, R. S.,** Experience using EMYCIN, in *Rule-Based Expert Systems,* Buchanan, B. and Shortliffe, E., Eds. Addison-Wesley, Reading, MA, 1984, 314.
7. **Bonissone, P. P. and Brown, A. L., Jr.,** Expanding the Horizons of Expert Systems, Proc. Conf. Expert Syst. Knowledge Eng., Gottlieb Duttwailer Institute, Zurich, Switzerland, April, 1985.
8. **Boose, J. H.,** Personal Construct Theory and the Transfer of Human Expertise, Proc. Natl. Conf. Artif. Intelligence, Austin, Texas, 1984, 27.
9. **Buckley, J. J., Siler, W., and Tucker, D.,** A fuzzy expert system, *Fuzzy Sets Syst.,* 20(1), 1, 1986.
10. **Buckley, J. J., Siler, W., and Tucker, D.,** FLOPS, A Fuzzy Expert System: Applications and Perspectives, in *Fuzzy Logics in Knowledge Engineering,* Negoita, C. V. and Prade, H., Eds., Verlag TUV Rheinland, Germany, 1986.
11. **Davis, R. and Lenat, D. B.,** *Knowledge-Based Systems in A.I.,* McGraw-Hill, New York, 1982.
12. **Dempster, A. P.,** Upper and lower probabilities induced by a multivalued mapping, *Ann. Math. Stat.,* 38, 325, 1967.
13. **Gaines, B. R.,** Foundations of fuzzy reasoning, in *Fuzzy Automata and Decision Processes,* Gupta, M. M., et al., Eds., North-Holland, New York, 1977.
14. **Gaschnig, J.,** PROSPECTOR: An expert system for mineral exploration, in *Introductory Readings in Expert Systems,* Michie, D., Ed., Science Publishers New York, 1982.
15. **Erman, L. D., Fennel, R. D., and Reddy, D. R.,** System organizations for speech understanding: implications for network and multiprocessor computer architectures for A.I., *IEEE Trans. Comput.,* C-25(4), 414, 1976.
16. **Erman, L. D.,** The HEARSAY-II speech-understanding system: integrating knowledge to resolve uncertainties, *Computing Surv.,* 12(2), 213, 1980.
17. **Hall, L. O. and Kandel, A.,** *Designing Fuzzy Expert Systems,* Verlag TÜV Rheinland, Germany, 1986.
18. **Hayes-Roth, F.,** The knowledge-based expert system: A tutorial, *Computer,* 17(9), 11, 1984.
19. **Kandel, A., Ed.,** Special issue on expert systems, *Inf. Sci.,* 37(1—3), 1985.
20. **Kandel, A.,** *Fuzzy Mathematical Techniques with Applications,* Addison-Wesley, Reading, MA, 1986.
21. **Kacprzyk, J. and Yager, R.,** Emergency-oriented expert systems: a fuzzy approach, *Technical Report MII-213/247,* Machine Intelligence Institute, Iona College, New Rochelle, NY, 1982.
22. **Keene, S. E.,** *Object-oriented Programming in Common LISP,* Addison-Wesley, Reading, MA, 1988.
23. **Leung, K. S. and Lam W.,** Fuzzy concepts in expert systems, *Computer,* 43, 1988.
24. **Mamdani, E. H.,** Fuzzy Logic Controllers with Industrial Applications, Proc. JACC, San Francisco, 1980.
25. **McDermott, J.,** Domain knowledge and the design process, *Design Stud.,* 3(1), 1982.
26. **Miyamoto, S. and Yasunobu, S.,** Predictive Fuzzy Control and its Application to Automatic Train Operation Systems, Fuzzy Inf. Proc. Soc. Conf., 1984.
27. **Ogawa, H., Fu, K. S., and Yao, J. P. T.,** Knowledge Representation, and Inference Control of SPERILL-II, ACM Conf. Proc., November 1984.
28. **Politakis, P. and Weiss, S. M.,** Using empirical analysis to refine expert system knowledge bases, *Artif. Intelligence,* 22, 23, 1984.
29. **Pople, H. E., Jr.,** Knowledge-based expert systems: the buy or build decision, in *Artificial Intelligence Applications for Business,* Reitman, W., Ed., Ablex, Norwood, NJ, 1984.
30. **Reddy, D. R., Erman, L. D., Fennel, R. D., and Neely, R. B.,** The HEARSAY Speech Understanding System: An Example of the Recognition Process, Proc. IJ-CAI 3, 1973, 185.
31. **Shafer, G.,** *A Mathematical Theory of Evidence,* Princeton University Press, New Jersey, 1976.
32. **Shortliffe, E. H.,** *Computer-Based Medical Consultation: MYCIN,* Elsevier/North-Holland, NY, 1976.
33. **Valverde, L. and Trillas, E.,** On Modus Ponens in Fuzzy Logic, 15th Int. Symp. Multiple-Valued Logic, Kingston, Ontario, 1985.
34. **van Melle, W.,** A Domain-independent Production-rule System for Consultation Programs, Proc. IJCAI-79, 1979, 923.
35. **Waterman, D. A.,** *A Guide to Expert Systems,* Addison-Wesley, Reading, MA, 1986.
36. **Weiss, S. M., Kulikowski, C. A., Amarel, S., and Safir, A.,** A model-based method of computer-aided medical decision-making, *Artif. Intelligence,* 11, 145, 1978.
37. **Weiss, S. M. and Kulikowski, C. A.,** Representation of expert knowledge for consultation: the CASNET and EXPERT projects, in *Artificial Intelligence in Medicine,* Szolovits, P., Ed., *AAAS Symp. Series,* Westview Press, Boulder, CO, 1982, 21.
38. **Wenstop, F.,** Applications of Linguistic Variables in the Analysis of Organizations, Ph.D. thesis, University of California, Berkley, 1975.
39. **Whalen, T. and Schott, B.,** Goal-directed approximate reasoning in a fuzzy production system, in *Approximate Reasoning in Expert Systems,* Gupta et al., Eds., North-Holland, New York 1985.
40. **Yager, R. R.,** Robot planning with fuzzy sets, *Robotica,* 1, 41, 1983.

41. **Zadeh, L. A.,** The role of fuzzy logic in the management of uncertainty in expert systems, *Fuzzy Sets Syst.,* 11, 199, 1983.
42. **Zadeh, L. A.,** Common sense knowledge representation based on fuzzy logic, *Computer,* 16(10), 61, 1983.
43. **Zadeh, L. A.,** Review of books: a mathematical theory of evidence, *AI Mag.,* 5(3), 81, 1984.
44. **Zadeh, L. A.,** A simple view of the Dempster-Shafer theory of evidence, and its implication for the rule of combination, *AI Mag.,* 7(2), 85, 1986.
45. **Zadeh, L. A.,** Outline of a theory of usuality based on fuzzy logic, Berkeley Cognitive Science Report Series, University of California Berkeley, CA, 1986.

Chapter 2

GENERAL PURPOSE FUZZY EXPERT SYSTEMS

Mordechay Schneider and Abraham Kandel

TABLE OF CONTENTS

I. INTRODUCTION

The expert system we describe here can be logically divided into three main parts:

1. The Front-end Compiler (FC)
2. The Inference Engine (IE)
3. The Question-Answer Program (QAP)

The Front-End Compiler is a program that receives input data from the user, or any other connected software, and compiles it. During compilation the data is transformed into a format that is understandable to the inference engine. Following compilation the reformatted data is passed to the inference engine. The inference engine uses this data in conjunction with a knowledge base in the form of production rules.

The QAP has only one function — to trace a certain conclusion and show the user how this conclusion was reached (what rules and data participated in contributing to the drawing of that conclusion).

To see how an expert system might be made fuzzy it is first necessary to understand the properties of fuzzy logic and fuzzy sets which are applicable here. This work is divided into four parts. First we explain the concepts of the fuzzy expected value and the fuzzy expected interval and their properties, and then we describe the fuzzy expert system.

II. ON FUZZY LOGIC

Fuzzy logic was introduced by Zadeh[1] in 1965. The basic idea was to extend the classical logic (the Boolean logic) in order to relax the harsh constraint that everything that can be said about anything is either absolutely true or absolutely false. Zadeh suggests that it is possible to understand a statement as being 0.75 true (which could be interpreted as *not really true*) or 0.5 true (may be true), etc. The use of fuzzy numbers enable us to use fuzzy variables (which will be described in the next section). In 1978 a paper by Kandel and Byatt[2] described a new concept for fuzzy average called the Fuzzy Expected Value (FEV). A description of the fuzzy expected value follows.

A. THE FUZZY EXPECTED VALUE

Let χ_A by a B-measurable function such that $\chi_A \in [0,1]$. The fuzzy expected value (FEV) of χ_A over the set A, with respect to the fuzzy measure $\mu(\bullet)$, is defined as

$$FEV(\chi_A) = \mathop{SUP}_{T \in [0,1]} \{min[T, \mu(\xi_T)]\} \tag{1}$$

where $\xi_T = \{x | \chi_A(x) \geq T\}$.

Now, $\mu\{x | \chi_A(x) \geq T\} = f_A(T)$ is a function of the threshold T.[3] The function μ maps ξ into the interval [0,1].

The actual calculation of $FEV(\chi_A)$ consists of finding the intersection of the curves T = $f_A(T)$, which will be at a value T = H, so that $FEV(\chi_A) = H \in [0,1]$.[4]

1. Example 1

Using the base variable *old*, assume a given population and a given subjective compatibility curve such that

10 people are 20 years old → χ = 0.20
15 people are 30 years old → χ = 0.30
25 people are 45 years old → χ = 0.45
30 people are 55 years old → χ = 0.55
20 people are 60 years old → χ = 0.60

As can be seen, we have five different thresholds (0.20, 0.30, 0.45, 0.55, 0.60). The first step in the process is to check how many people are above each threshold (in percentage terms). As can be seen, 100 people are above or equal to 0.2; 90 people are above or equal to 0.3; 75 people are above or equal to 0.45; 50 people are above or equal to 0.55 and 20 people are above or equal to 0.60.

Pairing the data and rearranging it by increasing order of the measure of belief, χs, we obtain the following five [χ,μ] pairs (see Equation 1):

(0.20, 1.00)
(0.30, 0.90)
(0.45, 0.75)
(0.55, 0.50)
(0.60, 0.20)

Now, the minimum value of each pair is

min(0.20, 1.00) = 0.20
min(0.30, 0.90) = 0.30
min(0.45, 0.75) = 0.45
min(0.55, 0.50) = 0.50
min(0.60, 0.20) = 0.20

and, therefore, following Equation 1, the FEV, which is the maximum of all these minima, is

max(0.20, 0.30, 0.45, 0.50, 0.20) = 0.50

Thus, the FEV is 0.50. From this result we can state that the fuzzy expected age is 50.
Next, we discuss the properties of the fuzzy expected interval.

B. THE FUZZY EXPECTED INTERVAL
The Fuzzy Expected Interval (FEI) was designed to handle cases where the FEV is not applicable. The application of FEI becomes useful when the data provided by the user to the expert system is fuzzy or incomplete. The following example illustrates the idea.

1. Example 2
Assume the following information has been provided by the user:

more or less 20 people are between the ages of 20 and 30
20 to 25 people are 15 years old
25 people are almost 40 years old

What is the typical age of the group of people described above?

Clearly the FEV is not applicable here, since the calculation of FEV requires complete information about the distribution of the population and their grade of membership (the χ's). The FEI was developed to solve these kinds of problems.

The solution to the problem in Example 2 will be provided in the following section. Here we describe some properties of the FEI.

In principle, the evaluation of FEI is performed using Equation 1. According to Equation 1, in order to find FEV it is necessary to find all MINs for each row and then the MAX over all MINs. The evaluation of FEI is performed in the same way. The following theorems will provide a method for evaluating MINs and MAXs among intervals.[5]

Theorem 1. Let S and R be two intervals such that (a) $S = \{s_1 \ldots s_n\}$ (b) $R = \{r_1 \ldots r_m\}$, and (c) $R \cap S = \varnothing$. Then the MAX of the two intervals is

$$MAX(S,R) = \begin{cases} R \text{ if } r_1 > s_n \\ S \text{ if } s_1 > r_m \end{cases} \qquad (2)$$

Theorem 2. Let S and R be two intervals such that (a) $S = \{s_1 \ldots s_n\}$ (b) $R = \{r_1 \ldots r_m\}$ and (c) $R \cap S = \varnothing$. Then the MIN of the two intervals is

$$MIN(S,R) = \begin{cases} R \text{ if } r_m < s_1 \\ S \text{ if } s_n < r_1 \end{cases} \qquad (3)$$

Theorem 3. Let S and R be two intervals such that (a) $S = \{s_1 \ldots s_n\}$ (b) $R = \{r_1 \ldots r_m\}$ (c) $R \cap S \neq \varnothing$ (d) $S \notin R$ and $R \notin S$. Then the MAX of the two intervals is

$$MAX(S,R) = \begin{cases} R \text{ if } r_m > s_n \\ S \text{ if } s_n > r_m \end{cases} \qquad (4)$$

Theorem 4. Let S and R be two intervals such that (a) $S = \{s_1 \ldots s_n\}$ (b) $R = \{r_1 \ldots r_m\}$ (c) $R \cap S \neq \varnothing$ (d) $S \notin R$ and $R \notin S$. Then the MIN of the two intervals is

$$MIN(S,R) = \begin{cases} R \text{ if } r_m < s_n \\ S \text{ if } s_n < r_m \end{cases} \qquad (5)$$

Theorem 5. Let S and R be two intervals such that $S = \{s_1 \ldots s_n\}$ and $R = \{r_1 \ldots r_m\}$, and $R \subseteq S$. Then the MAX of the two intervals is

$$MAX(S,R) = [r_1 \ldots s_n] \qquad (6)$$

Theorem 6. Let S and R be two intervals such that $S = \{s_1 \ldots s_n\}$ and $R = \{r_1 \ldots r_m\}$, and $R \subseteq S$. Then the MIN of the two intervals is

$$MIN(S,R) = [s_1 \ldots r_m] \qquad (7)$$

Definition 1. Let α and β be intervals. Then we say that α is *higher* than β if the upper bound of α is greater than the upper bound of β.

The following 2 equations are developed to handle fuzzy distribution of population.[6]

$$\underset{j}{UB} = \frac{\sum_{i=j}^{n} MAX(pi_1, pi_2)}{\sum_{i=j}^{n} MAX(pi_1, pi_2) + \sum_{i=1}^{j-1} MIN(pi_1, pi_2)} \qquad (8)$$

where pi_1 is the lower bound of group i and pi_2 is the upper bound of group i.

The process of finding the lower bound of any μ_i (or $\underset{i}{LB}$) is given by the following equation:

$$\underset{j}{LB} = \frac{\sum_{i=j}^{n} MIN(pi_1, pi_2)}{\sum_{i=j}^{n} MIN(pi_1, pi_2) + \sum_{i=1}^{j-1} MAX(pi_1, pi_2)} \qquad (9)$$

This concludes the description of the fuzzy expected interval. Next we describe the front-end compiler and show how it uses the fuzzy expected intervals in the process of compiling fuzzy data.

III. THE FRONT-END COMPILER

The Front-end Compiler (FC) is a program which receives data from the user (usually) and converts it to a form which can be understood by the inference engine. The inference engine places the transformed data on a blackboard to use in the evaluation of the knowledge base.

The task of compiling the data involves grouping the data according to some criteria, ordering each group by its characteristics, and, then, finding the fuzzy expected value or the fuzzy expected interval for some of the groups.

The blackboard is a global data structure (i.e., the contents of the blackboard is reachable from any part of the software) that contains important information about the data. Each line on the blackboard contains:

1. 2 or 3 key words
2. Upper and lower bounds of the value described by the object of the sentence
3. Certainty Factor (FC), which is used for the evaluation of the conclusion(s). (The evaluation process will be discussed in the next section.)
4. Rule Number (RN), which, if the sentence comes from user-supplied data, then the rule number is 0; otherwise, it is the same as the number of the rule which was fired and its conclusion is placed on the blackboard.

TABLE 1
Description of the
Information Stored on the
Blackboard

Name	Notation
key word 1	K1
key word 2	K2
key word 3	K3
lower bound 1	LB1
upper bound 1	UB1
lower bound 2	LB2
upper bound 2	LB2
certainty factor	CF
rule number	RN

The CF of any data provided by the user is always 1; the reason being that it is logical to trust the user to provide the right data. Even if the data is vague (for example, "almost 30 people are between the ages of 20 and 30"), the user is certain that this data is the only data available. Therefore, the expert system should treat the data as an absolute truth even if the evaluation will show otherwise.

In order to see how the compiled data is stored on the blackboard it is necessary to define the legal structure of a sentence. There are two types of sentences recognized by the FC:

1.   Sentences which start with the key word "THE"
2.   Sentences which do not start with the key word "THE"

A. EVALUATION OF SENTENCES OF TYPE 1

The general structure of a sentence of Type 1 is

THE A of [the] B is [not] [adjective] C

where A and B are key words and C is either a key word, a number, or a range of numbers describing B. The words in the brackets are optional.

As mentioned before, each line on the blackboard has important information:

The evaluation of a sentence depends on the value of the adjective, C, and whether or not a negation is used in the sentence (*not* C). We explain each case separately.

1. Case 1

Suppose we have the following sentence:

The color of the product is green.

The 3 key words are *color, product* and *green*. The CF is 1 (since the data was provided by the user) and the RN is 0 (again, since the data is provided by the user). The lower and upper bounds are values which describe K3. Since K3 is not a number, LB1 = 1, UB1 = 1, LB2 = UNDEF and UB2 = UNDEF. Thus, the compiled sentence will be stored on the blackboard as

K1	K2	K3	LB1	UB1	LB2	UB2	CF	RN
color	product	green	0	0.99	UNDEF	UNDEF	1	0

2. Case 2

Assume we have the following sentence:

The color of the product is not green.

Here the key word *not* is present in the data. The negation creates two intervals such that:

$$\bar{x} = [0..x - \epsilon] \text{ and } [x + \epsilon..1] \tag{10}$$

Using Equation 10 we can evaluate the new LBs and UBs and compile the sentence to look like (let $\epsilon = 0.01$):

K1	K2	K3	LB1	UB1	LB2	UB2	CF	RN
color	product	green	0	0.99	UNDEF	UNDEF	1	0

We set LB2 and UB2 to be UNDEF since they fall outside the interval [0,1].

3. Case 3

Consider the following sentence:

The color of the product is almost green.

We really don't know the meaning of *almost green* but, mathematically, we can represent it. *Almost* is an adjective and, therefore, has a numerical range (see Appendix A). Since *green* is not a number we assign the value 1 to it. Now we can find the values for almost green

$$\text{LB1 (ALMOST GREEN)} = 0.9$$
$$\text{UB1 (ALMOST GREEN)} = 0.9$$

The upper bound of *almost green* is 0.9 because UB (ALMOST GREEN) = $1 - 1 = 0$. But since LB > UB (logically inconsistent) we establish the following:

$$\text{IF LB > UB THEN UB = LB} \tag{11}$$

Thus, the compiled data is:

K1	K2	K3	LB1	UB1	LB2	UB2	CF	RN
color	product	green	0.9	0.9	UNDEF	UNDEF	1	0

4. Case 4

Suppose we have the following data sentence:

The age of Tom is 20.

In this case we have only two key words and a number. First we translate the value *20 years old* by mapping it to a value in the range [0,1]. (see Appendix B). Thus, the value corresponding to *20 years old* is 0.16. Since K3 does not exist (empty) the FC understands that the LB and the UB are mappings from the actual numbers to the grade of membership. Thus, the compiled sentence will be:

K1	K2	K3	LB1	UB1	LB2	UB2	CF	RN
age	Tom		0.16	0.16	UNDEF	UNDEF	1	0

Again, the interpretation of the table above is: *We are certain* (CF = 1) *that the age of Tom is 20 years old.* The CF = 1 and the RN = 0 came from the fact that the data was provided by the user.

5. Case 5

Consider the following data line:

The age of Tom is more or less 20.

First we have to interpret *more or less 20 years old* (using Appendix A):

LB(more or less 20 years old) = 18 years old
UB(more or less 20 years old) = 22 years old

Now we can use Appendix B to map the actual numbers into the range [0,1].

LB(18) = 0.15
UB(22) = 0.18

Again, since K3 does not exist, the LB and the UB represent the mappings from the actual numbers to the range [0,1]. Thus, the compiled data is:

K1	K2	K3	LB1	UB1	LB2	UB2	CF	RN
age	Tom		0.15	0.18	UNDEF	UNDEF	1	0

6. Case 6

In the next case we establish the rules of interpretation about the key word *not*. Assume we have the following data sentence:

The age of Tom is not 18.

Using Appendix B we can evaluate the grade of membership of *age 18* (which is 0.15). Thus, using Equation 10 we have

K1	K2	K3	LB1	UB1	LB2	UB2	CF	RN
age	Tom		0	0.14	0.16	1	1	0

7. Case 7

The last case combines all cases discussed above. Consider the following data line:

The age of Tom is not more or less 20.

The first step is to find the range of *more or less 20* (using Appendix A):

LB(MORE OR LESS 20 YEARS OLD) = 18 YEARS OLD
UB(MORE OR LESS 20 YEARS OLD) = 22 YEARS OLD

Next we use Appendix B to find the grade of membership of the ages found above:

$$\chi(18) = 0.15$$
$$\chi(22) = 0.18$$

Therefore the compiled data will be:

K1	K2	K3	LB1	UB1	LB2	UB2	CF	RN
age	Tom		0	0.14	0.19	1	1	0

Thus, for any Type 1 sentence the compiler performs the evaluation as described above, placing on the blackboard the results from that evaluation. Next we discuss the evaluation of sentences of Type 2.

B. EVALUATION OF SENTENCES OF TYPE 2

The general structure of a sentence of Type 2 is

[adjective] [number to] number K1 C,

where words inside the brackets are optional and C is a general statement that contains a key word and number (or a range of numbers), and may contain the key word *not* and any adjectives (from the list in Appendix A). From the general structure of a sentence of type 2 it can be seen that the first part of the sentence (excluding C) includes a key word (K1) and a range of numbers (in case there is only one number, we make LB = UB). C has a similar structure. Thus, the sentence:

Almost 20 people are between the ages 25 to 30

will be translated to:

LB1	UB1	K1	K2	LB2	UB2
18	19	people	age	25	30

Thus, the first range provides information about K1 and the second range provides information about K2.

After all sentences have been transformed to a form described previously, we perform the next step in the evaluation of sentences of type 2.

The next step involves grouping the relevant representations of the sentences together. A comparison is performed using K1 and K2 and their synonyms. Thus, in the stated example, we search for the sentences describing population of people and their ages and place them together in one group.

Next it is necessary to order each group by the second range which is the range LB2 — UB2 (according to Equation 1 and Definition 1).

After groups have been formed and ordered properly, we find the fuzzy expected interval for each group. The algorithm for the evaluation of the fuzzy expected value is presented below.

Algorithm 1.

1. Use the proper characteristic function to find all χ's.
2. Use Equations 8 and 9 to find all μ's.
3. Use Equation 3, 5, or 7 to find the minimum for each line.
4. Use Equation 2, 4, or 6 to find the maximum over the minima found in step 3.

When the four steps have been completed, the compiled data is placed on the blackboard. In the next example we show how the FC evaluates the data.

1. Example 3

Assume we have the following data:

more or less 20 people are between the ages of 20 and 30
20 to 25 people are 15 years old
the color of the house is green
10 people are 6 feet tall
the weight of John is 120 pounds
25 people are almost 40 years old
almost 15 people are 5 feet and 10 inches tall

First, we compile the two sentences which start with the key work *the*. Using the methods formerly stated, we compile the two sentences and place the compiled data on the blackboard:

K1	K2	K3	LB	UB	CF	RN
color	house	blue	0.9	0.9	1	0
weight	John		0.54	0.54	1	0

Next we group the sentences that describe ages and people and order them by their increasing age:

1. 20 to 25 people are 15 years old
2. more or less 20 people are between the ages of 20 and 30
3. 25 people are almost 40 years old

Using the algorithm for finding the fuzzy expected interval we follow the four steps:

STEP 1 — Find all χ's

RAW DATA	χ
20 to 25 people	0.125 - 0.125
more or less 20 people	0.16 - 0.25
25 people	0.3 - 0.325

Step 2 — Find all μ's.

μ	χ
1 - 1	0.125 - 0.125
0.63 - 0.70	0.16 - 0.25
0.34 - 0.39	0.3 - 0.325

Step 3 — Find the minimum for each pair.

$$\text{MIN}(1,[0.125 - 0.125]) = [0.125 - 0.125]$$
$$\text{MIN}([0.63 - 0.7],[0.16 - 0.25]) = [0.16 - 0.25]$$
$$\text{MIN}[0.34 - 0.39],[0.3 - 0.325]) = [0.3 - 0.325]$$

Step 4 — Find the maximum over the minima (found in Step 3):

$$\text{MAX}([0.125 - 0.125],[0.16 - 0.25],[0.3 - 0.325]) = [0.3 - 0.325]$$

The fuzzy expected interval for the last group is evaluated similarly and the compiled data is placed on the blackboard:

K1	K2	K3	LB	UB	CF	RN
color	house	blue	1	1	1	0
weight	John		0.54	0.54	1	0
people	age		0.3	0.325	1	0
people	weight		0.83	0.83	1	0

It is interesting to note that, in spite of the fact that we had fuzzy information about the height of the people, the result obtained was a single number (0.83) to show that there are cases in which the fuzzy expected interval can be reduced to the fuzzy expected value. The compiled data is used by the inference engine in the process of evaluating the knowledge base. Next we explain the features of the inference engine, and show how it uses the compiled data in the evaluation of the knowledge base.

IV. THE INFERENCE ENGINE

An inference engine is an expert system tool built for the evaluation of the knowledge base.[7,8,9,10] The inputs to the inference engine are the compiled data (provided by the FC) and the knowledge base. The output is the conclusions of rules that have been fired.

First, we explain the structure of the knowledge base and then we discuss the evaluation process.

A. THE KNOWLEDGE BASE

The knowledge base is a data structure which contains knowledge about the problem domain. A knowledge base can be represented by several different knowledge representation methods such as

1. Frames
2. Semantic nets
3. Production rules

The expert system described here uses production rules as a knowledge representation method.

The general structure of a production rule is

$$\text{IF P THEN C} \tag{12}$$

where P is the premise of the rule and C is the conclusion. The general structure of P is (using the BNF notation)

P :: = S |(P) |NOT P |S or P |S and P |
S :: = A sentence of type 1 (as was explained in the previous section).

The general structure of the knowledge base is

$$R \quad 1$$
$$\text{IF } P_1 \text{ then } C_1$$
$$R \quad 2$$
$$\text{IF } P_2 \text{ then } C_2$$
$$\cdot$$
$$\cdot$$
$$R \quad n$$
$$\text{IF } P_n \text{ then } C_n$$

The knowledge base is associated with a special matrix called Bit Matrix (BM). BM is a an NxN matrix where N represents the number of rules in the knowledge base, such that

$$BM[i,j] = \begin{cases} 1 \text{ if the conclusion of rule i can be} \\ \text{ involved in the evaluation of rule j} \\ 0 \text{ otherwise} \end{cases} \tag{13}$$

and,

$$\text{if} \quad \underset{j = 1 \text{ to } N}{BM[i,j]} = 0 \text{ then rule i is a conclusion} \tag{14}$$

The bit matrix has 2 functions:

1. It is used by the Question-answer Program (QAP) to trace the inference path when describing how a conclusion was reached.
2. It helps the inference engine to guide the user to reach a conclusion when the data which was provided by the user was not enough.

The bit matrix is created when the inference engine is activated. Next we discuss the evaluation process.

B. THE INFERENCE PROCEDURE

In order to understand the inference procedure, it is necessary to describe new data structures that are part of the inference mechanism. These data structure are two linked lists defined as:

1. A list of fired rules (LR)
2. A list of conclusions (LC)

The general algorithm for the evaluation process is

```
DONE := FALSE;

WHILE NOT DONE DO

BEGIN

    DONE := TRUE;

    I := 1;
```

```
FOR I := 1 TO N DO

IF NOT_IN_LR(I) THEN      { check only the rules which are not in LR }

BEGIN

    GET_PREMISE(ST,I);      { get premise of rule I }

    COMPILE_PREMISE(ST);

    COMPARE(ST,BB,MATCH);      { compare the premise against the blackboard }

    IF MATCH > 0.5 THEN

    BEGIN                              { if a match is found between BB and ST }

        GET_CONCLUSION(CONC,I);        { get conclusion of rule I }

        COMPILE_CONCLUSION(CONC);

        PLACE_CONCLUSION(CONC);      { place conclusion on the blackboard }

        ADD_TO_LR(I);      { add the rule number to LR }

        ADD_TO_LC(I);      { if I is a concluding rule then add I to LC }

        DONE : = FALSE;

    END;

END;

END;
```

In other words, the algorithm as stated allows us to fire as many rules as possible in one cycle. The process will continue until no rules can be fired in one complete pass through the knowledge base. In step 4 the determination is made as to whether or not a rule can be fired. The following is an explanation of the matching process.

The matching process is divided into two steps. First it is necessary to match the key words (the key words are either identical or are synonyms). If there is a match between the key words of the premise and the key words of the compiled data then the evaluation process continues.

Let LB1 and UB1 be the upper and the lower bounds of key word 3 in a sentence from the compiled data, and LB2 and UB2 be the lower and the upper bounds of key word 3 in the compiled premise. Then let M be the result of the matching between the compiled data and the compiled premise.

In order to see how well the compiled data matches the compiled premise it is necessary to examine the degree of overlap between the compiled data and the compiled premise. For convenience we force the result to be in the range [0,1].

There are several cases to consider.

1. If LB1 = UB1 and LB2 = UB2 then we are considering points matching. Thus if the points are identical (LB1 = LB2 or UB1 = UB2) then M = 1; otherwise, M = 0.

2. If LB1 $\neq$ UB1 and LB2 = UB2 then we compare a point to a range. The point is the premise and range is the data. Thus, if the point is not in the range [LB1 − LB2] then M = 0; otherwise, we divide the point by the range. But what will be the value of the point? If we choose the value of the point to be LB2 then we may have

$$LB2 > (UB1 - LB1)$$

and the result of the matching may be greater than 1. Thus, the solution will be to choose one point out of the range of possible points (the range can be found in the characteristic function). So, for example, if the characteristic function of old varies in the range [1..120], then the value of the point we choose will be $\frac{1}{120}$; and if the characteristic function of the variable *weight* is in the range [1..280], then we choose the value of the point to be $\frac{1}{280}$, and so on. But, if there is no characteristic function for the variable in the premise, we choose the value of the point to be $\frac{1}{100}$. Thus, the value of the point is

$$P = \frac{1}{\text{largest number the variable can accept*}}$$

and the result of the matching will be

$$M = \frac{P}{UB1 - LB1}$$

3. If LB1 = UB1 and LB2 $\neq$ UB2, then we compare a point with a range. This time the point is the data and the range is the premise. If the point is in the range [LB2..UB2] then M = 1 otherwise M = 0.

4. If LB1 $\neq$ UB1 and LB2 $\neq$ UB2 (the common case) then:
   If the ranges do not intersect then M = 0.
   If LB1 $\geq$ LB2 and UB2 $\geq$ UB1 (the range of the compiled data is included in the range of the compiled premise) then M = 1.
   If LB2 $\geq$ LB1 and UB1 $\geq$ UB2 (the range of the compiled premise is included in the range of the compiled data) then

$$M = \frac{UB2 - LB2}{UB1 - LB1}$$

   If UB1 $\geq$ UB2 and LB1 $\geq$ LB2 (the ranges intersect and the range of the compiled data is *higher*** than the range of the compiled premise) then

$$M = \frac{UB2 - LB1}{UB1 - LB1}$$

* This number can be found in the proper characteristic function.

** See Definition 1.

Here we divide the intersection of the two ranges by the range of the compiled data. That means that M is an indication of how much from the compiled data is matched against the range of the compiled premise.

IF $UB2 \geq UB1$ and $LB2 \geq LB1$ (the ranges intersect and the range of the compiled premise is *higher** than the range of the compiled data) then

$$M = \frac{UB1 - LB2}{UB1 - LB1}$$

Again, by dividing the intersection with the range of the compiled data, we find the degree to which the compiled data matches the compiled premise.

As mentioned earlier, the premise may contain the key words *not, and,* or *or*. This means that the premise may be composed of more than one part (clause). In this case the matching is done separately for each part and the results are combined such that:

$$M(A) \text{ OR } M(B) = MAX(M(A) , M(B)) \tag{15}$$

$$M(A) \text{ AND } M(B) = MIN(M(A) , M(B)) \tag{16}$$

and

$$NOT (M(A)) = 1 - M(A) \tag{17}$$

When the premise is evaluated (for each rule) then

$$CF = M \tag{18}$$

Let α be some threshold then if $CF > \alpha$ then rule I is fired and the compiled premise is placed on the blackboard.

K1	K2	K3	LB	UB	CF	RN
Key word 1	Key word 2	Key word 3	LB1	UB1	CF	I

When the inference procedure has finished (no more rules can be fired), the conclusion(s) is printed along with its certainty factor.

Once all inferences have been made and all conclusions stated, the user may elect to ask questions of the expert system. The most important question that the user can ask is, ''How did you arrive at conclusion I?.'' The question-answer dialogue is conducted via the Question-Answer Program (QAP) and is explained in the next section.

---

*   See Definition 1.

V. THE QUESTION-ANSWER PROGRAM

The Question-Answer Program (QAP) is used to retrieve the chain of firings in the inference procedure and display all rules and data that were involved in the firing of that rule.

The user may invoke the program by typing:

explain rule n

where n is any number. First the QAP checks to ensure that rule n is in the list of fired rules (LR). If the rule was not found in the LR, then the program is terminated (after an error message is displayed to the user). If the rule number is found in the list of fired rules, then the process continues.

The QAP uses the bit matrix (BM) to retrieve rules and data. As was mentioned (Equation 13),

$$BM\ [I,J]\ =\ 1$$

means that rule I may contribute to the firing of rule J. But it also means that rule J can be fired from rule I. Using this logic, we can derive a general recursive algorithm for retrieving the rules and data which were involved in the firing of the rule in question. The algorithm is presented in a Pascal-like language.

```
PROCEDURE conclusion (k); {k is the rule number to be retrieved}
FOR I = 1 to n do   {n is the number of rules in the knowledge base}
IF BM[I,K] = '1' THEN
BEGIN
  find out if rule I is in LR;   {if rule I was used in firing rule K}
  if so then
  BEGIN
    print the conclusion of rule I;
    activate procedure → conclusion(I);
  END;
END;
```

If nothing is found then conclude that the system has inferred that conclusion K came from user-supplied data.

Since the preceding procedure is activated recursively, it is guaranteed to find all the rules and data which were involved in the firing of rule K.

VI. CONCLUSION

The expert system discussed here is a general purpose expert system. In other words, the expert system is not domain dependent. The user can provide knowledge about any domain (in the form discussed in Section IV.A.) and the expert system should perform as described. The main achievements in this work are:

1.   The ability to interpret mathematically fuzzy variables (and, thus, use natural language in the front-end of the expert system)
2.   The use of heuristics in pattern matching in the inference procedure
3.   The use of the bit matrix eliminating the need for employing forward and backward reasoning

The application program was written in Pascal and we feel that it fully satisfied our goals in writing the expert system.

APPENDIX A.

Example of mapping table

for the variable PEOPLE		
ADJECTIVE	LB	UB
almost	x - 10%	x - 1
more or less	x - 10%	x + 10%
over	x + 1	x + 10%
much more then	2x	∞

APPENDIX B.

Examples of characteristic function

For the variable *Old*

$$\chi(x) = \begin{cases} 0 & \text{if } x \leq 0 \\ \dfrac{x}{120} & \text{if } x \leq 120 \\ 1 & \text{otherwise} \end{cases}$$

For the variable *Weight*

$$\chi(x) = \begin{cases} 0 & \text{if } x \leq 0 \\ \dfrac{x}{250} & \text{if } x \leq 250 \\ 1 & \text{otherwise} \end{cases}$$

REFERENCES

1. **Zadeh, L. A.,** Fuzzy sets, *Inf. Control,* 8, 338, 1965.
2. **Kandel, A. and Byatt, W. J.,** Fuzzy sets, fuzzy algebra, and fuzzy statistics, Proc. IEEE, 66 (12), 1619, 1978.
3. **Kandel, A.,** *Fuzzy Techniques in Pattern Recognition,* Wiley-Interscience, New York, 1982.
4. **Kandel, A.** *Fuzzy Mathematical Techniques With Application,* Addison-Wesley, Reading, MA, 1986.
5. **Schneider, M., and Kandel, A.,** Properties of the Fuzzy Expected Value and the Fuzzy Expected Interval, *Int. J. Fuzzy Sets Syst.,* 26, 1988.
6. **Schneider, M. and Kandel, A.,** Properties of the Fuzzy Expected Value and the Fuzzy Expected Interval in Fuzzy Environment, *Int. J. Fuzzy Sets Syst.,* 28, 1988.
7. **Negoita, C. V.,** *Expert System and Fuzzy Systems,* Benjamin/Cummings, Menlo Park, CA, 1985.
8. **Weiss, S. M. and Kulikowski, C. A.,** *A Practical Guide to Designing Expert Systems,* Rowman and Allanheld, Totowa, NJ, 1984.
9. **Hayes-Roth, F., Waterman, D. A. and Lenat, D. B.,** *Building Expert Systems,* Addison-Wesley, Reading, MA, 1983.
10. **Waterman, D. A.,** *A Guide to Expert Systems,* Addison-Wesley, Reading, MA, 1985.

Chapter 3

INFERENCES WITH INACCURACY AND UNCERTAINTY IN EXPERT SYSTEMS

Bernadette Bouchon-Meunier

TABLE OF CONTENTS

I. INTRODUCTION

The data base of a knowledge-based system contains inaccuracies and uncertainties which are inherent in the description of the rules given by the expert. They are due to the difficulty of representing the facts involved in the antecedents and the consequents of the inference rules, which are expressed in most cases by ambiguous characterizations (such as a color or an age described as "young"), or by imprecise data, (for instance, a length described as "approximately equal to 15 meters"). Uncertainties appear, particularly when the expert is not certain of the validity of the rule in any cases. Another problem stems from the utilization of these rules when the observed facts are not identical with the condition expressed in their premises, but are not too different from them. In a process using classical logic, the rule would not work in this case, but the fact is sometimes so close to the characterization indicated in the premise of a rule that it seems interesting to obtain a deduced fact, even if we must restrict its validity through a coefficient of uncertainty. Let us consider the following example:

Rule 1 (R1): IF the size of the object is small THEN the price is high.

An ambiguity stems from the meaning of *small* and *high,* which may change from an expert to the other one and which is rather difficult to determine precisely, and from the fact that the size may be evaluated with an inaccuracy. Another linguistic problem comes from modifiers such as *rather* or *very* which can be introduced in such a way that the fact "the size is rather small", for instance, will not be considered as sufficient to make the rule (R1) work directly. A coefficient of uncertainty comes from the doubt of the expert giving the rule about its validity.

Various techniques allow us to cope with the uncertainties or inaccuracies inherent in the evidence. We can think of using the semantic meaning of the words by means of natural languages. We can also use nonmonotonic logic that allows us to make hypotheses and to infer conclusions from incomplete information. The particular case of modal logic takes into account expressions such as "it is possible that the price is high", leading to noncertain, but plausible, conclusions. A coefficient may also be associated with the fact, expressing the certainty of the observer with regard to this evidence.

A simple method for dealing with ambiguities in the knowledge base consists of splitting the information contained in the imprecise rule into several crisp ones, in order to decompose the possible cases summarized in the imprecise rule. For example, R1 can be replaced by several crisp rules, such as:

1. If the size is less than 1 cm, the price is more than 30 units.
2. If the size is between 1 and 3 cm, the price is between 20 and 30 units.
3. If the size is more than 3 cm, the price is less than 20 units.

It is obvious that such a splitting of the intervals of size and price is somewhat arbitrary and the conclusion which is deduced at the limit points of the intervals can be discussed. It is possible to attenuate these crisp transitions by using the description of linguistic variables.[13]

The only way to manage both uncertainties and inaccuracies is by the utilization of possibility theory and fuzzy logic. We introduce, in the next section, the main methods for managing uncertainties and inaccuracies in such a way, but this domain is very rich and research is intensive.[7]

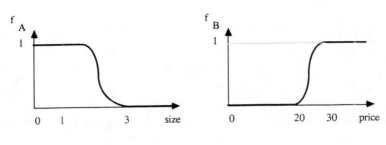

FIGURE 1.

II. FUZZY IMPLICATIONS

A. FUZZY INFERENCES

We consider propositions such as "X is A" or "Y is B", where X and Y are linguistic variables (such as the size, the color, the temperature, the nature, . . . of an object or an element) defined on given universes U_X and U_Y. A and B, respectively, are characterizations (for instance, short, blue, warm, mammalian, . . .) of the variable X (and Y, respectively) represented by a fuzzy subset of U_X (and U_Y, respectively) with membership function f_A (and f_B, respectively) defined on U_X (and U_Y, respectively) and lying in [0,1], as indicated in Figure 1. The rules describe fuzzy relations between two such propositions, and they take the form "if X is A then Y is B".

In Boolean logic, propositions "X is A" and "Y is B" would be considered as false or true, depending on the value 0 or 1 associated with them, and the implication expressed in the rule would take the value corresponding to $\neg$(X is A) $\lor$ (Y is B).

In the case of fuzzy logic, the propositions are considered satisfied with a degree $f_A(x)$ and $f_B(y)$ in every point x and y of the universes. They are not satisfied if the value is 0, completely satisfied if it equals 1, and intermediate values of satisfaction are possible. A possibility distribution can be associated with any X when characterized by A, by assigning to every element x of U_X a degree of possibility $p_X(x)$ equal to $f_A(x)$.[14]

The implication is defined, in the case of fuzzy logic, in several ways which lead to various definitions of the value r(x,y) representing the strength of the inference that involves the facts "X is A" and "Y is B", for any pair (x,y) of elements of the universes U_X and U_Y. These definitions present analogous properties and they can be compared through the behavior of r(x,y) with regard to x or y, according to probable forms of f_A or f_B. They present both advantages and difficulties and, in most cases, the user of the knowledge-based system must make a choice in the probable properties of the fuzzy implications, indicating which ones he considers essential or of minor importance.

The value of the strength of the deduction rule is principally useful to determine the fact which will be obtained as a conclusion of such a rule when the observation which enables us to use it is somewhat different from the premise of the rule, as will be explained in Section III.

B. DEFINITIONS OF THE MAIN FUZZY IMPLICATIONS

There exist several ways of introducing fuzzy implications, and general classes of such quantities may be exhibited.[1,8-10] The most classical forms, which have been used in the practical realization of knowledge-based systems, are indicated here, through the value r(x,y) of the implication involving premise "X is A" and conclusion "Y is B", $\forall$ x $\in$ U_X, $\forall$ y $\in$ U_Y:

$$r^1(x,y) = 1 - f_A(x) + f_A(x) f_B(y) \qquad \text{(Reichenbach implication)}$$
$$r^2(x,y) = \max(1 - f_A(x), \min(f_A(x),f_B(y))) \qquad \text{(Willmott implication)}$$

$r^3(x,y) = \min (f_A(x), f_B(y))$ (Mamdani implication)

$r^4(x,y) = 1$ *if* $f_A(x) \leq f_B(y)$ *and* 0 *otherwise*   (Rescher-Gaines implication)

$r^5(x,y) = \max (1 - f_A(x), f_B(y))$ (Kleene-Dienes implication)

$r^6(x,y) = 1$ *if* $f_A(x) \leq f_B(y)$ *and* $f_B(y)$ *otherwise*   (Brouwer-Gödel implication)

$r^7(x,y) = \min (f_A(y)/f_B(x), 1)$ *if* $f_A(x) \neq 0$ *and* 1 *otherwise*   (Goguen implication)

$r^8(x,y) = \min (1 - f_A(x) + f_B(y), 1)$ (Lukasiewicz implication)

$r^N(x,y) = \max (N(f_A(x)), f_B(y))$

For a strong negation N defined by $N(u) = f^{(-1)}(f(0) - f(u))$, where f is the additive generator of a triangular norm F,[6,10] which can be written:

$$F(u,v) = f^{(-1)} (f(u) + f(v)), u \in [0,1], v \in [0,1]$$

with regard to the continuous, strictly decreasing function f defined on [0,1] and lying on R^+, so that $f(1) = 0$ and $f(0) < +\infty$, with $f^{(-1)}(u)$ equal to $f^{-1}(u)$ if $0 \leq u \leq f(0)$, to 0 if $u \geq f(0)$, and to 1 if $u \leq 0$. A particular case corresponds to $i = 5$, with $N(u) = 1 - u$, $f(u) = 1 - u$.

$$r^F(x,y) = f(f_A(x), f_B(y))$$

where *f* is the quasi-inverse of a t-norm F, defined by:

$$f(u,v) = \max \{w \in [0,1], F(u,w) \leq v\}$$

Particular cases correspond to $i = 6$ with $F(u,v) = \min (u,v)$, $i = 7$ with $F(u,v) = uv$, $f(u) = -\log u$, and $i = 8$ with $F(u,v) = \max (u + v - 1, 0)$, $f(u) = 1 - u$.

It must be noted that r^3 does not satisfy the same properties as the other ones and cannot be considered as a real fuzzy implication, but it has been included in the list because it has actually been used to quantify the strength of fuzzy inference rules.

In order to compare some of their properties with those of the Boolean implication we denote this one as follows:

$r^B(x,y) = 1$ *if* $f_A(x) = 0$ or $f_B(y) = 1$, *and* 0 *if* $f_A(x) = 1$ and $f_B(y) = 0$ with $U_X = \{x\}$ and $U_Y = \{y\}$ and f_A and f_B ranging in $\{0,1\}$ in this case.

C. PROPERTIES OF THE IMPLICATIONS IN CRISP SITUATIONS

The previous definitions give more or less equivalent results, but some differences appear when we consider simple situations, which can be interpreted in different ways by the specialist constructing the knowledge base.[2,3,11] His choice of one of them will essentially depend on the kind of situations he comes upon and the behavior of the implication he thinks interesting.

1. Unambiguous (Crisp) Premise ''X is A''

A is characterized by a membership function f_A equal to 1 at some point x_1 of U_X and 0 elsewhere. We can consider the following example:

Rule 2 (R2): IF the size is 1 cm, THEN the price is high.

Almost all the fuzzy implications present the same behavior, analogous to those in Boolean logic:

- They give a value of $r^i(x_1,y)$ equal to $f_B(y)$, in every case including the Boolean one, except $i = 4$.

- If x is different from x_1, the value of $r^i(x,y)$ is generally 1 and the only exception concerns the case i = 3, yielding 0.

This situation is important when we consider a fact matching more or less the premise of the rule. It means that no conclusion will be available when the observation is not identical with A. Further remarks will be given in Section III.

2. Unambiguous (Or Crisp) Conclusion "X is B"
The membership function f_B equals 1 at one point y_0 of U_Y and 0 elsewhere. Such a situation corresponds, for instance, to the following rule, with a very precise conclusion:

Rule 3 (R3): IF the size is small, THEN the price is 50 units.

The preceding fuzzy implications present behavior similar to those in Boolean logic, except for i = 2, 3.
The strength $r^i(x,y_0)$ of the rule equals 1 for almost all the implications, as it would be the case in Boolean logic. The only exception concerns i = 2 and 3 which yield values depending on the characterization A of the premise, privileging particular points of U_X.
In any y different from y_0, we obtain several types of values:

- The value of $r^i(x,y)$ is a nonincreasing function of $f_A(x)$, as in the Boolean case, for i = 1, 2, 5, 8.
- For i = 3, 4, 6, 7, $r^i(x,y)$ has a different behavior and, in particular, equals 0 for any x in U_X such that $f_A(x) \neq 0$.

This means that such points will not be taken into account any more in further utilizations of the conclusion of the rule.

3. Crisp Values of the Antecedent "X is A" and the Consequent "Y is B" of the Rule
To conclude this part, we can remark that in the crisp situation defined for Boolean implication ($U_X = \{x\}$ and $U_Y = \{y\}$ and f_A and f_B ranging in $\{0,1\}$), all the indicated implications behave in the same way as the Boolean one, according to the results given in Table 1, except i = 3. In the latter case, the implication holds in the only case where both antecedent and consequent hold.

D. PROPERTIES OF THE FUZZY IMPLICATIONS FOR FUZZY SITUATIONS
In the general case where antecedent and conclusion are not crisp, a discontinuity appears for certain values of U_X and U_Y, as a consequence of the noncontinuity existing in the definitions of the implications themselves.
More precisely, the values of the fuzzy implications defined by i = 4, 6 or 7 change suddenly from 0 to 1 when $f_A(x)$ tends to 0, if $f_B(y) = 0$. Other cases of discontinuity may be remarked on Table 2, for instance, for some fuzzy implications defined as r^F, according to the t-norm F.
Extending the remarks concerning the case where antecedent and consequent are crisp, we examine the conditions equivalent to a value of the fuzzy implication equal to 0 or 1, which will be important in a generalization of the classical modus ponens in Section III.
We observe three types of conditions, indicated in Table 3. The first type is identical with the Boolean case (for i = 1, 5, N). The second type is an extension of the previous one (for i = 2, 4, 6, 7, 8, F), and the last type (for i = 3) is quite different from those obtained with B.

TABLE 1
Values of the Fuzzy Implications for Any $x \ \varepsilon \ U_X$, $y \ \varepsilon \ U_Y$, in Crisp Situations

i	A is crisp		B is crisp		A and B crisp
	$r(x_1, y)$	$r(x,y)$ $x \neq x_1$	$r(x,y_0)$	$r(x,y)$ $y \neq y_0$	$r(x,y)$
B,1,5,8	$f_B(y)$	1	1		
2			$\max (f_A(x), 1 - f_A(x))$	$1-f_A(x)$	0 if $f_A(x) = 1$ and $f_B(y) = 0$
4	1 If $f_B(y) = 1$ 0 Otherwise		1	1 If $f_A(x) = 0$ 0 Otherwise	1 if $f_A(x) = 0$ or $f_B(y) = 1$
6, 7	$f_B(y)$				
N, F				$N(f_A(x))$	
3		0	$f_A(x)$	0	0 if $f_A(x)$ or $f_B(y) = 0$ 1 if $f_A(x) = f_B(y) = 1$

TABLE 2
Cases of Discontinuity of $r^i(x,y)$ for Any $x \ \varepsilon \ U_X$, $y \ \varepsilon \ U_Y$

$i = 4, 6, 7, F$ (possibly)	$f_A(x) \rightarrow 0$ and $f_B(y) = 0$
$i = 4, 6, F$ (possibly)	$f_B(y) \neq 1$ and $f_A(x) \rightarrow f_B(y)^+$
$i = 4$	$f_B(y) \rightarrow 1$ and $f_A(x) = 1$

To conclude, we can say that the specialist must adopt a definitive position to choose a fuzzy implication in a knowledge-based system. Either he prefers fuzzy logic that has the same behavior as Boolean logic with regard to the criteria we have presented before (for instance, $i = 1, 5, N$) or he prefers a fuzzy implication that generalizes Boolean properties, which seem understandable since fuzzy logic involves less strict data than the classical one (for instance, $i = 4, 6, 7, 8, F$). Otherwise he prefers a strength of the fuzzy implication monotonous with regard to the value of f_A; this is the case for $i = 3$.[3]

III. COMPATIBILITY OF FACTS WITH THE PREMISE OF A RULE

Let us consider, for instance, rule (R1) which supposes a characterization of *small* for the size to be directly used. In the case of classical logic, an observation of the size expressed as "the size is 1.5 cm" (F1) or *the size is rather small* (F2), described in Figure 2, would not allow us to get a deduction from this rule. This is because the inference process is

TABLE 3
Necessary and Sufficient Conditions to Obtain a
Value of the Fuzzy Implication r(x,y) Equal to 0
or 1, for Any x ε U_X, y ε U_Y

i	Value = 0	Value = 1
B,1,5,N	$f_A(x) = 1$ and $f_B(y) = 0$	$f_A(x) = 0$ or $f_B(y) = 1$
2		$f_A(x) = 0$ or $f_A(x) = f_B(y) = 1$
8		$f_A(x) \leq f_B(y)$
4	$f_A(x) > f_B(y)$	
6,7,F	$f_A(x) \neq 0$ and $f_B(y) = 0$	
3	$f_A(x) = 0$ or $f_B(y) = 0$	$f_A(x) = f_B(y) = 1$

generally based on a modus ponens working in such a way that "if X is A then Y is B" and "X is A" holds, implies that "Y is B" holds.

In the case of fuzzy logic, a generalized modus ponens has been introduced[15] and takes into account both the rule "if X is A then Y is B" and an evidence "X is A'", where A' is identical to A or not, yielding a conclusion "Y is B'" with a characterization B' which can be different from B. The observation is described by means of a membership function $g_{A'}$ defined on U_X and lying in [0,1]. The obtained conclusion is given through a membership function $g_{B'}$ defined on U_Y and also lying in [0,1], taking into account the strength of the fuzzy implication r and the description of the observation through $g_{B'}$.

Obviously, if A' is too different from A, the given rule cannot provide any interesting result, and the conclusion "Y is B'", which is deduced, is weighted with a coefficient of uncertainty. This coefficient equals 1 when the doubt is absolute and the obtained result of the inference cannot be used any more to match the premise of any other rule, even in a generalized modus ponens process.

A combination law is necessary to determine $g_{B'}$ from r and $g_{B'}$. Once more, several possibilities exist, determined in a general way for a given operation G defined and ranging on [0,1] as follows:

$$g_{B'}(y) = \max (G(g_{A'}(x), r^i(x,y))), x \in U_X \quad \text{for any } y \in U_Y$$

It is classical to consider a t-norm for G, because of its satisfying properties.[6] The most commonly used is the minimum. Associated with the fuzzy implication of Lukasiewicz r^8, it was proposed to define a generalized modus ponens process,[15] but it must be noted that this expression may give a result $g_{B'}(y)$ different from $f_B(y)$ in particular cases, even if the

TABLE 4

Some Combination Operators G Defining a Generalized Modus Ponens Compatible with the Modus Ponens, for Every Fuzzy Inference r^i, and Value of the Corresponding Indetermination

i	G	$g_{A'} \leq f$	$g_{A'} \geq f$	A' crisp / A crisp ≠ A'	A' crisp / A fuzzy
1,2	$G(a,b) = \max(a + b - 1, 0)$	$g_{B'} = f_B$	$g_{B'} \geq f_B$	$g_{B'}(y) = 1 \wedge y$	$H = 1 - f_A(x'_1)$
3	$G(a,b) = \min(a,b)$		$g_{B'} = f_B$	$g_{B'}(y) = 0 \wedge y$	$H = 0$
5,8	$G(a,b) = \max(a + b - 1, 0)$		$g_{B'} \geq f_B$	$g_{B'}(y) = 1 \wedge y$	$H = 1 - f_A(x'_1)$
6	$G(a,b) = \min(a,b)$				$H = 1$ if $f_A(x'_1) = 0$; 0 otherwise
4	$G(a,b) = a\, b$	$g_{B'} \leq f_B$			
7	$G(a,b) = a\, b$	$g_{B'} = f_B$			
N,F	The associated t-norm F				$H = N(f_A(x'_1))$

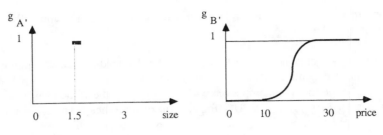

FIGURE 2.

observation is exactly described as in the antecedent of the rule. This is the case, for instance, if U_X is not continuous and there exists x in U_X so that $f_A(x) > f_B(y)$ and $f_A(x) \neq 1$.

A property (P) which seems reasonable to require from the combination law is the preservation of the conclusion of the rule when the observation coincides with its premise. It can be proved[2] that for G the choice of the t-norm F associated with the definition of the fuzzy implication (r^N or r^F, for instance, and consequently r^i, for $5 \leq i \leq 8$) ensures that requirement (P) is satisfied. Obviously, the choice of G is not unique and several combination operators may satisfy (P) for a given fuzzy implication r^i. Similarly, it is possible to exhibit various implications compatible with any combination law associated with an operation G.

It can be stated[3] that the combination law involving the t-norm G defined by $G(a,b) =$ max $(a + b - 1,0)$ for a and b in $[0,1]$, satisfies (P) for any previously-mentioned fuzzy implication r^i for every $i = 1, 2, 3, 5, 6, 7, 8$:

$$g_{B'}(y) = \sup (\max (g_{A'}(x) + r^i(x,y) - 1,0)), x \in U_X \quad \text{for any } y \in U_Y$$

The two other best known t-norms defined by $G(a,b) = $ min (a,b) and $G(a,b) = a,b$, respectively, for a and b in $[0,1]$ are also satisfying with regard to (P), with $i = 3, 6$ and $i = 3, 6, 7$, respectively.

Reasonable arguments are necessary to choose a combination law, according to the characterization of the conclusion when the observation (F1 or F2, for example) is not ambiguous or less precise than the premise of the rule.

The conclusion "Y is B' " depends on the compatibility of the premise "X is A" and the observation "X is A' ". The lowest value of the membership function $g_{B'}$ should be almost equal to 1 if the uncertainty is important, which means that A' is very different from A. On the contrary, $g_{B'}$ must be almost identical with f_B if A' and A are very similar.

The following properties hold with the examples of t-norms G satisfying requirement (P) indicated.

A. A' IS MORE SPECIFIC THAN A
This means that $g_{A'}(x) \leq f_A(x) \; \forall \; x \in U_X$. Then B' is generally also more specific than B: $g_{B'}(y) \leq f_B(y), \; \forall \; y \in U_Y$. Further, if we choose $i = 1, 2, 3, 5, 6, 7, 8$, or $i = N$ or F and their associated t-norm for G, then we get B' identical with B.

B. A' IS LESS SPECIFIC THAN A
This means that $g_{A'}(x) \geq f_A(x) \; \forall \; x \in U_1$. Then B' is also less specific than B: $g_{B'}(y) \geq f_B(y), \; \forall \; y \in U_Y$ if $i = 1, 2, 5, 6, 7, 8$, F, and $i = 3$ gives B itself as a conclusion.

C. A' IS NOT AMBIGUOUS
Its membership function g_A is 0 in every point of the universe U_X different from a given one, e.g., x'_1, where its value is 1 (example F1). Then any t-norm G gives a conclusion defined by $g_2(y) = r^i(x'_1,y)$.

If the premise of the rule itself is crisp with $f_A(x_1) = 1$ (example R2), then two probabilities may occur:

- *Either* $x_1 = x'_1$ and, according to Table 1, the deduced evidence coincides with the conclusion of the rule in all the cases, except $i = 4$
- *or* $x_1 \neq x'_1$, the fact is incompatible with the rule, and the value of $g_{B'}(y)$ is generally 1 (except for $i = 3$), $\forall\, y \in U_Y$, for any fuzzy inference, expressing that the indetermination is maximum, and every element of U_Y is probable.

The choice of $i = 3$ entails that no element of U_Y can be taken into account in a further utilization of the conclusion of the rule, since $f_B(y) = 0$, $\forall\, y \in U_Y$.

Two points of view are then possible. Either we consider that the incompatibility of the observation and the rule entails a 0 probability associated with any point of U_Y, no inferred fact being available, even with a certain indetermination, or we express that this misfit of the observation and the premise of the rule entails a complete uncertainty on the conclusion and every point of U_Y remains probable. It is clear that the choice of an implication and a combination law depends on the point of view which is preferred.

If the premise of the rule is fuzzy (example R1), in the case where $f_A(x'_1) = 0$, the fact is incompatible with the rule, and the previous remarks are still valid. In the case where $f_{A'}(x_1) \neq 0$, we can observe any point y of U_Y with a 0 membership function $g_{B'}(y)$, which means that there is no probability of y being concerned in the conclusion of the rule.

D. INDETERMINATION

An indetermination H appears when the two following facts occur at the same time: some elements y of U_Y are not probable with regard to the characterization of Y expressed in the conclusion of the rule ($f_B(y) = 0$), and the observed evidence is not perfectly identical with the premise of the rule ($g_{A'} \neq f_A$). Then the values of $g_{B'}$ may admit a bottom different from 0, and its height indicates the least probability of an element of U_Y to be involved in the deduced evidence.

The value of this indetermination is 0 for the fuzzy inference r^3. It is

$$\max_{\{x \in U_X / f_A(x) = 0\}} g_{A'}(x) \text{ for } r^i, \text{ with } i = 4, 6, 7$$

It equals

$$\max_{x \in U_X} G(g_{A'}(x), 1 - f_A(x)) \text{ for } r^i, \text{ with } i = 1, 2, 5, 8$$

and

$$\max_{x \in U_X} F(\, g_A(x), N(f_A(x))\,), \text{ for } r^i, \text{ with } i = N \text{ or } F$$

Once more, a position must be adopted by the specialist. Either he chooses a combination law and a fuzzy implication so that an indetermination appears (expressed by the minimum value of $g_{B'}$), implying that the fact deduced from the inference can be used in a further rule if it matches its premise in a generalized modus ponens process; or he prefers a solution so that an element of the universe U_Y is not probable in the deduced evidence if it was not probable in the conclusion of the rule.

IV. MODIFIERS OF THE ATTRIBUTES

Of interest in the use of linguistic variables in the management of uncertainty is the probability of changing the intensity of the characterization of the variable by means of modifiers (very, moderately, . . .), as indicated previously. Their use generates a modification of the probability distributions associated with the attributes.[15] General classes of operations can be defined to construct the probability distribution g representing the modified attribute from the probability distribution f of the attribute itself. For instance, *very* may induce $g(x) = f(x)^2$, for any x in the universe corresponding to the concerned variable, in such a way that g represents *very high* if f is associated with *high*.

An important criterion to choose a fuzzy implication and a combination law for the generalized modus ponens deals with the behavior of the modifiers, when used in a rule. In other words, if an observation is different from the premise of the rule because of the introduction of a modifier in the characterization of the variable, it is interesting to compare the fact inferred by the generalized modus ponens process with the consequent of the rule. We can express some remarks about two classes of modifiers, as follows.

First of all, expansive modifiers, such as *very, strongly, really,* which reinforce the attribute, can be represented by transformations of the possibility distributions so that the characterization A' of the observation is more specific than the characterization A of the variable in the premise of the rule, according to the indications given in Section III. For example, if A is *small* in the rule and A' is *really small* for the observation, the conclusion is generally *high* as indicated by R1, but could be more specific ($g_B \leq f_B$) for i = 4.

On the contrary, restrictive modifiers, such as *not very, moderately, somewhat,* can be represented by transformations of the possibility distribution describing the attribute they modify, so that A' is less specific than A. For instance, if A is *small* and A' is *not very small,* we deduce a conclusion from R1 telling that A' is *pretty high* if we choose i ≠ 3, and B' is B *(high)* otherwise. It seems that a discussion on the interest of a behavior or the other one depends on semantic reasons and also on the context of rule and observation. The expert must again express his preferences for a given knowledge-based system. The study of such properties is not yet finished, and only certain kinds of modifiers have been considered.

V. CONCLUSION

It appears that the management of uncertainty and imprecision is fundamental in knowledge-based systems, and the classical probabilistic methods are not always sufficient, because of the various types of ambiguity involved in rules and observations. Fuzzy logic provides new tools to cope with these problems, but a careful choice of the appropriate implication and combination law for the generalized modus ponens is necessary to avoid unacceptable conclusions.

REFERENCES

1. **Bouchon, B.,** Fuzzy inferences and conditional possibility distributions, *Fuzzy Sets Syst.,* 23, 33, 1987.
2. **Bouchon, B.,** On the forms of reasoning in expert systems, in *Approximate Reasoning in Expert Systems,* Gupta, M. M. et al., Eds., North Holland, Amsterdam, 1985.
3. **Bouchon, B. and Despres, S.,** Propagation of uncertainties and inaccuracies in an expert system, in *Uncertainty in knowledge-based systems,* Lecture Notes in Computer Science no. 286, Springer Verlag, 1987.

4. **Eshragh, F. and Mamdani, E. H.,** A general approach to linguistic approximation, in *Fuzzy Reasoning and Its Applications,* Mamdani, E. H. and Gaines, B. R., Eds., Academic Press, New York, 1981.
5. **Farinas del Cerro, L.,** Resolution modal logic, *Log. Anal.,* 110, 153, 1985.
6. **Klement, E. P.,** Operations on Fuzzy Sets and Fuzzy Numbers Related to Triangular Norms, 11th Int. Symp. Multiple-Valued Logic, Oklahoma, 1981.
7. PROC. *Int. Conf. Inf. Process. Manage. Uncertainty Knowledge-Based Syst.* (Conference IPMU), Paris (1986), Urbino (1988), Paris (1990). Selected papers in Lecture Notes in Computer Science no. 286 (1987) and 313, (1988), Springer-Verlag.
8. **Trillas, E. and Valverde, L.,** On indistinguishability and implication, memorandum 82/04, Universitat Politecnica de Barcelona, 1982.
9. **Valverde, L.,** On the structure of F-indistinguishability operators, *Fuzzy Sets Syst.,* 17(3), 313, 1985.
10. **Weber, S.,** A general concept of fuzzy connectives, negations and implications based on t-norms and t-conorms, *Fuzzy Sets Syst.,* 11, 115, 1983.
11. **Whalen, T. and Schott, B.,** Issues in fuzzy production systems, *Intern. J. Man-Machine Stud.,* 19, 57, 1983.
12. **Willmott, R.,** Two fuzzier implication operators in the theory of fuzzy power sets, *Fuzzy Sets Syst.,* 4, 31, 1980.
13. **Zadeh, L. A.,** The concept of linguistic variable and its application to approximate reasoning, *Inf. Sci.,* 8, part I: 199 and part II: 301, 1975.
14. **Zadeh, L. A.,** Fuzzy sets as a basis for a theory of possibilities, *Fuzzy Sets Syst.,* 1, 3, 1978.
15. **Zadeh, L. A.,** The role of fuzzy logic in the management of uncertainty in expert systems, *Fuzzy Sets Syst.,* 11, 199, 1983.

Chapter 4

ON THE REPRESENTATION OF RELATIONAL PRODUCTION RULES IN EXPERT SYSTEMS

Ronald R. Yager

TABLE OF CONTENTS

I. INTRODUCTION

Rule-based expert systems have shown themselves to be a powerful framework for building knowledge systems.[1] The primary knowledge structure in these systems is the production rule which is typified by:[2] If the tracing pattern is asymmetric gamma and the gamma quantity is normal then the concentration of gammaglobulin is within the normal range.

More formally, the structure of these rules is

$$\text{If } V_1 \text{ is } A_1 \text{ and } V_2 \text{ is } A_2 \ldots \text{ and } V_n \text{ is } A_n \text{ then U is B}$$

The typical antecedent element in these rules consists of a variable V_1 and an established, or fixed value, A_1 which must be achieved in order for the rule to fire. We shall call the term V_1 is A_1 a fixed value antecedent condition.

It appears that in trying to model some expert knowledge we may be faced with situations in which we have an antecedent element that requires some relationship between two or more variables be satisfied. That is, rather than simply requiring a variable to attain some established or fixed value, the success of the rule depends upon the attainment of some relationship between variables in the system. Examples of these kinds of situations are

If *the outer pressure is about the same as the inner pressure*

then lower the heat

$$\text{If } V_1 \text{ is } 2 \times V_2 \text{ and } V_3 \text{ is A then U is B}$$

In these examples the underlined portions exemplify the kinds of conditions which we have just mentioned. We shall call these types of conditions relational or variable antecedent elements.

Yager[3] has shown the potential of the theory of approximate reasoning (AR) as a representational scheme for building expert systems. In particular he has indicated its power in dealing with partial matching as necessitated by imprecise values of the A_is as well as imprecision in knowledge about the V_i's. He also indicated its ability to represent complex combining of antecedent conditions in particular cases in which we have to satisfy most of a collection of antecedent conditions. Gupta et al.[4] contains a number of applications of approximate reasoning to expert systems.

In this paper we extend the usefulness of the AR approach by investigating its ability to handle situations in which the antecedent requires the satisfaction of some relationship between variables.

II. APPROXIMATE REASONING

In this section we briefly review some aspects of the theory of approximate reasoning.[3]

Assume V is a variable which takes as its value an element in the set X, called the base set of V. A canonical statement is of the form:

$$\text{V is A}$$

where A is a fuzzy subset of X. The intent of this statement is to indicate that the value of V is an element of the set A. More formally, this statement induces a possibility distribution Π_v on X so that:

$$\Pi_v(x) = A(x)$$

In this statement $\Pi_v(x)$ indicates the possibility that V has assumed the value x.

In the framework of the theory of approximate reasoning a production rule such as:

$$\text{If V is A then U is B}$$

where U has base set Y gets translated into a joint possibility distribution $\Pi_{v,u}$ on $X \times Y$ so that:

$$\Pi_{v,u}(x,y) = 1 \wedge (1 - A(x) + B(y))$$

An alternative translation of these production rules is

$$\Pi_{v,u}(x,y) = (1 - A(x) \vee B(y))$$

A very lively literature exists discussing the relative merits of different interpretations.[5]

More generally, a rule:

$$\text{If } V_1 \text{ is } A_1 \text{ and } V_2 \text{ is } A_2 \ldots V_n \text{ is } A_n \text{ then U is B}$$

gets translated into a possibility distribution:

$$\Pi_{v1,v2,v3, \ldots vn,u} \text{ on } X_1 \times X_2 \ldots \times X_n \times Y$$

so that:

$$\Pi_{v1,v2,v3,..vn,u}(x_1,x_2,..x_n,y) = 1 \wedge (1 - (A_1(x_1) \wedge A_2(x_2)..\wedge A_n(x_n)) + B(y))$$

Assume $P_1, \ldots P_q$ is a collection of propositions which get translated into the possibility distributions $\Pi_1, \ldots \Pi_q$, respectively. The combined effect of all these propositions is a possibility distribution Π on the space Z, which is the cross product of base sets of all the variables appearing in the propositions, so that:

$$\Pi(z) = \text{Min } \Pi_i(z)$$
$$i = 1,\ldots n$$

Assume V is a variable which appears in some of the propositions, and we are interested in using $\Pi(z)$ to obtain the knowledge this system affords us about V. The process used to obtain this information is projection. In particular, if $Z = Z_1 \times Z_2 \times \ldots Z_n$ and Z_1 is the base set of V, then our inferred value to V is

$$\text{V is F}$$

where for any $z_1 \in Z$

$$F(z_1) = \text{Max } \Pi(z)$$
$$z \in \{z_1\} \times Z_2 \times Z_3...Z_n$$

Assume we have a knowledge base consisting of:

$$P_1: \text{if V is A then U is B}$$
$$P_2: \text{V is D}$$

P_1 translates into:

$$\Pi_1(x,y) = A^-(x) \vee B(y)$$

P_2 translates into $\Pi_2(x) = D(x)$
then:

$$\Pi(x,y) = (A^-(x) \vee B(y)) \wedge D(x) = (A^-(x) \wedge D(x)) \vee (B(y) \wedge D(x))$$

In order to find the inferred value of U we get:

$$\Pi(y) = \text{Max}_x \, [(A^-(x) \wedge D(x)) \vee (B(y) \wedge D(x))]$$
$$= \text{Max}_x \, [A^-(x) \wedge D(x)] \vee \text{Max}_x \, [B(y) \wedge D(x)]$$

If D is normal, it has at least one element with membership one:

$$\Pi(y) = \text{Max}_x \, [A^-(x) \wedge D(x)] \vee B(y)$$

We shall denote:

$$\text{Poss}[A^-/D] = \text{Max}_x \, [A^-(x) \wedge D(x)]$$

thus,

$$\Pi(y) = \text{Poss}[A^-/D] \vee B(y)$$

It should be noted that the spirit of the theory of approximate reasoning is very much like that of mathematical programming.

Assume V is a variable which takes its value in X. Let A and B be two normal subsets of X. If $A \subset B$, then the statement **V is A** is more informative than the statement **V is B.** This can be easily seen in the situation in which:

$$A = \{x_1, x_2, x_3\}$$

$$B = \{x_1, x_2, x_3, x_4, x_5\}$$

In the first case we know that V is one of three elements, while in the second it is one of five.

Yager[3] has discussed a measure of specificity to indicate how much information is contained in a set. In particular,

$$Sp(A) = \int_0^1 1/\text{Card} \, (A_\alpha) \, d\alpha$$

where

$$A_\alpha = \{x/A(x) \geq \alpha\} \text{ and Card } (A_\alpha) \text{ is the cardinality of } A_\alpha$$

III. RELATIONS IN PRODUCTION RULES

Assume V, W, and U are three variables taking their values in the sets X, Y, and Z, respectively. Consider the rule:

$$\text{If } \mathbf{V} \text{ is } \mathbf{W} \text{ then } \mathbf{U} \text{ is } \mathbf{B}$$

where we shall assume the base sets of V and W, X, and Y are the same.

In trying to represent this rule in a form amenable to application of the theory of approximate reasoning, the crucial issue becomes that of representing the relational antecedent condition **V is W**. The approach we shall take to representing this type of nonfixed requirement on variables is to replace the condition V is W by a fixed joint canonical statement of the form:

$$(V, W) \text{ is } R$$

where R is a subset on the set $X \times Y$ capturing the desired relationship between V and W. We shall call R the tolerance relationship.

As a first attempt to develop an R to capture the desired relationship we let R be defined as:

$$R(x,y) = 1 \text{ if } x = y$$
$$R(x,y) = 0 \text{ if } x \neq y$$

Thus, our rule can be written as:

$$\text{If } (V,W) \text{ is } R \text{ then } U \text{ is } B$$

This rule translates into a possibility distribution Π_1 with respect to V,W,U over the set $X \times Y \times Z$ so that:

$$\Pi_1(x,y,z) = 1 - R(x,y) \vee B(z)$$

If we have the additional knowledge that:

$$V \text{ is } D \text{ and } W \text{ is } E$$

then we get two more constraining possibility distributions:

$$\Pi_2(x) = D(x)$$
$$\Pi_3(y) = E(y)$$

Conjuncting these three distributions we get the overall distribution

$$\Pi(x,y,z) = D(x) \wedge E(y) \wedge (R^-(x,y) \vee B(y))$$

To get the inferred value of U, which we shall denote as F, we take the projection of Π onto Z, thus:

$$F(z) = \underset{x,y}{\text{Max}} \ \Pi(x,y,z)$$

thus,

$$F(z) = \underset{x,y}{Max} [D(x) \wedge E(y) \wedge R^-(x,y)] \vee B(z)$$

when D and E are assumed normal.
Thus,

$$F(z) = Poss[R^-/(D \cap E)] \vee B(z)$$

What becomes evident in this formulation is that the condition of equality of the two sets D and E does not imply that $F(z) = B(z)$.

Consider the simple case in which:

$$E = D = \{x_1,x_2\}$$

First, we note that:

$$Poss[(R^-/(D \cap E)] = Max_{x,y} [R^-(x,y) \wedge D(x) \wedge E(y)]$$

and that $R^-(x,y) = 1$ if $x \neq y$. Since $x_1 \neq x_2$ and $D(x_1) = 1$ and $E(x_2) = 1$, then:

$$Poss[R^-/D \cap E] = 1$$

thus $F(z) = 1$ for all z and we infer that U is unknown rather equal to B. At first this seems anti-intuitive, but a careful analysis makes it seem reasonable. The statement V is D indicates that V is either x_1 or x_2 but does not tell us which element it is. It just reduces the possible values of V to this smaller set. Similarly, W is E also just tells us that W is either X_1 or X_2. What is important to note is that $E = D$ does not imply that the value of V equals the value of W, but rather that the values of these elements are drawn from the same set. This fact becomes more obvious when we have the data $D = E = X$; here we see that the values of V and W are completely unknown. It can be easily shown that with the requirement of normality on D and E and the current choice for R, the only situation in which we will infer $F = B$, perfectly fires the rule, is if:

$$D = E = \{x\}$$

where x is any element of X. That is, only if D and E are the same singleton will we get B as our inferred value for U. Alternatively, this condition is seen to be that E and D are equal and have the highest specificity.

It can also be easily shown that if there exist two distinct elements x_1,x_2 so that $D(x_1) = 1$ and $E(x_2) = 1$ then we shall always infer $F = Z$; that is, U is unknown.

The choice of R as our tolerance relation is a very restrictive choice. It is highly specific in that it only perfectly fires the rule if E and D are the same singleton sets. We can say that this R is an intolerant tolerance relationship. In many cases of relational antecedent conditions we may prefer a more relaxed requirement. For example, in a control problem, if V and W are two pressures, we may desire that the antecedent condition be satisfied if both variables are very close to each other in value. In order to capture these less restrictive conditions we are forced to redefine the tolerance relationship R which determines the satisfaction of the antecedent condition.

Let R be a relationship on $X \times X$ we can indicate for each pair $x_1, x_2 \in X$, $R(x_1,x_2)$

$\in [0,1]$ as the degree to which x_1 and x_2 can tolerate each other as being possible values for V and W, respectively. The only requirement we make on our selection of R is that there exists at least one pair (x_1,x_2) so that $R(x_1,x_2) = 1$; that is, we require that R be normal. The reason for this requirement is that unless this is made it will not be possible for the rule to ever fire completely. We note that the larger $R(x_1,x_2)$, the more tolerant. The choice of membership of elements in R is a reflection of our desired relationship between V and W.

We recall that a rule:

$$\text{If } (V,W) \text{ is R then U is B}$$

with the data:

$$V \text{ is D and } W \text{ is E}$$

results in an inferred value for U of F where:

$$F(z) = \text{Poss}[R^-/D \cap E] \vee B(z)$$

in which:

$$\text{Poss}[R^-/D \cap E] = \text{Max}_{x,y} [R^-(x,y) \wedge D(x) \wedge E(y)]$$

We note the smaller Poss $[R^-/D \cap E]$, the more specifically we know F. That is, the rule has fired to a higher degree the smaller Poss $[R^-/D \cap E]$. We further note that if $R_1 \subset R_2$, $R_1(x,y) \leq R_2(x,y)$ for all x,y, then $R^-_1 \supset R^-_2$ and, hence;

$$\text{Poss}[R^-_1/D \cap E] \geq \text{Poss}[R^-_2/D \cap E]$$

Thus, we see that the larger the relationship R, the more the rules fired. We shall say R_1 is more tolerant than R_2 if $R_2 \subset R_1$. The most tolerant relationship is $R = X \times Y$. In this case $R^- = \phi$ and the rule always fires perfectly. The least tolerant case consists of the situation in which:

$$R(x^*,y^*) = 1$$
$$R(x,y) = 0$$

for all other. In this case, the rule perfectly fires only if V is $\{x^*\}$ and W is $\{y^*\}$.

While we have concentrated on the situation in which we are trying to capture V **R** W where R is a form of equals ("close", "nearly", etc.), we can also use the same ideas to capture any relationship of the form:

$$f_1(V) \textbf{ R } f_2(W)$$

where f_1 and f_2 are any functions. An example of this would be a situation in which we require:

$$3*V \textbf{ close to } W^2 + 5$$

In this case we would have (V,W) is R' where R' would reflect the degree of tolerance, as well as the structures of f_1 and f_2.

Again, assume we have the rule:

$$\text{If } (V,W) \text{ is } R \text{ then } U \text{ is } B$$

and we have the knowledge:

$$V \text{ is } D \text{ and } W \text{ is } E$$

This situation gives us the inference U is F where:

$$F(z) = \text{Poss}[R^-/D \cap E] \vee B(z)$$

We shall say that the rule has fired to at least degree α if:

$$\text{Poss}[R^-/D \cap E] \leq 1 - \alpha$$

We therefore see that a necessary and sufficient condition for the rule to fire to at least degree α is that for each pair $(x,y) \in X \times Y$

$$R^-(x,y) \wedge D(x) \wedge E(y) \leq 1 - \alpha$$

One question we may ask is, given a value V is D, what are the allowable values for W, so that our rule fires to at least level α?

In order to answer this question we introduce a family of subsets on Y, one for each x, denoted U_x, so that for each pair (x,y) $U_x(y)$ is the largest value of K for which:

$$R^-(x,y) \wedge K \wedge D(x) \leq 1 - \alpha$$

Then any subset E of Y so that:

$$E \subset \cap_x U_x$$

fires the rule to at least degree α.

We further note that if $D(x) \leq 1 - \alpha$, then $U_x(y)$ can be any value, and we still satisfy our requirement. Thus:

$$U_x = Y \text{ if } D(x) \leq 1 - \alpha$$

On the other hand, if $D(x) > 1 - \alpha$, we then require that for each y:

$$R^-(x,y) \wedge U_x(y) \leq 1 - \alpha$$

In this case, if $R(x,y) \geq \alpha$ then we have met the conditions regardless of the value of $U_x(y)$ and, thus, we can make $U_x(y) = 1$. If $R(x,y) < \alpha$, then we require that $U_x(y) \leq 1 - \alpha$. Thus, for $D(x) \geq 1 - \alpha$:

$$U_x(y) = 1 \qquad \text{if } R(x,y) \geq \alpha$$
$$U_x(y) = 1 - \alpha \text{ if } R(x,y) < \alpha$$

Thus, the largest possible value for W for the rule to fire to degree α is A, where:

$$A = \cap_x U_x$$

Since we require E to be normal, it is possible that given a D it may be impossible to fire the rule. This situation occurs if for each y there exists some $x \in X$ where $D(x) > 1 - \alpha$ and $R(x,y) < \alpha$. On the other hand, if there exists $y \in Y$ such that for all $x \in X$ either $D(x) \leq 1 - \alpha$ or $R(x,y) \geq \alpha$, then we can always find a normal W to fire the rule.

We further note that if $D_1 \subset D_2$ then if A_1 and A_2 are the respective maximal values for W for firing to degree of at least α, it follows that $A_2 \subset A_1$. Further, we easily show that if $R_1 \subset R_2$ then under a value V is D, R_2 requires a less specific value for W.

In the preceding situations we have assumed that the variable condition has occurred in the antecedent. In some cases this variable condition may occur in the consequent. Consider the situation in which we have the following rule:

$$\text{If V is A then U is R-related to W}$$

This translates to:

$$\text{If V is A then (U,W) is R}$$

where R is a relationship on $Y \times Z$. This translates to:

$$H = A^- \cup R$$

Assume we have the data:

$$\text{V is D}$$
$$\text{W is E}$$

and we are interested in finding F so that U is F.

In this case:

$$F(z) = \underset{x,y}{\text{Max}} \ [(A^-(x) \vee R(y,z)) \wedge D(x) \wedge E(y)]$$

$$F(z) = \text{Poss}[A^-/D] \vee \underset{y}{\text{Max}} \ [R(y,z) \wedge E(y)]$$

IV. ON THE STRUCTURE OF RELATIONAL RULES

In this section we look at the relationship between relational requirements and fixed requirements.

Assume V, W, and U are three variables taking their values in the sets X, Y, and Z, respectively. Consider the fixed antecedent rule:

$$\text{If V is A and W is B then U is C}$$

where A, B, and C are normal fuzzy subsets of X, Y, and Z, respectively. This rule translates into:

$$H_1 = A^- \cup B^- \cup C$$

If we have the data:

$$V \text{ is } D$$
$$W \text{ is } E$$

then we can infer U is F_1 where:

$$F_1(z) = \underset{x,y}{\text{Max}} \ [(A^-(x) \lor B(y)) \land D(x) \land E(y)] \lor C(z)$$

Consider a relational rule:

$$\text{If } (V,W) \text{ is } R \text{ then } U \text{ is } C$$

where R is a relation on $X \times Y$. This induces

$$H_2 = R^- \cup C$$

With the same data, V is D and W is E, we infer U is F_2 where:

$$F_z(z) = \underset{x,y}{\text{Max}} \ [R^-(x,y) \land D(x) \land E(y)] \lor C(z).$$

Consider the situation when the relation R is defined by:

$$R(x,y) = A(x) \land B(y)$$

This is always possible since the only requirement on R is that it is normal. In this case:

$$R^-(x,y) = A^-(x) \lor B^-(y)$$

and thus $F_2(z) = F_1(z)$.

Thus, we see that any rule involving a pair of fixed antecedent requirements can be represented as a relational rule where $R(x,y) = A(x) \land B(y)$. We note in the case when A = X and B = Y we get $R = X \times Y$ which is the most tolerant relationship. On the other hand, if A and B are singletons:

$$A = \{x^*\}$$

$$B = \{y^*\}$$

then R is the most specific tolerance relation:

$$R = \{(x^*,y^*)\}$$

Next consider the situation in which we have two fixed antecedent rules implying the same conclusion:

$$P_1 \text{ if } V \text{ is } A_1 \text{ and } W \text{ is } B_1 \text{ then } U \text{ is } C$$
$$P_2 \text{ if } V \text{ is } A_2 \text{ and } W \text{ is } B_2 \text{ then } U \text{ is } C$$

Then P_1 induces:

$$H_1 = A_1^- \cup B_1^- \cup C$$

and P_2 induces:

$$H_2 = A_2^- \cup B_1^- \cup C$$

The combined effect of these two rules is

$$H = H_1 \cap H_2$$
$$= (A_1^- \cup B_1^- \cap (A_2^- \cup B_2^-) \cup C$$
$$= \overline{(A_1 \cap B_1)} \cap \overline{(A_2 \cap B_2)} \cup C.$$

We can effectively represent this as:

$$\text{If } (V,W) \text{ is R then U is C}$$

where

$$R = (A_1 \cap B_1) \cup (A_2 \cap B_2)$$

More generally, if we have a collection of n rules, implying the same consequent, of the form:

$$\text{If V is } A_i \text{ and W is } B_i \text{ then U is C}$$

we can represent this in one relational rule:

$$\text{If } (V,W) \text{ is R then U is C}$$

where

$$R = \bigcup_{i=1}^{n} (A_i \cap B_i)$$

$$\text{i.e., } R(x,y) = \text{Max } [A_i(x) \wedge B_i(y)]$$

$$i = 1,\dots n$$

Thus we see that the use of relational type rules allows us to capture in one rule the information contained in a whole family of fixed antecedent rules inferring the same implication. In many cases, when the knowledge about a situation is expressed by an expert in terms of a relational rule, such as "if V is close to W then U is B"; this rule is in reality the expert summarization of a whole family of fixed rules. In many regards this idea of using relational type rules to capture this multiplicity of information typifies the approach of experts in that these rules characterize heuristics useful to analyzing complex situations.

While any combination of normal fixed rules can be represented in terms of a relational type rule, it is not always the case that a normal relational type rule can be expressed in terms of fixed type rules.

Consider the relation:

$$R = \begin{array}{c} \\ x_1 \\ x_2 \\ x_3 \end{array} \begin{array}{ccc} y_1 & y_2 & y_3 \\ \left[\begin{array}{ccc} \alpha_1 & \alpha_2 & \alpha_3 \\ 0 & 1 & 0 \\ 0 & 0 & 0 \end{array} \right] \end{array}$$

so that $1 > \alpha_1 > \alpha_2 > \alpha_3 > 0$.

We desire to find a family of A_is and B_is, each of which is normal, so that:

$$R = \bigcup_{i=1}^{n} (A_i \cap B_i)$$

In order to capture this R we need at least one pair A, B, so that

$$A(x_1) \wedge B(y_1) = \alpha_1$$

Then either $A(x_1) = \alpha_1$ or $B(y_1) = \alpha_1$. Assume $A(x_1) = \alpha_1$ then $B(y_2) \geq \alpha_1$. Since each A must be normal, then either $A(x_2) = 1$ or $A(x_3) = 1$; but if $A(x_2) = 1$, then $R(x_2,y_1) \geq A(x_2) \wedge B(y_1) \geq \alpha_1 > 0$. Similarly for $A(x_3) = 1$. Assume $B(y_1) = \alpha_1$ then $A(x_1) \geq \alpha_1$. Since B must be normal, either $B(y_2)$ or $B(y_3) = 1$. If $B(x_2) = 1$, then $R(x_1,y_2) \geq A(x_1) \wedge B(y_2) \geq \alpha_1 > \alpha_2$; similarly for $B(x_3) = 1$. Thus there exist no A and B which can give us $R(x_1,y_1)$ and, hence, we cannot model R.

Thus we see that the use of relational antecedents allows us to capture a wider range of functions than using a set of fixed rules. On the other hand, there are classes of Rs that can always be modeled as conjunctions of fixed rules.

Theorem. If each x has at least one y such that $R(x,y) = 1$, then there exists a collection of rules of the form:

$$\text{If V is } A_i \text{ and W is } B_i \text{ then C}$$

that is equivalent to:

$$\text{If (V,W) is R then C}$$

Proof. Consider $A_i = \{1/x_i\}$ and $B_i = \{1/y^*, R_{i,j}/y_j\}$ where y^* is such that $R(X_i, Y^*) = 1$. Then $A_i \cap B_i = \{1/(x_i,y^*), R_{i,j}/(x_i,y_j)\}$

$$\bigcup_{i=1}^{n \times n} (A_i \cap B_i) > R$$

Acknowledgment

Research is supported in part by NSF grants IST8503841 and DCR8513044.

REFERENCES

1. **Waterman, D. A.,** *Expert Systems,* Addison-Wesley, Reading, MA, 1986.
2. **Weiss, S. M. and Kulikowski, C. A.,** *A Practical Guide to Designing Expert Systems,* Rowman and Allanheld; Totowa, NJ, 1984.
3. **Yager, R. R.,** *Approximate reasoning as a basis for rule based expert systems,* IEEE Trans. Sys., Man, Cybern., 14, 636, 1984.
4. **Gupta, M., Kandel, A., Bandler, W., and Kiszka, J. B.,** *Approximate Reasoning in Expert Systems,* North-Holland, Amsterdam, 1985.
5. **Yager, R. R.,** An approach to inference in approximate reasoning, *Int. J. Man-Mach. Stud.* 13, 323, 1980.

Chapter 5

REDUCTION PROCEDURES FOR RULE-BASED EXPERT SYSTEMS AS A TOOL FOR STUDIES OF PROPERTIES OF EXPERT'S KNOWLEDGE

Antonio Di Nola, Witold Pedrycz, and Salvatore Sessa

TABLE OF CONTENTS

I. INTRODUCTION

The actual state of the art in designing expert systems, viewed both from methodological as well as applicational points of view, indicates a rapid development of new techniques. It has been observed in new tools applied for knowledge representation schemes (various data and control structures), and in extensive streams of investigations on coping with imprecision being, however, in its infancy. Here we can easily report several approaches diverse in their nature.[1,2] Fuzzy sets are one among them and they are well suited for processing a linguistic form of information; for some examples, we refer again to Reference 1.

Following the way in which fuzzy sets are discussed, we will present one of the problems which arises, while a knowledge base is arranged, and a useful and concise form of any query to a rule base, when a mode of forward reasoning is created.

It is an evident fact that the process of knowledge acquisition structured in *if-then* statements (condition-action format) is an error prone one and it requires a lot of effort to reach a satisfactory status by introducing some modification mechanisms (e.g., conflict detecting).

The actual tendency is to enlarge the size of the knowledge bases, in the sense of the number of the rules, as well as the number of subconditions forming the condition part of the rules. The possibility that an extra effort has to be made to improve properties of a collection of raw empirical data in the rules increases significantly.

It is not surprising that any attempt to make the process of knowledge acquisition more efficient from a conceptual and a numerical point of view deserves special attention.

On the other hand, too many subconditions of the condition part can influence a reduction of the user-friendly property of the expert system and impose a significant computational effort. Since the user is forced to feed the system with information dealing with every condition, this stage, if not well arranged, may lead to the user's conservative behavior. This fact strongly implies studying every subcondition part with respect to its importance for the knowledge base and with respect to its appropriate arrangement to increase the performance of the reasoning mechanisms.

In this chapter we focus our attention on the rule-based expert systems in which the production rules:

$$\text{if } A_i \text{ then } B_i$$

$i = 1,2,\ldots,N$, contain a fuzzy condition, as well as a control (action part). Thus, A_i and B_i are represented as fuzzy relations (or fuzzy sets) in appropriate spaces, and the reasoning mechanism is realized with the use of fuzzy relation equations of the form:

$$B_i = A_i \circ R$$

$i = 1,2,\ldots,N$, where R is constituted by means of the facts contained in the rules. More precisely, R can be obtained by solving this set of fuzzy relation equations formed on the basis of facts collected in the rules. For more details, we refer to reference papers 3—6.

Now it is instructive to have a closer look at the rule-based expert systems working with fuzzy information. As usual, one deals with many subcondition variables on the basis of which the action is performed. Thus, any suitable relationship (given by R) between the condition parts and action parts forms a core (here a fuzzy relation) of the expert system. It means that in practice, for the reasoning stage, the values, fuzzy or nonfuzzy, of all the subconditions have to be provided to generate a hypothesis on the action realized.

It happens, however, that certain subconditions are difficult to evaluate by the user and/

or the reliability of this information is rather low. Therefore, it is reasonable to reduce the condition space by taking only a few subconditions (features) to form a condition subspace. These are the most "significant" ones considered for an action point of view. Nevertheless, such a procedure of reduction may lead to a slight modification of fuzzy actions, but it is an essential price to pay in order to neglect some subconditions. The main point, while the most irrelevant subconditions parts are eliminated, is to achieve a certain balance between imposed changes of the action parts and an achieved reasonable size of the condition space. Notice, however, that the fuzzy relation equation of the reasoning scheme should be modified with regard to the original one, where the entire original conditions space has been utilized. The problem, described here in a very concise form, will be called a reduction problem of the knowledge base.

After the presentation of some underlying ideas, we will then formulate precisely the statement of the problem in terms of fuzzy relation equations and tackle it with suitable reduction techniques. A reconstruction problem will then be discussed to give an overall picture of how the information coming from the reduced knowledge base can be combined, bearing in mind the influence of different levels of difficulty to get reliable results. Moreover, we will present a way in which the fuzzy relation R may not be computed. The evaluation, however, as to whether the reduction is necessary, is realized on the basis of the same set of rules.

An illustrative numerical example forms an additional clarification of the entire presentation. Despite the fact that the entire discussion will be realized in a framework of fuzzy sets, the reader interested in some general concepts shown here may follow the chapter replacing a concept of fuzzy sets and relations by crisp sets and relations. Fortunately, the results derived are also valid.

II. REDUCTION PROBLEM FOR FUZZY RELATION EQUATIONS

Formulating the problem we are going to work on, we first specify the universes of discourse. As mentioned in Section I., the model has inputs (conditions) defined in input spaces (subcondition spaces) denoted by A_1, A_2...,A_N. We assume that they have finite cardinality and denoted by:

$$A_1 = \{a_{11}, a_{12}, \ldots, a_{1n_1}\}$$

$$A_2 = \{a_{21}, a_{22}, \ldots, a_{2n_2}\}$$

.

.

.

$$A_N = \{a_{N1}, a_{N2}, \ldots, a_{Nn_N}\}$$

The same holds with the output space (action space) denoted by B and consisting of a finite collection of elements of cardinality m, namely $B = \{b_1, b_2, \ldots, b_m\}$. All the fuzzy sets A_j: $A_j \rightarrow [0,1]$, $j = 1,2\ldots,N$, describing inputs and output, will be denoted using capital letters and the relationship between them are modeled by a fuzzy relation R defined in the Cartesian product of the set A_j and B, R: $A_1 \times A_2 \times \ldots \times A_N \times B \rightarrow [0,1]$. Then the output and inputs are interrelated, and these ties are given in the form of the fuzzy relation equation:[3]

$$B = A_1 \circ A_2 \circ \ldots \circ A_N \circ R$$

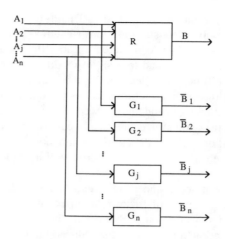

FIGURE 1.   The Fuzzy Relation

that, in terms of membership functions, read accordingly:

$$B(b_k) = \bigvee_{i_1=1}^{n_1} \bigvee_{i_2=1}^{n_2} \ldots \bigvee_{i_N=1}^{n_N} [A_1(a_{1i_1}) \wedge \ldots \wedge A_N(a_{Ni_N}) \wedge R(a_{1i_1}, a_{2i_2}, \ldots, a_{Ni_N}, b_k)] \quad (1)$$

for any k = 1,2,...,m, where $\wedge$ and $\vee$ are the min and max operators of [0,1].

A reduced form of the model Equation 1 is put down as a formula for single input-single output model. For the jth input variable, j = 1,2,...,N, this is given by:

$$\bar{B}_j = A_j \circ G \quad (2)$$

where G is a fuzzy relation defined in $A_j \times B$, G: $A_j \times B \to [0,1]$, cf. Figure 1.

Here we denote the output by $\bar{B}_j$ instead of B underlying that the reduced form of the model may not, and usually does not, generate the same output, namely B. Of course, we intend to determine such fuzzy relation G which allows to obtain the closest results to B. In an ideal situation, we can ask to satisfy an equality B = $\bar{B}$ despite which values of inputs are considered. In more formal fashion, choose such G that the equality:

$$A_j \circ G = A_1 \circ A_2 \circ \ldots A_{j-1} \circ A_j \circ A_{j+1} \circ \ldots \circ A_N \circ R \quad (3)$$

is satisifed for any j = 1,2...,N and for any fuzzy set A_j.

Unfortunately, the previous requirement is too restrictive and the equality that has to be fulfilled for every fuzzy set A_j is even difficult to verify. Therefore, it is reasonable to speak about the model equivalence in Equations 1 and 3 in the sense of an assigned class of inputs. More clearly, we say that the fuzzy models are equivalent with respect to the class of inputs X if the equality of the latter equation holds for any fuzzy set A_j, j = 1,2,...,N, of X, X being an assigned class of inputs.

In the sequel, for simplicity, we put $a_{jij} = a_{ij}$ so that no misunderstanding can arise. Two classes X of inputs are of a significant interest:

(i)   Inputs in the universe A_j are considered as singletons, while the inputs defined in the remaining universes are modeled as "unknown," i.e., their membership functions are equal identically to 1.0 in the entire universe of discourse.

(ii) Similarly, as before, n_j inputs in the universe A_j are viewed as singletons, and the inputs in the remaining universes are also singletons.

Of course, in both cases, Equation 2 is viewed as a system of n_j fuzzy equations, whose solution set is assumed nonempty.

Denoting by $\underline{1}$ the fuzzy set with the membership function equal identically to 1 and by $\overline{G}$, and $\underline{G}$, the fuzzy relations resulting in (i) and (ii), the formulas for the membership functions of $\overline{G}$ and $\underline{G}$ are contained in the following propositions, respectively.

Proposition 1—If the inputs of the Equation 1 are specified as follows:

if $A_i = 1$ for any $i \neq j$, $i,j = 1,2,\ldots,N$,
if n_j input fuzzy sets A_i^t, $t = 1,2,\ldots,n_j$, defined in A_j^t are singletons and such that they are disjoint and completely cover A_j (i.e., their fuzzy union has the membership function equal identically to 1.0),
then the fuzzy relation $\overline{G}$ of the reduced model (2), defined in $A_j \times B$, is given by the membership function:

$$\overline{G}(a_{i_j}, b_k) = \overset{n_i}{\underset{i_1 = 1}{\vee}} \ldots \overset{n_{j-1}}{\underset{i_{j-1} = 1}{\vee}} \overset{n_{j+1}}{\underset{i_{j+1} = 1}{\vee}} \ldots \overset{n_N}{\underset{i_N = 1}{\vee}} R(a_{i_1}, \ldots a_{i_j}, \ldots a_{i_N}, b_k)$$

for any $i_j \in A_j, b_k \in B, i_j = 1,2,\ldots,n_j$

Proof—According to the assumptions made in the statement, we have the following membership function for the output:

$$B(b_k) = \overset{n_1}{\underset{i_1 = 1}{\vee}} \ldots \overset{n_N}{\underset{i_N = 1}{\vee}} \{A_j^t(a_{i_j}) \wedge [A_1(a_{i_1}) \wedge \ldots \wedge A_{j-1}(a_{i_{j-1}})$$

$$\wedge A_{j+1}(a_{i_{j+1}}) \wedge \ldots \wedge A_N(a_{i_N}) \wedge R(a_{i_1}, \ldots a_{i_j}, \ldots a_{i_N}, b_k)]\}$$

$$= \overset{n_j}{\underset{i_j = 1}{\vee}} \{A_j^t(a_i) \wedge [\overset{n_1}{\underset{i_1 = 1}{\vee}} \ldots \overset{n_{j-1}}{\underset{i_{j-1} = 1}{\vee}} \overset{n_{j+1}}{\underset{i_{j+1} = 1}{\vee}} \ldots \overset{n_N}{\underset{i_N = 1}{\vee}}$$

$$R(a_{i_1}, \ldots a_{i_j}, \ldots a_{i_N}, b_k)]\}$$

Since A_j^t, $t = 1,2,\ldots n_j$, are n_j singleton fuzzy sets, i.e., we have $A_j^t(a_{i_j}) = 1$ if $i_j = t$ and $A_j^t(a_{i_j}) = 0$ otherwise, then the membership function of B_t simplifies into:

$$B_t(b_k) = \begin{cases} \overset{n_1}{\underset{i_1 = 1}{\vee}} \ldots \overset{n_{j-1}}{\underset{i_{j-1} = 1}{\vee}} \overset{n_{j+1}}{\underset{i_{j+1} = 1}{\vee}} \overset{n_N}{\underset{i_N = 1}{\vee}} R(a_{i_1}, \ldots a_{i_j} \ldots a_{i_N}, b_k) & \text{if } i_j = t \\ \\ 0 & \text{otherwise} \end{cases} \quad (4)$$

for any $t = 1,2,\ldots,n_j$, $k = 1,2\ldots,m$. Bearing in mind the properties of the fuzzy sets defined in A_j^t and performing computations for any $t = 1,2,\ldots,n_j$, we have the following input-output fuzzy sets of the reduced model:

$$A_j^1 = [1 \ 0 \ \ldots \ 0] \text{ and } B_1$$

$$A_j^2 = [0 \ 1 \ \ldots \ 0] \text{ and } B_2$$

$$\cdot \qquad \qquad \cdot$$

$$\cdot \qquad \qquad \cdot$$

$$\cdot \qquad \qquad \cdot$$

$$A_j^{n_j} = [0 \ 0 \ \ldots \ 1] \text{ and } Bk_{n_j}$$

In accordance to Reference 4, a solution (the greatest) of the system in Equation 2 is obtained realizing the ⓐ-composition[7] of these fuzzy sets and, hence, their fuzzy intersection. Thus, we have:

$$\overline{G} = \bigwedge_{t=1}^{n_j} [A_j^t \ \text{ⓐ} \ B_1]$$

$$= ([1 \ 0\ldots0] \ \text{ⓐ} \ B_1) \wedge ([0 \ 1\ldots0] \ \text{ⓐ} \ B_2) \wedge\ldots\wedge ([0 \ 0\ldots1] \ \text{ⓐ} \ B_{n_j})$$

$$= \begin{bmatrix} B_1 \\ \underline{1} \\ \vdots \\ \underline{1} \end{bmatrix} \wedge \begin{bmatrix} \underline{1} \\ B_2 \\ \vdots \\ \underline{1} \end{bmatrix} \wedge\ldots\wedge \begin{bmatrix} \underline{1} \\ \underline{1} \\ \vdots \\ B_{n_j} \end{bmatrix} = \begin{bmatrix} B_1 \\ B_2 \\ \vdots \\ B_{n_j} \end{bmatrix}$$

Here $\underline{1}$ denotes the membership function equal identically to 1.0 in the entire space B. The result formulated in the proposition follows of course from Equation 4.

Similarly, it is proved that:

Proposition 2—If the n_j input fuzzy sets A_j^t, $t = 1,2,\ldots,n_j$, are singletons and such that in every universe of discourse A_j, $j = 1,2,\ldots,N$, they are pairwise disjointed and completely cover A_j, (in the sense of proposition 1), then the fuzzy relation $\underline{G}$ of the reduced model (2) is given by:

$$\underline{G}(a_{i_j},b_k) = \bigwedge_{i_1=1}^{n_1} \ldots \bigwedge_{i_{j-1}=1}^{n_{j-1}} \bigwedge_{i_{j+1}=1}^{n_{j+1}} \ldots \bigwedge_{i_N=1}^{n_N} R(a_{i_1},\ldots a_{i_j},\ldots a_{i_N},b_k)$$

for any $i_j \in A_j$, $b_k \in B$, $i_j = 1,2,\ldots,n_j$.

III. GENERALIZATION OF PRODUCTION RULES

In this section we will present how, without any calculations of the fuzzy relation of the knowledge base, one can detect a situation in which this reduction procedure may be applied.

As before, we discuss the rule-based system where a knowledge is of a format of condition-action statements. For clarity, we consider two of the rules, the ith and the jth one, $i,j = 1,2,\ldots,N$, $i \neq j$ and we discuss four cases:

1. Both A_i and A_j are very similar (close to each other), (the similarity notion or closeness of fuzzy notions will be explained later); the same holds for actions, e.g., B_i and B_j. Thus, we can say that these two rules are consistent and there is no necessity to modify them.

2. The same holds A_i and A_j, as well as B_i, B_j, are very different.

3. Here $A_i = A_j$ while B_i and B_j differ significantly. We are faced with a case where one has to extend the condition space to get consistency of this pair of rules. In other words, this situation requires a particularization of the rules, as it is shown in Reference 8.

4. This case is widely discussed here: A_i and A_j are different, yielding the same or very similar actions B_i and B_j. Thus, if the same elements of the space of conditions were eliminated, A_i and A_j, redefined in this new reduced space of conditions, would become closer to each other. The phenomenon, already needed to resolve this inconsistency, will be called rule generalization.

Thus, we need a mechanism that allows us to detect the situation 4. At the beginning one has to express precisely how *nearness* of the fuzzy relations (conditions) and fuzzy sets (actions) can be evaluated. Here we may give a lot of approaches; all of them generate indices taking their maximal values for identical fuzzy sets, i.e., equal (but not necessarily) to 1.0, approaching to the minimum (i.e., zero) for completely different fuzzy sets. An example of this index, borrowed by multivalued logic, is the following. We recall that for truth values of two propositions "p" and "q", $p, q \in [0,1]$, a degree, to which extent they are equal to each other, is given by:

$$(p \equiv q) = (p \to q) \,\&\, (q \to p)$$

where "$\to$" is treated as a Gödelian implication "α", namely, $(p \,\alpha\, q) = 1$ if $p \leq q$, $(p \,\alpha\, q) = q$ if $p > q$ and "&" is treated as minimum. Then this expression reads as:

$$(p \equiv q) = (p \,\alpha\, q) \wedge (q \,\alpha\, p)$$

For two fuzzy sets (relations) of a universe of discourse A, we may speak about a degree to which the membership functions of A_i, A_j at some elements $a \in A$ are equal to each other:

$$[A_i(a) \equiv A_j(a)] = [A_i(a) \,@\, A_j(a)] \wedge [A_j(a) \,@\, A_i(a)]$$

In order to obtain a global overview to which extent A_i is equal to A_j, we calculate the values of $[A_i(a) \equiv A_j(a)]$ for all $a \in A$ building an empirical probability distribution function defined as:

$$F_{A_i \equiv A_j}(w) = \operatorname{card} \{a \in A : [A_i(a) \equiv A_j(a)] \leq w\}/\operatorname{card} A$$

that one can utilize, for instance, to calculate:

- an averaged value,

$$[A_i \equiv A_j] = \int_0^1 w \cdot dF_{A_i = A_j}(w)$$

• a model value, i.e., the value of $w \in [0,1]$ for which a probability density function:

$$\frac{dF_{A_i \equiv A_j}}{dw}$$

attains its maximum.

The same computations are performed for the fuzzy sets B_i and B_j. It is easily seen that an expression:

$$\gamma_{ij} = [B_i \equiv B_j] \, \textcircled{a} \, [A_i \equiv A_j]$$

can serve as a suitable index for detecting whether the rules have to be generalized. Note that this index takes quite high values for situations 1—3, and reaches values nearly zero for case 4. Performing the calculations for all the pairs of indices, we arrange the results in a matrix form $\Gamma = [\gamma_{ij}]$, $i = 1,2,...,N$. Obviously, $\gamma_{ii} = 1.0$, $\gamma_{ij} = \gamma_{ji}$ but $N(N - 1)/2$ elements of Γ must be calculated.

Again, to get a global view to which extent the rules viewed as the entire collection should be generalized, i.e., the condition space reduced, we may take any synthetic index, perhaps an average value of γ_{ij}' seems to be a reasonable choice:

$$\gamma_{av} = \frac{1}{N^2} \cdot \sum_{i,j=1}^{N} \gamma_{ij}$$

This value may indicate a need to reduce the condition space. The proposal just presented, also has been studied extensively in Reference 8.

At this point both of the methods given so far can be conveniently treated as complementary ones. The test for rule generalization enables us to detect a situation where a requirement is to reduce the condition space without any explicit knowledge of the fuzzy relation R. Moreover, the second method described here points out which particular rules are responsible for such a status of the knowledge base. Nevertheless, this approach does not indicate clearly which subspaces of the conditions space should be formed. The information necessary for this purpose comes from applying the results of Section II.

IV. RECONSTRUCTION PROBLEM

First, we observe that the obvious inequality $\underline{G} \le \overline{G}$ holds, where it is understood that $\underline{G}(a_{ij}, b_k) \le \overline{G}(a_{ij}, b_k)$ for any $j = 1,2,...,N$ and $k = 1,2,...,m$. For simplicity, we put $\underline{G}(a_{ij}, b_k) = \underline{G}_j$ and $\overline{G}(a_{ij}, b_k) = \overline{G}_j$ for any $j = 1,2,...,N$.

Having at our disposal a collection of reduced models determined, with the aid of Propositions 1 and 2, we may formulate the following problem. Assume that a subset K of indices of the set $\{1,2,...,N\}$ and related inputs are given. Therefore, the following system is known:

$$\begin{cases} A_j \circ \underline{G}_j = \underline{B}_j \\ \\ A_j \circ \overline{G}_j = \overline{B}_j \end{cases}$$

where $j \in K$. Note that $\underline{B}_j \leq \overline{B}_j$ for any $j \in K$ and we deal, not with single membership function B_j, but with an interval-valued fuzzy set, the so-called Φ fuzzy set[9]. Instead of the two preceding equations, we can introduce a shorthand notation:

$$A_j \circ [\underline{G}_j, \overline{G}_j] = [\underline{B}_j, \overline{B}_j] \qquad j \in K$$

It is clear that the reduced fuzzy model generates the interval-valued fuzzy set of output. Then, if several input fuzzy sets are available, the relevant outputs forming Φ fuzzy sets should be combined to get an overall result.

A reasonable concept to aggregate the results of the reduced models from the subset K would work as follows. For any $b_k \in B$, calculate the number of Φ fuzzy sets contributing to the previously fixed values ranging between zero and card K.

By a simple normalization, we reach a fuzzy set of the second order B; this means that to every $b_k \in B$, a fuzzy set defined in $[0,1]$ has been established. More precisely, the value of the membership function B (before normalization) is equal to:

$$B(b_k, w) = \frac{\text{card } \{j \in K: \underline{B}_j(b_k) \leq w \leq \overline{B}_j(b_k)\}}{\text{card } K}$$

for any $w \in [0,1]$, $b_k \in B$.

Remembering the main idea behind the concept of model reduction that relies on diverse grades of difficulty and precision while the fuzzy sets of inputs are determined, the fuzzy set B may be modified to take it into account. A level of difficulty to get reliable data of the corresponding input in A_j is expressed by u_j, $j \in K$; u_j is in $[0,1]$. The higher the value of u_j, the more severe difficulties are manifested which deal with input determination. The results obtained are less reliable and, in consequence, should not significantly contribute to the final result. The weights u_j, $j \in K$, can be easily obtained, e.g., by the well-known Saaty's priority method.

This weight factor is useful to control a width of the interval-valued fuzzy set $[\underline{B}_j, \overline{B}_j]$. Now, instead of the original one, namely, $[\underline{B}_j, \overline{B}_j]$, one has:

$$[\underline{B}^{\mu_j}, \overline{B}^{\mu_i}],$$

where $b_k \in \mathbf{B}$. We illustrate these stated considerations with the help of a numerical example.

V. AN ILLUSTRATIVE EXAMPLE

We consider a two-input, single-output fuzzy model $A_1 \circ A_2 \circ R = B$, where $n_1 = n_2 = m = 3$ is described by the following fuzzy relation:

$$R = \begin{array}{c} \\ a_{11} \\ a_{12} \\ a_{13} \end{array} \underbrace{\begin{bmatrix} a_{21} & a_{22} & a_{23} \\ 0.2 & 0.7 & 0.8 \\ 0.5 & 0.6 & 0.4 \\ 0.9 & 0.0 & 0.7 \end{bmatrix}}_{b_1} \underbrace{\begin{bmatrix} a_{21} & a_{22} & a_{23} \\ 0.4 & 0.3 & 0.9 \\ 0.3 & 1.0 & 0.5 \\ 0.7 & 1.0 & 0.6 \end{bmatrix}}_{b_2} \underbrace{\begin{bmatrix} a_{21} & a_{22} & a_{23} \\ 1.0 & 0.3 & 0.7 \\ 0.4 & 1.0 & 0.2 \\ 0.5 & 0.3 & 0.9 \end{bmatrix}}_{b_3}$$

Construct two reduced fuzzy models making use of the families of the fuzzy sets introduced earlier. Let A_1 be singleton and A_2 be "unknown".

Then we have for the first and second model, respectively:

$$
\overline{G}_1 = \begin{array}{c} a_{11} \\ a_{12} \\ a_{13} \end{array}
\begin{array}{ccc} b_1 & b_2 & b_3 \\ \left[\begin{array}{ccc} 0.8 & 0.9 & 1.0 \\ 0.6 & 1.0 & 1.0 \\ 0.9 & 1.0 & 0.9 \end{array} \right] \end{array}
\qquad
\underline{G}_1 = \begin{array}{c} a_{11} \\ a_{12} \\ a_{13} \end{array}
\begin{array}{ccc} b_1 & b_2 & b_3 \\ \left[\begin{array}{ccc} 0.2 & 0.3 & 0.3 \\ 0.3 & 0.6 & 0.2 \\ 0.5 & 0.0 & 0.6 \end{array} \right] \end{array}
$$

$$
\overline{G}_2 = \begin{array}{c} a_{21} \\ a_{22} \\ a_{23} \end{array}
\begin{array}{ccc} b_1 & b_2 & b_3 \\ \left[\begin{array}{ccc} 0.9 & 0.7 & 1.0 \\ 0.7 & 1.0 & 1.0 \\ 0.8 & 0.9 & 0.9 \end{array} \right] \end{array}
\qquad
\underline{G}_2 = \begin{array}{c} a_{21} \\ a_{22} \\ a_{23} \end{array}
\begin{array}{ccc} b_1 & b_2 & b_3 \\ \left[\begin{array}{ccc} 0.2 & 0.3 & 0.5 \\ 0.3 & 0.6 & 0.0 \\ 0.7 & 0.2 & 0.6 \end{array} \right] \end{array}
$$

To visualize how the reduced model works, we consider two input data A_1' and A_2' so that:

$$
A_1' = \begin{bmatrix} 0.7 & 0.5 & 0.2 \end{bmatrix}
\qquad
A_2' = \begin{bmatrix} 0.3 & 1.0 & 0.4 \end{bmatrix}
$$

The first reduced model gives an interval fuzzy set:

$$\{[0.3\ 0.5\ 0.3],\ [0.7\ 0.7\ 0.7]\}$$

while the second yields:

$$\{[0.3\ 0.6\ 0.4],\ [0.7\ 1.0\ 1.0]\}$$

Their combination, applying the results of Section IV. for $\mu_j = 1$, $j = 1,2$, lead to the fuzzy set of the second order $\widetilde{B}$, with membership function given by:

$$
\widetilde{B}(b_1, w) = \begin{cases} 1 & \text{if } w \in [0.3, 0.7] \\ 0 & \text{otherwise} \end{cases}
$$

$$
\widetilde{B}(b_2, w) = \begin{cases} 0.5 & \text{if } w \in [0.5, 0.6] \\ 1 & \text{if } w \in [0.6, 0.7] \\ 0 & \text{otherwise} \end{cases}
$$

$$
\widetilde{B}(b_3, w) = \begin{cases} 0 & \text{if } w \in [0, 0.3] \\ 0.5 & \text{if } w \in [0.3, 0.4] \\ 1 & \text{if } w \in [0.4, 0.7] \\ 0 & \text{if } w \in [0.7, 1] \end{cases}
$$

For comparison, the fuzzy set resulting from the complete model equation $A_1 \circ A_2 \circ R$ is equal to $[0.7, 0.5, 0.5]$ and is covered by the fuzzy set B.

VI. CONCLUSIONS

The problem discussed, at length, in this chapter has been addressed to a class of those tasks of verification and improvement properties of the knowledge bases for rule-based expert systems. It has been indicated how the fuzzy relation of the knowledge base can be modified to reach a state of more compact shape yet producing acceptable changes in the control part of the rules. Two general lines of this construction have been underlined, namely, saving reasoning time and preserving the user-friendly property of the expert systems.

It is also of interest that the algorithms provided here can be applied to pattern recognition schemes, e.g., References 10—12. In this second case the reduction already performed corresponds to a feature selection problem: the changes of control actions find their analogies in an increment of the classification error.

Further studies in this direction can be concentrated for the solution of the reduction problem in the case that the fuzzy relation of the knowledge base is given at a certain precision (i.e., the set of the corresponding fuzzy relation equations has only an approximate solution). In such a situation it would be interesting to discover a set of subcondition candidates that might be eliminated.

REFERENCES

1. **Gupta, M. M., Kandel, A., Bandler, W., and Kiszka, J. B., Eds.,** *Approximate Reasoning in Expert Systems,* North-Holland, Amsterdam, 1985.
2. **Zadeh, L. A.,** The role of fuzzy logic in the management of uncertainty in expert systems, *Fuzzy Sets Syst.,* 11, 199, 1983.
3. **Di Nola, A., Pedrycz, W., and Sessa, S.,** Fuzzy relation equations and algorithms of inference mechanism in expert systems, in *Approximate Reasoning in Expert Systems,* Gupta, M. M., Kandel, A., Bandler, W., and Kiszka, J. B., Eds., North-Holland, Amsterdam, 1985, 355.
4. **Di Nola, A., Pedrycz, W., Sessa, S., and Wang, P. Z.,** Fuzzy relation equations under a class of triangular norms: a survey and new results, *Stochastica,* 2, 99, 1984.
5. **Higashi M., and Klir, G. J.,** Identification of fuzzy relation systems, *IEEE Trans. Syst. Man Cybern.,* 2, 349, 1984.
6. **Pedrycz, W.,** Applications of fuzzy relational equations for methods of reasoning in presence of fuzzy data, *Fuzzy Sets Syst.,* 16, 163, 1985.
7. **Sanchez, E.,** Resolution of composite fuzzy relation equations, *Inf. Control,* 30, 38, 1976.
8. **Pedrycz, W.,** Generalization and particularization of production rules in expert systems, in *Cybernetics and Systems '86,* Trapple, R., Ed., D. Reidel, Dordrecht, Netherlands, 1986, 783.
9. **Sambuc, R.** Functions Φ-flous: Application a l'Aide en Diagnostic in Pathologie Thyroidienne, Ph.D. thesis, Marseille, France, 1975.
10. **Cayrol, M., Farreny, H., and Prade, H.,** Fuzzy pattern matching, *Kybernetes,* 11, 103, 1982.
11. **Kandel, A.,** Fuzzy Techniques in Pattern Recognition, Addison-Wesley, 1982.
12. **Pedrycz, W.,** Techniques of supervised and unsupervised pattern recognition with the aid of fuzzy set theory, in *Progress in Pattern Recognition II,* Gelsema, E. S. and Kanal, L. N., Eds., North-Holland, Amsterdam, 1986, 439.

Chapter 6

THE PHYSIOLOGY OF THE EXPERT SYSTEM

A. F. Rocha, F. Giorno, B. Leäo, and M. Theoto

TABLE OF CONTENTS

I. INTRODUCTION

Fifth generation computer systems present new challenges that go beyond the realms of hardware and software to the process of human reasoning and its underlying logical and philosophical foundations.[13]

This new generation of computers will have two main distinctive features:[34] they will be based on knowledge processing rather than on information processing, and they will offer natural human-computer interfaces. The success of the expert systems endorses these objectives.[13,38]

This new approach puts human reasoning in the focus of both the design and application of fifth generation computer systems, and triggers two new questions of major importance:[13]

1.   Do we have an adequate theory of knowledge on which to base knowledge processing systems?
2.   Do we have an adequate theory of human psychology and brain physiology on which to base human-computer interface design?

For more than 2000 years man has tried to find the basic operation of thinking and has privileged logic.[12,13] Man started to investigate the function of his brain 150 years ago and has discovered that learning is not only dependent on logic but also heuristic. Physiology and psychology have stressed the importance of knowledge acquisition based on the repeated observation of the facts, and since Pavlov and Skinner, they have been studying this mechanism[20,28,32,43]

Expert systems use logic to handle human knowledge, without trying to understand and to model both heuristic and its acquisition. Because of this, the development of an expert system requires a tremendous effort in generating the heuristic of the human expert. It is also a consensus that human expertise is founded over a great number of production rules and/or frames, so that high performance for expert systems can be achieved only on mainframe machines.

Conditioning is the best studied approach for heuristic information acquisition in the case of animals and humans. Besides conditioning, biologists investigate many other processes of behavior modification introduced by repetitive data observation, such as sensory adaptation and message transmission attenuation.[28,29,40,41-43]

Behavioral modification induced by information repetition will be called here inductive learning, in contrast to deductive learning founded over the process of logical inference.

Behavioral research on conditioning and electrophysiological studies about distribution of the information throughout the brain have demonstrated that learning results in segregation of the information inside reduced brain circuits.[2,43] The idea is that well-learned activities are processed by small neural nets or frames operating a reduced number of almost context-free production rules.[15,43] These conclusions are in conflict with the popular ideas on the AI field, assuming that any reasonable expert machine has to handle at least some hundreds or thousands of production rules and/or frames.

Despite its tremendous success on approaching the process of heuristic acquisition or conditioning, biology is only on its initial steps toward a serious investigation of the deductive or logic process of learning. It has been only recently that the technique to correlate brain activity with complex task processing has been improved.[14,44,45] This technique correlates the brain activity recorded in the electroencephalogram with the understanding and performance volunteers have on cognitive tasks being performed during the recording.[4,17,18,40,49]

The initial results from these new studies seem to confirm the disposition of man to use a small amount of information to process highly complex tasks such as speech understanding. They do point, however, to fuzzy logic[40,43-45] and expectancy[44] as the important tools improving the human performance of dealing with a particular activity. Expectancy uses previous knowledge to guide goal-oriented processing by favoring previous successful relations.[15,19,21,24,25,33,47]

General rules may be deduced and formalized from all this knowledge acquired by biology, and they may be used in the development of fuzzy expert systems that will be, at least, much more user friendly, and, for sure, more efficient and compact, too. This chapter will focus its attention on the subject. The formalization summarized here was developed in a series of papers published since 1980. The reader interested in deeper details is referred to these papers, since no attention is presently paid to provide a full description of proofs and concepts.

II. TODAY'S EXPERT SYSTEMS

An expert system is a computer program based on AI techniques; it uses knowledge about a particular domain and reasoning techniques to support those activities performed by human experts.[7,22,48,51] Any person gains expertise through special education and, more important, through special training. Expert systems are believed to obtain this capacity from its interaction with the human expert.

Fuzzy expert systems are proposed to behave more humanlike because they handle concepts and reasoning in the same fuzzy and natural way as the human being does.[46,51,54]

The simple architecture of expert systems includes a data base, a knowledge base, and an inference machine.[7] The data base contains an organized collection of permanent and specific data concerned with the particular problem to be solved. The knowledge base is the repository of a codified form of the knowledge about the problem. The inference machine controls and executes all reasoning over the data base and the knowledge base towards, achieving the defined goals.

Recent discussions about the architecture of expert systems[8,22,51] have stressed the necessity to introduce descriptions at different levels of abstraction and to represent different kinds of knowledge in order to accurately model a knowledge domain.[51]

More attention has now been paid to provide systems with more structured organizations, by partitioning knowledge according to its use and by subdividing the reasoning process into steps.[8,23] This has stimulated the development of hybrid systems in which different formalisms coexist or are integrated together.[1,11,22,23] For instance, framelike formalism may be used for representing typical characteristics (prototypical knowledge) and properties of the entities to be modeled, whereas production rules may be selected in formalizing control knowledge.[22,23,51]

No matter what approach is to be used, the principle task of the knowledge engineer in the development of any expert system is to obtain and to represent the specialized knowledge to be processed by his particular system.

III. NEURAL NETS

Animal behavior is accepted as dependent on the distribution that messages have inside the brain. Knowledge representation is obtained through association of neurons, and it is dependent on both the structure of this association as well as on individual properties of each neuron.[40] Neural nets are shaped and reshaped by many biological processes depending on how facts are associated.

A neural net may be considered as a fuzzy graph or simplex having the strength of its arcs dependent on the strength of the coupling between the neurons they represent.[6,40] The behavior of individual neurons may be assumed equivalent to that of a fuzzy automaton having its family of state membership dependent on both the history of the received stimulation as well as on internal states.[40,43]

Briefly, any neuron N is assumed to be a fuzzy automaton:

$$N = (S, SC, Q, m, a)$$

to which S is the set of inputs or stimuli; SC is the set of outputs or classes of neural responses; Q is the set of neural states; a is the output function:

$$a: S \times SC \rightarrow [0,1]$$

and m is the family of state membership (u(q,q)) matrices

$$u\,(q,q) : Q \times Q \times S \rightarrow [0,1]$$
$$u\,(q,q) = u\,(j,t) \cdot (r) \cdot u\,(s,s) \cdot u(sc,sc)$$

That is to say, u(q,q) depends on:

1. The time j the neuron is able to hold information — The function u(j,t) controls the state transition depending on the relation between the neural memory and the time of the stimulus occurrence. States of high entropy are favored by repeated application of excitatory (inhibitory) stimuli during (outside) the retention period j.
2. The number of stimuli received in the retention time j — States of high entropy are favored by frequent (infrequent) excitatory (inhibitory) stimuli.
3. The stimulus membership (u(s,s)) — This function controls the state transition depending on the relations held by the stimuli. States of high entropy are favored by frequent paired stimuli.
4. The control (u(sc,sc)) exercised over the neuron by other neural structures.

Two neurons Na, Nj are functionally coupled if the output of one of them (i.e., Na) is the input to the other (i.e., Nj)

$$Na = (SCa, SCj, Qj, mj, aj)$$

The coupling (u (Na, Nj) between Na and Nj depends on the capacity Na have to trigger different responses on Nj. Because of this u(Na, Nj) is dependent on both the entropy h(SCa, j) transmitted from Na to Nj, as well as on the entropy (h(SCj)) of Nj:

$$u\,(Na,Nj): SCa \times SCj \longrightarrow [0,1]$$

$$u\,(Na,Nj) = h\,(SCa,j)/h(SCj)$$

$$SCa,j = aj\,(q,SCa)$$

The traffic of messages (c (Na, Nj)) between any pair of neurons Na,Nj is determined by the balance between inhibitory (I) and excitatory (E) actions exercised over them by the incoming messages:

$$c\,(Na,Nj) = u\,(E,Nj) - u\,(I,Nj)$$

The neural nets NG are organized in the nervous systems (NS) depending on the neural coupling:

$$NG = (\{SCj\}_{NS}, \{u\,(Na,NJ)\}_{NS})$$

The message distribution on these NGs are dependent on u(Na, Nj); that is to say on the message spreads in operational nets OG defined over NG:

$$OG = (\{SCj\}_{NS}, \{c(Na,Nj)\}_{NS})$$

The coupling power of these nets are calculated as:

$$cp\ (NG) = \underset{P}{M}\ \underset{s}{m}\ u(Na,Nj)$$

$$cp\ (NG) = \underset{P}{M}\ \underset{s}{m}\ c(Na,Nj)$$

where M and m stand for the operations realized over parallel (p) and serial (s) pathways, respectively. M could be a max-operation, and m could be a min-operation as preconized by classical fuzzy set theory in the case of logic operators **or** and **and**, respectively. However, other functions are more likely to replace these max and min, in order to turn formalization closer to reality.[17,22]

The fuzzy grammars realized over such nets are fuzzy relation grammars, with productions of the type:

$$SCa \longrightarrow SCj$$
$$f\ (p)$$

or

$$f\ (p) = u\ (SC_{a,a+1}, SC_{a+1,j})$$
$$f\ (p) = c\ (SC_{a,a+1}, SC_{a+1,j})$$

where f(p) expresses the dependence on the context.

Let FG denote NG and OG. The fuzzy languages L(FG) defined by these grammars are composed by productions of the type:

$$L\ (FG)\ (n) = M\ ((p1)\ m\ (p2),\ldots,m\ (pk))$$

M being taken over all derivation chains from the point basis to the node n.[40]

IV. THE INDUCTIVE LEARNING

Conditioning is the best-known paradigm to explain the dynamics of neuronal association induced by frequent message pairings. Data association that are frequent strengthen useful neural couplings, inhibit antagonistic actions, and disfacilitate unrelated neural linkages.[40,43] Through these processes, the animals acquire their heuristic knowledge about the external environment, knowledge which is composed of those sets of actions providing the best animal-environment adaptation. Because of this conditioning, it is governed not only by data pairing but also by the success in reaching desired goals.

The stages of high entropy of excitatory neural automata are favored in nonhomogeneous and nonregular environments[43] since the state membership u(q, qh) is strengthened in such conditions. The opposite is true in the case of inhibitory neural automata. Because of this, specific neural nets FG are modeled in the nervous system (NS) by repeated associations of facts. This modeling depends on the relations between the characteristics (j, (r), etc.) of the neurons in NS and the patterns of information present in the specified environment. Goal-oriented development of neural nets is provided by u(sc, sc), since it formalizes the control exercised by NS over both the neural coupling and neural traffic.

$$cp\ (FG) \longrightarrow 1$$

for some neural nets, because environmental stimulation approaches the states of their neurons to states of high entropy.[43]

Heuristic acquisition is governed by the frequency of and the relations established between repeated paired stimuli. In other words, it is guided by the information patterns existing the surrounding environment, which model the nervous systems changing the neuronal coupling. This process is responsible for the heuristic knowledge imprinting during the training of the expert.

Experiments on conditioning, sensory adaptation, and message transmission attenuation, etc.[2,28,32,45,46,49] provide the data necessary to shape (u(t, j), Å (r) and u(s, s) and to model inductive learning. Inductive knowledge acquisition, however, is believed to be guided, too, by goal-oriented processes to provide the best adaptation of the animal to the surrounding environment. Experiments on operative and aversive conditioning may provide information about u(sc, sc) necessary to model goal-oriented inductive learning.

It has been demonstrated[45] that as inductive learning progresses:

1.   The number of nodes of FG are reduced
2.   FG is more likely to be a tree
3.   Context-free languages are to be handled on FG
4.   Processing time is reduced

because in this condition cp(FG) → 1.

Heuristic acquisition is accompanied by the strengthening of neural coupling on specific neural nets[2] and the segregation of information inside reduced neural circuits, all of this resulting in brain specialization.[4,21,36,37,44]

V. KNOWLEDGE MODELING

Learning is the process of abstracting the key features or evidence E related to a proposed question in order to develop a model M to guide the behavior of the system posing the question. The model is a set of relation R (logic or not) between these evidences E:

$$M = \{R = M \mid R = E\}$$

and has a neural net NG guaranteeing its existence. The functional description F of M is provided by the fuzzy language L (NG) defined over NG. Thus,

$$M = \{FG, F = L(FG)\}$$

Learned models M must:

1.   Have autonomy (self-reproduction): M must have the capacity to be maintained or to be re-created in NS, providing the substrate for comprehension.
2.   Have mobility (exogenous-reproduction): M must have the capacity to be re-created into other neural systems NS, providing the substrate for communication.

3. Be generative: M must have the capacity to generate expectancies and predictions about the observed environment E.
4. Have plasticity: M must have the capacity to generate (to be a germ for) new learnings by posing new questions to the system; the new questions being the result of discrepancies between expectancies or predictions and observations on E.

Well-learned or expert models are those realized over neural nets FG for which:

$$cp \ (FG) \longrightarrow 1$$

Expert models will exhibit low descriptive and high functional complexities[26,39,43,45] since, on the one hand, their FG tends to be trees with a reduced number of nodes and, on the other hand, neurons will be operating at or near their high entropy states. Thus, expert models have strong autonomy, mobility, and plasticity. Also, they have an important capacity of prediction.

If new information is in disagreement with a model Mj, it may reduce cp(FGj) and may increase the coupling power of other models Mk in the complementor net:

$$cp(FG)$$

In such a condition, disagreement and error may trigger a procedure to determine whether a new and more complete model may be developed from Mj or from Mk, which is able to accommodate the new knowledge.[45] This procedure will search for i-cut level sets[35] at FGj or FGj able to maintain the highest possible coupling power and the smallest disagreement with the new data.[45] If this i-cut level set exists, it will be the germ for the development of a new expert model.

Expert models may furnish strong germs for development of new learnings, since their cp(FG)s tend to 1, permitting the search for i-cut level sets enjoying also strong coupling powers.

Expert neural net or the model it represents is built around a basic and highly structured unit or "germ" formed by the most representative and consolidated atoms of the respective knowledge.[44] Other "subnets" are connected to this germ, forming "halos" of weaker and variable associations.

One of the results of the message imprinting promoted by inductive learning is that part of the surrounding environment E is represented inside the brain. However, it is not possible to speak about an isomorphism for this representation, even in the case of sensory systems.[44] The principal fact against the idea of an isomorphic representation is the observation that the same information may be represented many times in the nervous system, each time enjoying a different neighborhood and, thus, different relations with other information.[9,44]

Expert models will be represented many times in the brain, each neural net enjoying a different neighborhood and establishing different associations between models. This cerebral organization optimizes association of information and generation of new knowledge, by creating the background for the establishment of unobserved associations between old models.

VI. AN EXPERIMENTAL APPROACH FOR DEDUCTIVE LEARNING

Fuzzy set theory now provides the background to bring logic under experimentation at the laboratories of physiology. This approach has been used by us to investigate the brain and logic mechanisms underlying speech decoding.[15-18,44,50]

The brain activity evoked by a particular task may be recorded in the electroencephalogram (EEG). Its statistical analysis furnishes the key elements to the understanding of the involved cerebral processing if the results are correlated with behavioral components of the task under investigation. In the case of language perception, this means that it is necessary to correlate the EEG activity during the decoding of the verbal information with its comprehension. A good picture of verbal comprehension may be obtained from its recall. Hence, it is necessary to correlate the statistical analysis of the EEG recorded during the verbal decoding with a statistical analysis of the recall of this verbal information.

The recall of a listened or read text may be described by fuzzy graphs having the recalled phrases as their terminal nodes and their nodes associated in the same way the subject organizes the text information to conclude for its main subject or theme.[15,16,18,50] The mean graph associated to the "mean decodification" of the text in a given population is dependent on the frequency of phrases, nodes, and nodes associations in the recalled graphs of this population.

The logic structure of the text decodification is obtained asking the volunteer to assign to each node of the graph the logic operator he assumes to be the best one used for the deduction of the chosen theme.

There is a clear distinction between language as a set of symbols and meanings shared by a population, and speech as the use of this language to vehiculate personal "senses" in this society.[27] As a matter of fact, each text may be viewed as a partially closed structure[47] depending, on the one hand, on meaning, guaranteeing a minimum coherence among the symbols used and, on the other hand, on sense permitting people to speak about individual ideas and models. Each piece of information has a correlation with the structure of the text used to convey the desired information or sense, and it triggers a confidence on it at the receiver.

Logic operators at each node, therefore, have to handle two different sets of truth.[17] One of them relates the information represented at the node to the structure of the text itself, and the other associates the same information with the set of beliefs of the receiver. The first of these values will be called correlation and the other will be referred to as confidence.

The concept of "importance", "relevance", or "correlation" was introduced in the expert system Centaur,[1] and it is used by others[22,51] to differently weigh the contributions of each piece of information in the evaluation of the frame or diagnostic. The concept of "confidence", "uncertainty", etc. was used by Mycin.[48] and it is introduced in many expert systems to equip them with a computational capability to analyze the transmission of uncertainty from the premises to the conclusion and associate the conclusion with what is commonly called a certainty factor.[54]

To investigate both the concept of correlation and confidence, the volunteers are divided in two experimental groups, each one of them working one of these fuzzy values. The rest of the experimental paradigm is maintained the same. A recorded text is played to the subject both with and without having his brain activity registered in the electroencephalograph. After this listening session, the volunteer is asked to recall the text in written form and to point out its theme. The next step is to order the phrases of the recalled text according to the importance they have to the definition of the chosen theme. These ordered phrases are considered as the terminal nodes of the decoding tree, and the subject is asked to join them in the same way he believes he joined the same pieces of information to conclude the theme. After the graph is completed, the volunteer assigns to each node his evaluation of the confidence or correlation shared by the information represented at that node. The final step requires the subject to choose one of the logic operators *or, and, not,* and *if* as the operator used to combine information at the node.

Confidence and correlation are averaged for each node of the calculate mean tree as well as for its corresponding logic operator. Linear correlations between these true values and their position in the tree are investigated to assess any dependence of these values on the structure of the text. The correlation between the true values handled by each operator is studied for both confidence and correlation in order to investigate whether the physiological rules[16-18] are the same as the proposed classical ones.[52]

The association of the most frequent logic operators to the tree representing the ''mean comprehension'' of the text, and the assignment of the calculated confidence and correlation mean values to each node of this tree, complete the description of the text decodification in the studied population.

If the brain activity is to be studied, then the EEG is sampled by a microcomputer together with a filtered version of the sound. This permits the localization of the epochs of the EEG associated with each phrase listening. Amplitude averaging of phrase epochs and epoch averaging in the case of repeated words are calculated for each scalp site of EEG recording. Coefficients between left/right or anterior/posterior recording sites are calculated, and statistically correlated with the probability of occurrence of the phrases in the recalled text, with the confidence or correlation assigned to each phrase, etc. By this method, cerebral activity is correlated with the decoded information of a given population.[18]

VII. THE PHYSIOLOGY OF THE DEDUCTIVE PROCESS

The experimental data obtained thus far have shown that texts are fuzzy decoded in all the studied populations, the fuzziness depending not only on the structure of the text itself but also on the degree of knowledge shared by the receiver about the theme of the text.[16-18,50]

A. THE ''MEAN'' GRAPH

The calculated mean trees exhibited a common pattern in all the experiments. The probability of occurrence of their terminal nodes decreased from left to right. Also, the connectivity of the right nodes was fuzzier than that of the left ones. The correlation assigned to the nodes decreases in some experiments from left to right and from the root to the terminals,[17] whereas confidence did not show any defined pattern in any of the experimental groups.

The recalled phrases had a crisp distribution over the left nodes. There was no difficulty in selecting the most frequent phrase at each one of these nodes, which also enjoyed a crisp connectivity. The distribution of the phrases over the ''mean'' decoding graph paralleled the node connectivity once it became fuzzier for the right nodes, too.[50]

The general picture is that of a decoding graph being much more structured at its left than at its right part. This was reflected in the fact that it was easy to separate İ-cut subgraphs for probability thresholds as high as 60% to represent the germs of the text decoding in the chosen population. The germs were always a subgraph occupying the left portion of the decoding tree.

Training on the theme of the text[50] seems to be accompanied by a reduction of the fuzziness of the decoding. The decoding of untrained people was described by a family of fuzzy graphs, since it was impossible to decide for only one graph based on node probability, connectivity, and distribution of phrases. In contrast, the decoding graph of trained people could be represented by a İ-cut subgraph with a 70% threshold, exhibiting a crisp node connectivity and phrase distribution.

B. THE NATURAL RULES FOR LOGIC OPERATORS

All studied logic operations were of the type

$$a \ m \ b$$

with m standing for one of the connectives: **and, or,** and **if.**

Confidence and correlation were averaged for a, b, and each *m*. Also, multiple linear correlations were tested between correlation and confidence assigned to *m* and those values associated to a and b.

The results confirmed the use of the classical max rule in the case of the connective **or,** since confidence and correlation averaged for *m* was equal to the maximum averaged for a and b.

The same was not true for the connective **and**. In this case, confidence (cn) and correlation (cr) averaged for *m* were intermediate values for those averaged for a and b. As a matter of fact, the calculated value for *m* was a liner function of the values obtained for a and b:

$$\text{cr } (m) = .30 + 45\text{cr (a)} + 16\text{cr (b)}$$
$$\text{cn } (m) = .39\text{cn (a)} + .40 \text{ cn (b)}$$

the correlation coefficients (r) being around 70% and the probability of the results being obtained by chance ranging from .5 to 2%.

Surprising is the fact that the linear coefficient calculated for correlation is different from zero. This implies a correlation is different from zero for *m* even if it is zero for both a and b. However, correlation was also linearly correlated to the node position in the tree. This could mean that correlation depended not only on the values a and b, but also on the structure of the text itself. The correlation cr (N) of the nodes of level N was calculated as:

$$\text{cr (N)} = .79 + .8 \text{ N}$$
$$r = 81\% \text{ and } p = 1\%$$

This kind of dependence was not observed in the case of confidence.

It may be assumed that people tend to use powered means to assess the true values in the case of the connective and instead of the classical min rule.[22,30,53,55]

Correlation averaged for the operator *if* was greater than the maximum value calculated for a and b. Confidence, in turn, was equal to the calculated maximum for a and b. Despite this, both correlation and confidence for *m* was linearly related to the values assigned to a and b:

$$\text{cr } (m) = .32 + .35\text{cr (a)} + .25\text{cr (b)}$$
$$r = 63\% \text{ and } p = 2\%$$

$$\text{cn } (m) = .49\text{cr (a)} + .40\text{cr (b)}$$
$$r = 75\% \text{ and } p = .5\%$$

Once more the linear coefficient was different from zero only in the case of correlation, and data did not support the classical propositions.[31]

C. BRAIN ACTIVITY, DECODING, AND LOGIC

The brain activity recorded in the EEG while the subject was listening to the text was linearly correlated to the probability of the phrases in the recalled text, as well as with the confidence and correlation assigned later to these phrases.[18]

The brain activity was studied for each phrase of the text through the analysis of the mean amplitude of the EEG calculated for the epoch corresponding to the listening of the phrase and for each one of the six scalp recording sites: left and right frontal, parietal, and central areas. Coefficients of dominance were calculated by dividing the area averaged for the left and right sites (Cf, Cp, and Cv) or for each cerebral hemisphere, by dividing frontal and parietal areas (Cfp), frontal and central areas (Cfc), and central and parietal areas (Ccp).

The probability of phrases p (F) on the recalled texts was a linear function of the coefficient of dominance at the frontal, central, and parietal areas:

$$p(F) = 345 - 2Cf + .36Cc - .9Cp$$

The correlation coefficient (r) was .73 and the probability (p) that the calculated relation was explained by chance was of 3.3%.

The dependence on Cc is in accordance with the well-known dominance of the left central areas in the case of language processing.[5,36,37] However, the inverse dependence on Cf and Cp points also to a right hemisphere involvement on speech recognition.

Correlation assigned to each phrase cr(F) depended on the same parameters:

$$cr (F) = 126 + .24Cf + .86Cc - .16Cp$$
$$r = 92\% \text{ and } p = .01\%$$

It was dependent on a left dominance at frontal and central areas. However, the participation of the right hemisphere was stressed by the dependence of c(F) on Cfc, Cfp, and Ccp calculated for the right but not for the left sites:

$$cr (F) = 37 + .56Cfc + .35Cfp - .53 CP$$
$$r = 94\% \text{ and } p = .1\%$$

This function stressed the participation of the frontal areas on the organization of the text structure as proposed by Luria.[27]

Confidence on each phrase cn (F) was also dependent on the cerebral activity:

$$cn (F) = 316 - 6.35Cf + 1.85Cc - 1.87Cp$$
$$r = 74\% \text{ and } p = 7\%$$

$$cn (F) = 1586 - 12 Cfc + 8 Cfp - 10Ccp$$
$$r = .55 \text{ and } p = 5\%$$

Confidence has a more complicated and weak dependence on the EEG activity if compared to correlation.

These results,[18] although initial ones, clearly demonstrated that it is possible to bring logic under neurophysiological investigation by using fuzzy set theory. If biology was successful in handling inductive learning, now it might contribute to a better understanding of the process of logic inference. This interdisciplinary approach opens new frontiers toward a sound theory of human knowledge.

VIII. KNOWLEDGE INVESTIGATION

A new paradigm for knowledge investigation may be outlined from the above experiments, devoting its attention to disclose the basic rules used by human experts in dealing with his special domain of knowledge. The target on this new paradigm is not the isolated expert, but a whole population of them. Its main objective is to understand and to model the knowledge shared by a group of experts, which may represent the general way of thinking of a school of knowledge, in order to try to understand and to model the human expertise, instead of the human expert.

The investigation begins with the delimitation of its own purposes. It is necessary to outline the limits (goals, data and population) of the expertise to be investigated. For instance, in the case of medical expert systems, this implies definition of:

1.   The set of diagnoses (goals) to be modeled
2.   The set of generic (data) signals, symptoms, subsidiary tests, etc. to be used by the systems
3.   The population of experts (reasoning) to be studied

The study will involve the investigation of:

1.   The knowledge environment — composed of all the models shared by the experts
2.   The working environment — composed of the most important data experienced by the experts during their special training
3.   The expert environment — composed of the processes the experts use to handle knowledge at the working environment

Initial results obtained by applying this approach to model congenital heart diseases will be reported in the last section of this chapter.

A. THE KNOWLEDGE ENVIRONMENT

The first step of the investigation is to obtain the expert knowledge using the fuzzy graph theory to construct the structure of the used expert models.

The selected experts are provided with a list of generic data and asked to separate those elements of the list correlated with the specified goals (in our experiments, a clinical diagnosis). Generic data means the theoretical data associated with the chosen goals by the specialized literature.

The next step is to order this sublist according with the importance the experts assume the data have in relation to the chosen goals (diagnosis). These ordered sublists are considered as the sets of terminal nodes of the fuzzy graphs that will describe the expert model used to reach the defined goals or diagnosis.

The investigation proceeds by asking the experts to join the terminals on second order nodes, the second order nodes on third order nodes, etc. in the same way they assume the data have to be organized to support the chosen goals. This provides the experimenter with a family of graphs related to the set of goals or diagnosis under investigation.

Once the expert model graphs are built the experts are requested to assign the degree of correlation (in the closed interval [0,1]) that they assume the information at each node shares with the chosen goals. Also, they have to select the logic connectives used to join the information at each node. This completes the initial phase of knowledge acquisition.

In the sequence, the collected expert model graphs are submitted to a statistical analysis to generate mean expert model graphs for each of the selected goals or diagnosis. This is accomplished by selecting f-cut level graphs composed by those nodes exhibiting a frequency equal or greater than the specified f-threshold. If the node distribution is not crisp, a family of f-cut fuzzy graphs will be generated to describe the representation of the knowledge in the studied population.

This procedure may be applied to a set of f-threshold values or it may be generalized a p-level set[35,46] to represent each expert model in the population, by making:

$$\lim f \longrightarrow 0$$

The generic data are associated to the terminal nodes of the calculated mean expert model graphs according to their probability of occurrence in the studied population and according to their calculated distribution over the terminal nodes. If different data have the same probability to occupy a specified node, the most frequent datum used by the population will be preferred at this position.

There is another source of fuzziness at this point. If data distribution over the mean graph is not crisp, then a family of graphs have to be constructed to accommodate the most frequent data association.

The calculated mean for each node is then assigned to the nodes of the mean expert model graphs. The used connectives are distributed over the mean graphs with the aid of the same rules used to localize data at the terminal nodes.

This step ends the study of the knowledge environment. This phase provides the experimenter with a family of fuzzy expert models representing the specific domain of knowledge in the chosen population.

B. THE WORKING ENVIRONMENT

The purpose of this phase of the investigation is to provide the experimenter with the environment where the population of experts uses its knowledge. The expert applies his knowledge to a specific data population and his knowledge is modeled by the existing patterns in this working environment. This is the basic process of the inductive learning discussed before. For instance, in the case of medical expert systems, this means that it is necessary to know the existing surrounding patterns associated to the diagnosis being studied.

The same methodology discussed for characterizing the knowledge environment will be used here, the only difference being the fact that the experts will not be asked to organize the data related to the chosen goals, but to allow the experimenter to investigate their professional archives. The purpose of this is to disclose what the data are the experts are using and how they are being organized according to the goals they induced. For instance, in the case of medical expert systems, the experts are requested to allow the experimenters to investigate the profiles of their patients according with their diagnosis.

The same list of generic data (in our case, signal, symptoms, etc.) used in the first phase of the investigation will be applied here, too. The initial step is to translate the contents of the professional (in our case, the medical) record into the same jargon[47] used by the generic data list. If the knowledge engineer does not have a full capacity to understand this professional jargon, he has to be helped by an expert not included in the investigated population.

Once the professional records are translated into a common jargon used by the population of the studied experts, their contents may be described by means of graphs constructed in a similar way the expert model graphs were built in the previous phase of the investigation. The data are ordered according to their appearance in the record under study. The second, third, etc. order nodes, will join this information according to defined syntactic rules similar to the following ones:

1. All data included in a paragraph of the record are joined at a second order node numbered according to the chronological order of the paragraph in the record.
2. Second order nodes are joined to form the third order nodes according to the segmentation of the record into subsections.
3. Third order nodes are grouped according to the segmentation of the record into sections.

Once the record graphs are built they will be submitted to the same statistical procedure outlined before to generate the mean graphs. This will generate mean record graphs representing the profile of each chosen goal (diagnosis) in the working environment of the experts.

The comparison between the mean expert models and the mean record graphs will distinguish the theoretical from the expert knowledge in the studied population of experts. Besides this, and more importantly, the knowledge of the mean record graphs enables the knowledge engineer to stimulate well-controlled tests to study the expert environment. It permits the study of the behavior of the expert in an "almost real situation", since it provides a control of the complexity of the furnished information by controlling both data as well as their association on each of the planned tests.

C. THE EXPERT ENVIRONMENT

Once the working environment has been characterized, the expert environment can be studied through simulation of "real situations" requiring the expert to use his knowledge to reach the chosen goals (or diagnosis).

The mean record graphs will be used in constructing these real situations, because the experimenter can create "testing" records by supplying quantitative data to quantitative variables, as well as to deleting or adding qualitative data to the mean record graphs. Also, the complexity of the real situation can be controlled by modifying data association on the testing records. Task difficulty can be increased by partially merging the contents of two or more mean record graphs. The advantage of the present approach is to provide the knowledge engineer with a full control of the complexity of the experiment.

The testing records will be presented to the same population of experts studied on the previous phases of the investigation. They will be presented in a base of "case exercise"; that is to say, after receiving the information, the expert has to point out the goal supported by the data, as well as to pick the information he used out of the testing records, and to organize it as he assumed these pieces of knowledge need to be correlated to induce (trigger) the chosen goal (diagnosis). The same procedure used to construct the expert model and record graphs will be used here to generate "case" record graphs.

Once case record graphs are obtained for different goals and different experts, they undergo the same statistical treatment discussed before to generate mean case graphs.

The comparison between these mean case graphs and the corresponding mean expert model graphs will furnish more information to differentiate theoretical from expert knowledge in the studied population. More importantly, it will provide the knowledge engineer with information about the possible relations shared by different expert models, both when antagonistic or cooperative models are concerned. The knowledge engineer will use the conclusions from this study to start the organization of the knowledge space. Here the different expert models are assembled to optimize the association of information and generation of new knowledge. At this very moment, the expert system will enhance its capacity for prediction.

If case reports are presented to the expert having its brain activity monitored by the electroencephalogram, the mean case report graphs may be correlated with this activity as described before. This kind of study opens the frontier to a sound investigation of a physiological and psychological theory of expert knowledge that could be the foundation for the development of a more humanlike software and hardware.

IX. CONCLUSION

The proposed paradigm is now being applied to model the expert knowledge on congenital heart diseases. Twenty experts and twenty general practitioners from different regions of Brazil are involved in the project. The general practitioners act as a control to characterize the change in information representation induced by expertise.

The generic data list was prepared from a list of signals, symptoms, and subsidiary tests being used at the Instituto de Cardiologia, Fundacão de Cardiologia do Rio Grande do Sul, to codify clinical records on cardiology. The experimenter acquiring the information is a physician trained in cardiology and medical informatics, but is not an expert on congenital cardiac diseases.

TABLE 1
The Frequent Data on Expert IAC Graphs

Signal, symptom, lab tests	Data prob[a]%	Corr[b]	Node/prob[c] (%)
Fixed and wide split of S2	100	.95	1/100
Ejection murmur of increased flow	80	.90	2/70
Mesodiastolic flow murmur (RV area)	77	.83	3/70
RBBB on ECG	57	.75	4/70
Increased pulmonary vascular markings on CXR	52	.75	6/60
Recurrent chest infection	48	.62	5/50

[a]   Data prob — probability of data on the population.
[b]   Corr — correlation averaged for data.
[c]   Node/prob — teminal node and probability of association datum/node.

The first phase of the investigation or the characterization of the knowledge environment is now being concluded for noncyanotic pathologies. The results briefly presented below are those concerned with interatrial communication (IAC).

A. THE EXPERT MODEL OF IAC

The number of terminal nodes of the IAC expert model graphs averaged 6.3 nodes, this implying that the mean IAC expert graph is composed by six terminal nodes. The probability of these six nodes are from left to right: 100, 90, 85, 80, 78, and 70%, respectively. Second and third order nodes, averaged two and one nodes, respectively. Second order node probabilities (around 60%) are lower than the terminal nodes.

Experts tended to link the first two left terminal nodes almost directly to the root, saying that they would think about the diagnosis in the presence of only this information. The other terminal nodes are linked to the left data in order to improve certainty about the diagnosis.

The most frequent elements of the generic list on the IAC graph are found in Table 1.

The distribution of this data on the mean IAC expert model was crisp, as demonstrated by the distribution of data on terminal nodes (left column on Table 1). The results also pointed to an increase of fuzziness from the left toward the right nodes. Correlation evaluated by the expert parallel the probability of data on expert IAC graphs.

A striking feature on the expert IAC graphs was the multiple connections of at least node one with other right nodes, reflecting the opinion of the experts that this other information may increase the confidence of a diagnosis triggered by the information at node one. The information at node one could be taken almost as a ''synonym'' of its associated diagnosis. Because of this, the expert IAC graphs were, in general, a family of trees, each one representing one of the associations between the left and right nodes. The mean number of expert IAC trees was four.

B. THE GENERAL PRACTITIONER MODEL OF IAC

The mean number of nodes of the general IAC graphs was 13.5, implying that the mean general practitioner model IAC must have 13 terminal nodes. Second and third order nodes average three and one, respectively.

Besides the increased number of nodes, the mean general practitioner graph showed a very fuzzy structure, since it was both impossible to easily define node connectivity as well as characterize data distribution over these nodes.

The most frequent data on the general practitioner IAC graphs are found in Table 2.

The data stressed the following difference between the expert and the general practitioner handling IAC:

TABLE 2
The Most Frequent Data on General Practioners IAC Graphs

Signal, symptom, lab test	Data prob[a] (%)	Corr[b]
Dilatation of pulmonary artery trunk	100	.75
Fixed and wide split S2	100	.90
Increased pulmonary vascular markings on CXR	100	.68
RBBB on ECG	75	.80
RV hypertrophy on ECG	75	.64
RA enlargement on ECG	75	.67
Ejection murmur	50	.70

[a]   Data prob — probability of data on the population.
[b]   Corr — correlation averaged for data.

1.   The expert uses reduced and crisp models associated to the diagnosis, whereas the general practitioner is fuzzy and prolix in organizing his reasoning. This confirms the hypothesis that learning shrinks the knowledge representation. In other words, expertise reduces the amount of information and processing necessary to support decision.
2.   The expert uses some data as almost a "synonym" of the goal (diagnosis) to be reached (trigger). The other information is used to strengthen confidence in the decision. This is not the case in the nonexpert, since it always uses all the information available.
3.   Confidence paralleled use by the expert, but not by the general practitioner.
4.   Experts and general practitioners differed on the used information. The expert physician uses clinical data as the most important support for his decision, whereas the general practitioner prefers to base his diagnosis on laboratory tests, whose results are believed to be endorsed by an expert. Thus, the expert prefers to use his expertise on evaluating both the information provided by the patient and, more importantly, the clinical data he observes on physical examination. The general practitioner seems to prefer to use the knowledge of the results of laboratory tests.

   The present data, although supported by a reduced number of experiments, confirm the theoretical point of view that expertise reduces the amount of information to be processed, as well as the time required to reach the desired goal. If this is proved to be true for a great number of other experiments with other types of experts, working in different physical and cultural environments, then it will be possible to assume that expert systems could also be a job for a microcomputer.

   It seems possible to affirm that it is worthwhile any effort devoted to expand the present data, because of their theoretic importance and because of the new frontiers opened by the present approach.

REFERENCES

1. **Aikins, J. S.**, Prototypical knowledge for expert systems, *Artif. Intell.*, 20, 163, 1983.
2. **Barttlet, F. and John, E. R.**, Equipotentiality quantified: the anatomic distribution of the engram, *Sciente*, 182, 764, 1973.
3. **Brown, W., Marsh, J. T., and Smith, J. C.**, Contextual meaning effect on speech evoked potentials, *Behav. Biol.*, 9, 755, 1973.
4. **Brown, W., Marsh, J. T., and Smith, J. C.**, Principal component analysis of ERP differences related to the meaning of an ambiguous word, *Electroencephalogr. Clin. Neurophysiol.*, 46, 709, 1979.

5. **Butler, S. R. and Glass, A.,** Asymmetries in the electroencephalogram associate with cerebral dominance, *EEG Clin. and Neurophysiol.,* 36, 481, 1974.
6. **Burnstein, G., Nicu, M. D., and Balaceanu, C.,** Simplicial differential geometric theory for language cortical dynamics, *Fuzzy Sets Syst.,* 1986.
7. **Clancey, W. J.,** The epistemology of a rule-based expert systems — a framework for explanation, *Artif. Intell.* 20, 215, 1983.
8. **Clancey, W. J.,** Heuristic classification, *Artif. Intell.,* 27, 280, 1985.
9. **Dreyer, D. A., Schneider, R. J., Metz, C. B., and Whitsel, B. L.,** Differential contributions of spinal pathways to body representation in postcentral gyrus of Macaca mullata, *J. Neurophysiol.,* 35, 119, 1974.
10. **Dreyer, D. A., Loe, P. R., Metz, C. B., and Willy, W. D.,** Representation of head and face in postcentral gyrus of the macaque, *J. Neurophysiol.,* 38, 555, 1977.
11. **Fikes, R.,** The role of frame-based representation in reasoning, *Comm. of the ACM,* 28, 904, 1985.
12. **Fuchi, K., Sato, S., and Miller, E.,** Japanese approaches to high-technology R&D, *Computer,* 17, 14, 1984.
13. **Gaines, B. R. and Shaw, W. L. G.,** Systemic foundations for reasoning in expert systems, in *Approximate Reasoning in Expert Systems,* Gupta, M. M., Kandel, A., Blander, W., and Kiska, J. B., Eds., Elsevier, New York, 1985, 271.
14. **Glasser, E. M. and Ruchkin, D. S.,** *Principle of Neurobiological* Signal Analysis, Academic Press, New York, 1976.
15. **Greco, G., Ferreira, R. R., and Rocha, A. F.,** Language Neurophysiology: EEG and speech perception, Brazilian J. *Med. Biol. Res.,* 5, 464, 1983.
16. **Greco, G., Rocha, A. F., and Theoto, M.,** Fuzzy Logical Structure of a Text Decoding Proc. 6th Int. Cong. Cybern. Syst. Paris, 1984, 193.
17. **Greco, G. and Rocha, A. F.,** The fuzzy logic of a text understanding, *Fuzzy Sets Syst.,* 3, 347, 1987.
18. **Greco, G. and Rocha, A. F.,** Brain activity and fuzzy belief, in *Fuzzy Sets in Psychology,* Zétényi, T. Ed., Elsevier, Amsterdam, 1988, 297.
19. **Johnston, J. C. and McClelland, J. L.,** Perception of letters in words: seek not and ye shall find, *Science,* 84, 1192, 1974.
20. **Hernadez-Pon, R., Scherrer, H., and Jouvet, H.,** Modification of electric activity in cochlear nucleus during "attention" in unanesthetized cats, *Science,* 123, 331, 1956.
21. **Kutas, M. and Hillyard, S. A.,** Reading senseless sentences: brain potentials reflect semantic incongruity, *Science,* 207, 203, 1980.
22. **Lesmo, L., Saitta, L., and Torasso, P.,** Applications of fuzzy expert systems in hepatology, in press, 1986.
23. **Lesmo, L. and Torasso, P.,** Prototypical knowledge for interpreting fuzzy concept and quantifiers, *Fuzzy Sets Syst.,* in press, 1986.
24. **Lieberman, P.,** Some effects of semantic and grammatical context on the production and perception of speech, *Language Speech,* 6, 172, 1963.
25. **Lieberman, P.,** *Intonation, Perception and Language,* MIT Press, Cambridge, MA, 1968.
26. **Loefgren, L.,** Descriptive and interpretative complexities, *Int. J. Gen. Syst.,* 3, 26, 1976.
27. **Luria, A. R.,** Cerebro y lenguaje, *Paidos,* Buenos Aires, 1974.
28. **Mackintosh, N. J.,** The psychology of animal learning, Academic Press, New York, 1974.
29. **Magleby, K. L. and Zengel, J. E.,** A dual effect of repetitive stimulation on post-tetanic potentiation of transmitter release at the frog neuromuscular junction, *J. Physiology (London),* 245, 163, 1975.
30. **Mandami, E. H.,** Application of fuzzy logic to approximate reasoning using linguistic terms, IEEE Trans. Comput. C-26; 1182, 1977.
31. **Mizumoto, M.,** Fuzzy inference using max-composition in the compositional role of inference, in *Approximate Reasoning in Decision Analysis,* Gupta, M. M. and Sanchez, E., Eds., North-Holland, Amsterdam, 1982, 67.
32. **Marks, E. L.,** Sensory processes: *The New Psychophysics,* Academic Press, New York, 1974.
33. **Miller, G. A. and Isard, S.,** Some perceptual consequences of linguistic rules, *Verbal Learn. Verbal Behav.,* 2, 217, 1963.
34. **Moto-oka, T.,** *Fifth Generation Computer Systems,* North Holland, Amsterdam, 1982.
35. **Negoita, C. V. and Ralescu, D. A.,** Applications of Fuzzy Sets to Systems Analysis, Birkhause Verlage, 1975.
36. **Neville, H. J., Kutas, M., and Schmidt, A.,** Event related potential studies of cerebral specialization during reading. I. studies of normal adults, *Brain Language,* 16, 300, 1982.
37. **Neville, H. J., Kutas, M., and Schmidt, A.,** Event related potential studies of cerebral specialization during reading; II. Studies of congenitally deaf adults, *Brain Language,* 16, 316, 1982.
38. **Reitman, W.,** Artificial Intelligence Applications for Business, Alex, Nordwood, NJ, 1984.

39. **Reggia, J. A., Nau, D. S., Peng, Y., and Perricone, B.,** A theoretical foundation for abductive expert systems, in *Approximate Reasoning in Expert Systems,* Gupta, M. M., Kandel, A., Blander, W., and Kiska, J. B., Eds., Elsevier, New York, 1985, 549.
40. **Rocha, A. F., Francozo, E., Handler, M. I., and Balduino, M. A.,** Neural languages, *Fuzzy Sets Syst.,* 3, 11, 1980.
41. **Rocha, A. F. and Francozo, E.,** EEG activity during the speech perception, *Revue Phonetique Appliquee,* 55, 307, 1980.
42. **Rocha, A. F.,** Neural fuzzy point processes, *Fuzzy Sets Syst.,* 5, 127, 1981.
43. **Rocha, A. F.,** Basic properties of neural circuit, *Fuzzy Sets Syst.,* 7, 109, 1982.
44. **Rocha, A. F.,** Brain activity during language perception, in International Encyclopedia on *Information and Control,* M. Singh, Ed., Pergamon Press, Elmsford, NY, S30140-8, 1985, 115.
45. **Rocha, A. F.,** Toward a theoretical and experimental approach of fuzzy learning, in *Approximate Reasoning in Decision Analysis,* Gupta, M. M. and Sanchez, E., Eds., North-Holland, Amsterdam, 1982, 191.
46. **Rocha, A. F.,** Expert sensory systems: initial considerations, in *Approximate Reasoning in Expert Systems,* Gupta, M. M., Kandel, A., Blander, W. and Kiszka, J. B., Eds., Elsevier, New York, 1985, 549.
47. **Rocha, A. F. and Rocha, M. T.,** Specialized speech: a first prose for language expert systems, *Inf. Sci.* 37, 193, 1985.
48. **Schortliffe, E. H.,** *Computer-based medical consultations: Mycin,* Elsevier, New York, 1976.
49. **Sutton, S., Braren, M., Zubin, J., and John, E. R.,** Evoked potential correlates of stimulus uncertainty, *Science,* 150, 1187, 1965.
50. **Theoto, M. T., Santos, M. R., and Uchiyama, N.,** The fuzzy decodings of educative texts, *Fuzzy Sets Syst.,* 3, 315, 1987.
51. **Torasso, P. and Console, L.,** Knowledge organization and approximate reasoning in medical diagnostic expert systems, in press, 1986.
52. **Zadeh, L. A.,** Fuzzy sets as a basic for a theory of possibility, *Fuzzy Sets Syst.,* 1, 3, 1978.
53. **Zadeh, L. A.,** The role of fuzzy logic in the management of uncertainty in expert systems, *Fuzzy Sets Syst.,* 11, 199, 1983.
54. **Zadeh, L. A.,** The role of fuzzy logic in the management of uncertainty in expert systems, *in Approximate Reasoning in Expert Systems,* Gupta, M. M., Kandel, A., Blander, W., and Kiska, J. B., Eds., Elsevier, New York, 1985, 549.
55. **Zimmerman, H. J. and Zysno, P.,** Latent connectives in human decision making, *Fuzzy Sets and Systems,* 4, 37, 1980.

Chapter 7

ON THE PROCESSING OF IMPERFECT INFORMATION USING STRUCTURED FRAMEWORKS

Andrew P. Sage

TABLE OF CONTENTS

I. PRINCIPAL APPROACHES

Structured frameworks for human information processing, associated inference analysis based upon statistical and epistemic reasoning, and realistic algorithms for software realization of support systems were the three primary areas for investigation reported on here. There are many approaches to the processing of information that is, for any of a variety of reasons, imperfect. A recent paper by Stephanou and Sage[1] summarizes many of these, and a recent monograph[2] illustrates application of some of these approaches. While many of the recent efforts, including most that appear in this book, are based no fuzzy set theory, here we report on a somewhat different approach to representation and use of imperfect information. The research presented here is intended to integrate, especially in the areas of inference analysis, contemporary approaches in artificial intelligence for expert system construction and management science or systems engineering approaches for the design of decision support systems.

II. BACKGROUND FOR THE RESEARCH

Our research was concerned with the appropriate use of imperfect information in planning, problem solving, and decision making. By imperfect information, we mean information that is incomplete, imprecise, uncertain, inconsistent, unreliable, or some combination of these. We have been especially concerned in our research with design protocols for knowledge support systems that aid individuals in coping with imperfect information.

We have been especially concerned with the development of approaches that will cope with evidential or epistemic information as well as with statistical or aleatory information. The latter part of our efforts have concerned initial examination of distributed decision making and the need to acquire, represent, and use imperfect knowledge in forms that are appropriate for group use, where individual members of the distributed group represent a variety of stages of human experiential familiarity with the task at hand and may not necessarily all have access to the same information.

Our work has a basis in, but extends considerably, the decision-making problem with a finite set of possible alternatives in that it considers imperfect knowledge of information about the decision situation structural model that described the task at hand, the impacts of alternative courses of action, and the value perspectives of the decision makers.[3] Our assumptions of imperfect knowledge of information may refer to available information that may be imprecise, relative to the degree of refinement with which the assessment is made; information that may be inconsistent in the sense of being in disagreement with presumed principles or laws of an assumed decision situation model; or information that may be incomplete in that needed elements are missing. One of the principal purposes to which an operational aid based on these concepts could be put is in the interactive determination of appropriate information requirements for judgment and choice. We wish to deal with information that is statistical and imperfect,[4-6] as well as with information that is epistemic and imperfect.[8-13]

Several advances associated with the use of formal model-based management efforts are associated with this research. The central issues studied here lie at the interface between the human model and the formal model used to represent a particular judgment situation.[8,14,15] This interface concerns the human information processing activities required for the use of situation frameworks, schemata, or models, and enhancement of the ability of people to organize them in a way that results in an efficient and effective aid for judgment and choice activities, including situation understanding.[16,18] This research is concerned with support for formal knowledge-based reasoning as well as skill-based and rule-based inputs to judgment that are appropriate in situations in which there exists considerable experiential familiarity with a task at hand.

Most contemporary formal models for decision support depend, for their use, on a very organized process to adequately frame the knowledge surrounding an issue. At the representation level, this knowledge is required to be precise relative to the degree of refinement with which the assessment is made, consistent in agreement with principles and laws of the model, and complete in the sense of containing all the information necessary to obtain a linear preference ordering of the alternatives.[2] It is often in the real life situation, however, that various parameters of a model for these purposes are unknown. Immeasurableness, or perhaps partial knowledge of the parameters of the model, prevents the wide use of formal models for decision support. Even when sufficient information is gathered, it is not unusual to obtain a recommendation for a best alternative that is unacceptable to the decision maker. A preference order, supplied by a formal model, may not reflect some aspects of the decision maker's intuitive feelings concerning the situation at hand. In addition, information may be discovered to be self-contradictory after it is interpreted.[5] In this circumstance, some aspects of the semantic character of the knowledge structure and information contained in the model of the decision situation have been improperly assessed or represented.

Real life decision problems are so complex, unstructured, and poorly understood initially that a formal process for decision support must allow and encourage learning so that the decision maker gradually obtains increased understanding of the decision situation as well as the personal value perspectives which led to the decision, so as to be able to make a more informed and hence typically better decision. Thus, support to formal-based reasoning results in the learning of rules and ultimately the skills that augment, and perhaps even replace, formal reasoning in later performance of these same tasks.[8,18,19] This appears to firmly introduce skill and rule-based knowledge into the judgment situation and to make it very desirable that formal processes for model-based management and knowledge support incorporate these forms of knowledge. It is particularly important to recognize the possible information imperfections that may surround a decision situation when determining information requirements and requirement specifications for decisions and for designs that result from decisions.[20] In each of these cases, the ability to cope with imperfect information is essential.

III. RESEARCH ACCOMPLISHMENTS

An approach that accommodates both probabilistic (or aleatory) and logical (or evidential or epistemic) support, and that is able to cope with several types of imperfect information has been developed. It is important that alternatives to probabilistic representations of knowledge be considered as real information about specific situations that is often more evidential or epistemic than it is aleatory. The principal accomplishments of this research have been the development of a decision-aiding concept that allows processing of information that is imprecise, incomplete, inconsistent, and otherwise imperfect. A structural representation of knowledge is utilized so that the effects of information imperfection are available to the decision maker in the form of information presentation aids displaying the results of information that has been input to the support system up to the present.

The process results in interactive decision aiding in which the support system user, rather than the dictates of an analysis algorithm, guides the process.[5] At any point in time, evaluation of identified courses of action are available that are based on information obtained up to that point in time. This acts as a value of information guide in determining needs for additional information, and in encouraging the system user to not seek information of low saliency or relevancy, or of greater consistency than warranted by the task at hand. The approach appears very capable of implementation on current generation microcomputers, and a realistic effort at implementation, together with extensions to the distributed decision-making case, is suggested as an important, potentially high payoff topic for future research.

We have been concerned with interactive formulation of knowledge support processes as learning processes in which the decision maker, through interaction with the support system, is able to successively gain a better understanding about the decision situation and to adapt judgments accordingly.[8,12] This is accomplished in such a way as to enable selective resolution of inconsistencies in the knowledge base during the search for an appropriate schema representation of knowledge, perhaps in the presently instrumented form of a dominance structure among alternative courses of action, that is sufficient to enable judgment and choice.

This learning aspect of the research was complemented by research efforts associated with knowledge acquisition.[19] One object of knowledge acquisition is to enlarge the knowledge base of a support system with specific new knowledge, perhaps in the form of schemata or productions which might be expressed in terms of rules, contexts, and interpretations. Inconsistency and incompleteness in the knowledge base generally are detected only through the application of concepts that are often not known to the support system at the time the imperfect knowledge is first acquired. Thus, meta-knowledge is essential to construction of a support system. It can be used to prioritize the relevancy of the schema, scripts, or rules in the knowledge base so that those most likely to be useful, given a current state of knowledge on the part of the support system, are used first.

There are several possible criteria for evaluating the correctness, reliability, and operational utility of the results obtained in this research. In the absence of general quantitative criteria to rigorously assess the reliability and operational utility of the results of the proposed work, our evaluation procedure should be based heavily on simulated case studies and operational evaluation based on these. The main criteria that appear appropriate for evaluating the reliability of the conceptual knowledge-based support system developed here include

1. The system should be robust relative to performance in the face of missing rules and/ or evidence. The performance of the reasoning scheme should "degrade gracefully" by generating less specific conclusions as information is subtracted rather than qualitatively incorrect ones.
2. The sequential knowledge acquisition and information-gathering schemes should also detect significant inconsistencies in the rules and/or evidence, and warn the user of this and possibly other knowledge imperfections. Through this, it should also be possible to measure information and knowledge acquisition cost. This figure of merit reflects the (primarily personnel) costs associated with the knowledge acquisition step of the knowledge support system aiding process.
3. Enhancement of decision quality should be measurable in case study comparisons against expert, intermediate, and inexperienced decision makers. Decision quality is a function of the specificity and correctness of the conclusions, as well as of the cost of the information that was required to reach those conclusions.
4. Decision-making productivity enhancement should be measurable in case study comparisons.
5. A knowledge base updating flexibility measure reflects the cost of updating the knowledge base as rules are added, deleted, or substituted.

Apparent throughout this research were concerns relative to the structuring of inference and decision problems,[6] and the learning that is reflected by revisions of problem structure in the light of new knowledge as it evolves over time.[9-11] The principal research agenda was, in a very real sense, to determine how to design knowledge base representations in a way that facilitates adequate problem structuring. This is primarily accomplished through an interactive process in which the decision maker is able to evaluate the effects of new information upon judgment as this information is received. Research is, we believe, almost nonexistent on how people form structural representations or "cognitive models" of complex inference and decision problems, and how we revise such structure in order to incorporate what is learned. This study was undertaken in the belief that, if we are successful in developing useful strategies for organizing knowledge and structuring problems, we will also enhance the degree to which evidential and inferential subtleties can be captured and incorporated in operationally useful support processes that aid human inference and decision.[13]

IV. HIGHLIGHTS OF THE APPROACH

In the standard decision analysis paradigm, it is assumed that a set of feasible alternatives $A = (a_1,...,a_m)$ and a set $(X_1,...,X_n)$ of attributes or evaluators of the alternatives can be identified. Associated with each alternative course of action a in A, there is a corresponding consequence $X_1(a), X_2(a),...,X_n(a)$ in the n-dimensional consequence space $X = X_1X_2,...,X_n$.

The decision maker's problem is to choose an alternative a in A so that the maximum pleasure with the payoff or consequence, $(X_1(a),...,X_n(a))$, results. It is always possible to compare the values of each $X_i(a)$ for different alternatives; however, in most situations, the magnitudes of $X_i(a)$ and $X_j(a)$ for i not equal j cannot be meaningfully compared since they may be measured in totally different units. Thus, a scalar-valued function defined on the attributes $(X_1,...,X_n)$ is sought that will allow comparison of the alternatives across the attributes. The existence of the value function as a mechanism for representation and selection of alternatives in a utility space is based on the fundamental representation theorem of simple preferences; this states that under certain conditions of rational behavior there exists a real-valued utility function U so that alternative a_1 is preferred to alternative a_2 if and only if the utility of a_1, denoted by $U(a_1)$, is greater than $U(a_2)$, the utility of a_2.

A primary interest in multiattribute utility theory (MAUT) is to structure and assess a utility function of the form:

$$U[X_1(a),...,X_n(a)] = f\{U_1[X_1(a)],...,U_n[X_n(a)]\}$$

where U_i is a utility function over the single attribute X_1 and f aggregates the values of the single attribute utility functions so as to enable one to compute the scalar utility of the alternatives. The utility functions U and U_i are assumed to be continuous, monotonic, and bounded. Usually, they are scaled by $U(x^*) = 1$, $U(x^0) = 0$, $U_i(x^*_i) = 1$, and $U_i(x^0_i) = 0$ for all i. Here $x^* = (x^*_1, x^*_2,...x^*_n)$ designates the most desirable consequence, and the expression $x^0 = (x^0_1, x^0_2,...x^0_n)$ denotes the least desirable consequence. The symbols x^*_i and x^0_i refer to the best and worst consequence, respectively, for each attribute X_i, i.e., $x^*_i = X_i(a^*)$ where a^* is the best alternative for attribute i, and $x^0_i = X_i(a^0)$ where a^0 is the worst alternative for attribute i. In the simplest situation, additive independence of attributes[3,24] exists so that the MAUT function may be written as:

$$U(A_i) = w_1U_1(A_i) + w_2U_2(A_i) +...+ w_nU_n(A_i)$$

Here the w_j are the weights of the various attributes of the decision alternative A_i and the U_j are the attribute scores for the alternative.

The methodology of decision analysis, using multiattribute utility theory, is generally decomposed into the same generic steps we have outlined for the scalar utility case. The major difference is that it is, of course, necessary to elicit the multiple attribute utility function.

Even though the theory and procedures of multiattribute decision analysis are conceptually straightforward, there are other circumstances that make its implementation very complex. Putting the methodology into practice is much more involved than one might believe. Each of the decision analysis steps requires substantial interaction between the analyst and the decision maker. A very stressful thinking process is demanded of the decision maker while the analyst is in charge of coordinating a series of activities in order to facilitate this process. The analyst must obtain the minimum amount of relevant information about the decision problem to determine the various utility functions. Often redundant information should also be obtained in order to check for consistency. It is interesting that in most of the literature on decision analysis, there is little or no mention of the information system functions that need to be accomplished in order for a recommended decision to evolve. In effect, it is assumed that the decision maker is an expert with respect to knowledge of relevant information but is unable to aggregate this information in a proper fashion needed for an effective judgment or decision.

In the exercise of an effective decision-aiding process, much is required of the analyst. The analyst must be sensitive to biases and flawed heuristics that the decision maker may utilize; the analyst must be able to structure the decision problem regardless of the degree of complexity and, above all, retain the confidence of the decision maker with respect to the belief that a formal analysis of the problem will result in a more intelligent and informed decision.

The large amount and complexity of information required for complete specification of multiattribute utility functions and probabilities lead, especially in practice, to the use of simplified heuristics. Often these are flawed.[2,25] Even in prescriptive or normative situations, screening procedures will often be needed to reduce the time, stress, and effort demanded from decision makers. Screening methods are intended to identify and reduce the size of the nondominated set of alternatives, that is to say those that are not bettered by at least one other alternative on each and every one of the attributes of importance, through use of behaviorally relevant and easily available information. One of the first to develop a screening method for decision making with incomplete knowledge of probabilities was Fishburn.[26,27] He was concerned with the use of incomplete information on probabilities in comparing alternative strategies in a typical formulation of decision making under risk. The criterion of choice or strategy that should normatively be used is the principle of maximum expected utility, so that the decision maker seeks a strategy a* which is the maximum over i of the expected utility of alternatives a_i. The utility function $U_j(a_i)$ is a precisely assessed multiattribute utility function defined on the set of possible strategies $\{a_i\}$, and $p_j(\cdot)$ is a measure of the likelihood of the possible state of nature j given that a particular strategy was selected. The imprecise forms of the measure of probability that Fishburn considered are

1. No information about $p_j(\cdot)$
2. Ordinal measure: an ordering of p_j, (e.g., $p_1 \geq p_2 \geq p_3$)
3. Linear inequalities: an ordering of sums of p_j, (e.g., $p_1 + p_2 \geq p_3$)
4. Bounded interval measure: (e.g., $c_i \leq p_i \leq c_i + d_i$)

The search for the best alternative is, in this approach, performed by pairwise comparisons of expected utility among candidate strategies. Because of the restricted form of the available information on the p_j, the search for the best alternative can be put, in general, into a straightforward linear programming problem.

Sarin[28] proposed a screening procedure similar to that of Fishburn with the additional assumption of additive independence on the set of attributes. This assumption simplifies the search for dominance structures and results in a procedure which can be formulated as a mathematical programming problem. The parameters of the mathematical programming formulation include probabilities, importance weights, and single attribute utility functions. A simple procedure is then developed for the case when the probabilities and the importance weights are precisely known and utilities are stated in the form of linear inequalities, thereby resulting in simple linear programming formulations. Extensions to this research have been reported by Sage and White[5] who develop an **A**lternative **R**anking **I**nteractive **A**id based on **D**ominance structural **E**licitation (ARIADNE) concept. The general mathematical programming formulation of the search for dominance results in the interactive solution of a large number of relatively simple linear programs. Several cases have been considered:

1. The probabilities (p_j) are known precisely, and the importance weights (w_i) and utilities (U_i) are described by linear inequalities.
2. The importance weights (w_i) are known precisely, and the utilities (u_i) and probabilities (p_j) are described by linear inequalities.
3. The importance weights (w_i) and utilities (u_i) are known precisely, and the probabilities (p_j) are described by linear inequalities.

In cases 1 and 2 the solution results in a set of hierarchically organized linear programming problems. The simplest formulation is that of case 3, which is equivalent to the problems solved by Sarin and Fishburn, resulting in a set of simple linear programming problems. Other approaches have been developed, but they all rely on the existence of additive independence conditions to facilitate the computations.

Decision-making problems in which there are several conflicting objectives have been formulated in terms of multiobjective programming problems. The mathematical programming formulation of the multiple objective decision problem is

$$\max_a U[u_1(a),u_2(a),\ldots,u_n(a)]$$

where the u_i are the real-valued utility functions of the n objectives. The single attribute utility functions u_i are assumed known; the overall utility function U is unknown. Generally, the solution to this problem is not a unique alternative but a set of nondominated alternatives. Multiobjective programming techniques operate under the notion of dominance for generating the set of nondominated alternatives. Interactive algorithms to gradually gain knowledge about U and solve this problem uniquely have been proposed by many researchers.

Screening procedures can, in general, be made to be interactive. Interactive approaches of this type assume that the decision maker can provide preference information on simple, often hypothetical, alternatives. Initially, a reduction of the nondominated set is made with the available information. If the decision maker can select an alternative from the nondominated set, then the process is stopped. Otherwise, further information is requested. Often, but not always, very little guidance is provided by these procedures about the information needed. The decision maker is asked to provide further information when a single nondominated alternative is not present. This information may be redundant in that the decision model could have inferred it from previous information. In this case it could serve only as a check for consistency. The new information may be inconsistent with the existing information. Alternately, the decision maker may never recognize and provide information needed to reduce the nondominated set. All this may make decision support processes that are based on interactive multiple objective procedures ineffective and inefficient.

The flexibility that the interactive screening, or scanning, procedures provide potentially results in a more effective support process. However, this flexibility complicates, to a considerable extent, the task of the decision analyst as a facilitator. In order to make efficient use of these new screening procedures, it seems necessary to provide the analyst with a suitable dialogue generation and management system. Such a system must be designed with full knowledge of the particular data base and model-based management system used to allow for the interactive dominance-based scanning. At the very least, it is necessary to provide assistance to the analyst in determining what information is most needed, in terms of relevancy to the task at hand and with due consideration being given to information that is both important and cognitively easy to assess. The analyst should be aided in the evaluation of acquired information for consistency and in ways to avoid and resolve inconsistencies that do result. Finally, the information that is acquired must be represented and used within a model-based management system that is valid and appropriate for decision making.

The foundations of multiattribute expected utility theory provide useful models of normative behavior for decision making under risk. The assessment of precise utility functions in these models is mathematically justified by the existence of a real-valued utility function. Whether it is behaviorally justified is yet another question. In addition to the several practical difficulties encountered in assessing precise utilities, there are a number of semantic issues involved with the precise representations of preference judgments. One of the aims of contemporary research is to seek representation that incorporates in complete measurements of preference and risk attitude and to provide a behaviorally meaningful as well as rationally correct approach for decision support. Instead of assessing a real-valued utility function, various "fuzzy" kinds of imprecise and otherwise imperfect representations of utilities and probabilities are allowed.

Procedures for screening multiattribute alternatives with partial information about preference judgments appear to be more behaviorally attractive to decision makers than more precise approaches for decision support such as multiattribute utility theory (MAUT) based models. This is primarily because these procedures require less precise assessments of preference and often reduce the number of assessments needed in order to reach a decision. These characteristics of screening procedures reduce the time and effort demanded from decision makers.

Screening procedures are, in general, interactive. They assume that the decision maker can provide preference information on simple, often hypothetical alternatives. Initially, a reduction of the set of alternatives is sought given the available information. If the decision maker can select an alternative from the reduced set of alternatives, then the process is stopped. Otherwise, additional information is requested until a further reduction of the set of alternatives is achieved.

Very little guidance is provided by these procedures about the information needed. The decision maker is asked to provide any additional information he may know and wish to express about the problem. In some cases, this information may be redundant, meaning that the decision model could have inferred it from previous information; consequently, it could serve only as a check for consistency. It may be inconsistent with the existing information, or the decision maker may never recognize and provide the necessary information to reduce the set of alternatives. All of this may cause the decision process based on interactive screening procedures to be ineffective and inefficient.

The analyst will generally need aid to determine what information is best to be requested, in terms of being relevant to the task at hand and cognitively easy to assess. He must be aided in the evaluation of acquired information for consistency and in ways to avoid and resolve them. Finally, the information must be aggregated and interpreted within a valid model for decision making. In this chapter, we develop an inquiring system, aimed at directing and controlling the assessment step of screening procedures that addresses the needs of the decision analyst.

The problem we are dealing with here is identifying the minimal set of information required for typical screening procedures of decision support in order to make a recommendation for decision.

Let the set of feasible alternatives be represented by $A = (a_1, a_2, \ldots, a_m)$ and the attribute set by $X = (X_1, X_2, \ldots, X_n)$. Define X^+ as the extended consequence set of dimension n v m that contains all the possible consequences that may occur. For the case of m alternatives and n attributes, the extended consequence set is given by:

$$X^+ = (x_1^1, x_2^1, \ldots, x_n^1, x_1^2, x_2^2, \ldots, x_n^2, \ldots, x_1^m, x_2^m, \ldots, x_n^m)$$

which contains nvm elements $x_i^j = X_i(a_j)$. The subscripts correspond to attributes and the superscripts, to alternatives. The power set of X^+ is the family of all subsets of X^+, and F_X represents a subset of the power set of X^+. We denote the set of alternatives by F_X.

The set of alternatives F_X allows only simple preference comparisons. That is, given two alternatives A^i and A^j in F_X, we could only assess whether $A^i > A^j$, $A^j > A^i$, or $A^i \sim A^j$, where the symbol $>$ indicates "is at least as good or preferred to" and $\sim$ indicates "indifference". Consider a strength of preference notion that involves comparison of preference differences. Denote by F_X^* a nonempty subset of $F_X v F_X$ so that $(A^i, A^j) \epsilon F_X^*$ if and only if $A^i > A^j$ and not $A^j > A^i$. The elements in F_X^* are the ordered pairs (A^i, A^j) defined by the relation $>$ in F_X. They indicate that the strength of preference of A^i over A^j is positive. F_X^* contains the weak preference information provided by the decision maker excluding indifference relations. Define a relation $> *$ in F_X^* and interpret $(A^i, A^j) > *(A^k, A^l)$ to mean that the strength of preference of A^i over A^j is at least as much as the strength of preference of A^k over A^l. The notation $(A^i, A^j) \sim * (A^k, A^l)$ means both $(A^i, A^j) > * (A^k, A^l)$ and $(A^k, A^l) > * (A^i, A^j)$. Also, $(A^i, A^j) > * (A^k, A^l)$ means not $(A^k, A^l) > * (A^i, A^j)$.

When certain conditions of rational behavior are satisfied by the binary relations $>$ in $F_X F_X$ and $> *$ in $F_X^* F_X^*$, representation results stating the existence of interval-valued mappings Q and Q* can be derived. These representation results can be described by set inclusion inequalities. The problem becomes a simple linear programming problem. When some form of independence among the attributes exists, such as preferential independence or mutual preferential independence, the set of inequalities is obtained by explicit assessments of simple and strength of preference relations and from preference relations inferred by independence and explicit assessments.

A goal of the inquiring system considered here is to identify a set of minimial information necessary to induce a linear order in the set of candidate alternatives. This minimal set constitutes the inquiry pattern, or set of queries that the system will present to the decision maker in order to learn about personal value perspectives and relevant factual information about the decision situation. At any time, the decision maker may refuse to respond to any query. The system will then search for another inquiry pattern leading to another, possibly the same, linear order. This approach is directed at achieving effectiveness of the overall process. Effectiveness, in this case, clearly requires adaptation and learning.

This goal-seeking behavior can be formulated as a set of simple deterministic dynamic programming problems. At every stage $k = 0, 1, \ldots, N$ we have a set $\{R^k\}$ representing the possible orders of the alternatives. At the initial stage $0, R^0$ represents no order on the set of alternatives. At the final stage $N, \{R^N\}$ represents the set of all possible linear orders. At intermediate stages $k = 1, \ldots, N-1$, $\{R^k\}$ represents the set of partial orders.

The policy q_k is a query obtained from the set of admissible queries Q_k at each stage $k = 0, 1, \ldots, N - 1$. There are a large number of possible preference comparisons in F_X, some of which may be more susceptible than others to the effects of biases and flawed heuristics. Through use of procedures developed in subsequent sections, we are able to select from the set of possible queries at each stage those that are behaviorally meaningful to the decision maker and operationally relevant to our goal of inducing a linear order on the set of alternatives F_X.

The effort function $g_k[R^k, q_k(R^k)]$ represents the effort associated with responding to query q_k at a particular state R^k. There will generally exist queries that are more difficult to understand, and hence respond to, than others. There will also be some queries that lead to a final state faster than others. The effort function should reflect this. The system equation $f(R^k, q_k)$ is obtained from the decision model for interactive aiding developed in Reference 19, and just reviewed. It computes the next state (order) of the system $R^{k+1} = f(R^k, q_k)$

given the current state (order) and the response to query q_k. The objective is to minimize the total effort of the inquiring process through identification of the optimal inquiry pattern, or minimal set of information, required to provide a recommendation.

Before proceeding with some commentary concerning development of the dynamic programming algorithm for the inquiring system, it is important to clarify the underlying sequence of events envisioned in the decision support process. These are

1. Identify an initial state R^0 (no order on the set of alternatives) and initial stage k = 0.
2. The inquiring system recognizes the state R^k and presents query $q_k(R^k)$ to the decision maker.
3. If no response is given to the query, then go to step 2 (another query is presented at the same stage).
4. The next order (R^{k+1}) on the set of alternatives is computed according to the decision model equation $R^{k+1} = f[R^k, q_k(R^k)]$.
5. If we are at the final stage (k = N − 1), then stop. Otherwise, go to step 2.

After this process is completed, a sequence of queries $\{q_0, q_1, \ldots, q_{N-1}\}$ is obtained. We call this sequence an inquiry pattern, and it specifies the queries to be presented at each stage for every possible order that may occur. In other words, it tells which query to select after the decision maker responds to the currently unresolved query.

It is important to note that although this process evolves much like a deterministic dynamic programming problem, i.e., the order of the alternatives is fully determined by knowing the order and the response to the query at each stage, there is an element of uncertainty in that the response to the query is not known *a priori*. Thus, the possible orders of the alternatives at each stage is a stochastic variable determined by the order at the previous stage and the probability distribution of the possible responses to each query at that stage. The decision model equation then becomes $R^{k+1} = f[R^k, w_k(q_k(R^k))]$. This is a stochastic control problem. Hence, it is more difficult to solve than a simple deterministic problem. We also face the problem of identifying a probability measure for the set of responses to each query at each stage. Thus, implementation of the inquiring system is not necessarily a trivial task without the ability to rapidly solve the posed computations.

The inquiry space is determined by the set of all possible simple preference comparisons among elements on the set of alternatives F_X and also strength of preference judgments in F_X^*. A set of m feasible alternatives to be evaluated in terms of n attributes results in a set of alternatives F_X with cardinality m^n. The inquiry space is then a set containing $m^n(m^n - 1) \sim 2$ possible queries regarding the relation of preference among pairs of alternatives in F_X, and the number of strength of preference queries depends on previous preference judgments in F_X. For notational purposes, let [i,j] denote a query requesting the preference relation among alternatives A^i and A^j in F_X, and assuming (i,j) and (k,l) are in F_X^*, denote [(i,j),(k,l)] a query requesting the strength of preference relation among previous simple preference judgments.

The response space constitutes the element of uncertainty in the inquiring system. When the decision maker is presented with query [i,j], the possible responses could be that alternative A^i is as good or preferred to A^j, A^j is as good or preferred to A^i, there is indifference between the two, or the decision maker might even refuse to respond to that query. The response space can be described by the set of relations {(i,j),(j,i)}, with (i,j) meaning that $A^i \cdot A^j$. That is, not (i,j) and not (j,i) is equivalent to not responding to that query, (i,j) and not (j,i) that A^i is as good or preferred to A^j, and (i,j) and j,i) that the decision maker is indifferent between the two alternatives. Likewise, the response for query [(i,j),(k,l)] is described by the set of relations {((i,j),(k,l)),((k,l),(i,j))} with a similar interpretation. A probability measure defined on the set of responses for each query at each state could represent the uncertainty associated with the response. This probability measure may depend explicitly on the present order of the alternatives and the query presented.

There are at least two possible representations for the inquiry space and the response space. The representation of the inquiry space is highly interrelated with the representation of the response space. First, we can assume that the set of queries is independent of the state of the system. This means, for example, that even when the decision maker has already stated that $A^i \cdot A^j$, we can present again query [i,j]. However, in this case the probability of obtaining the same response should be one. The probability measure on the set of responses is dependent on the current order of the alternatives and hence to the responses of previous queries. The other approach is to let the set of queries be state dependent in the sense that queries already responded to become inadmissible in subsequent stages. In this case, the probability of the responses only depends on the query selected.

Although both representations allow for the resolution of our inquiry problem, the second approach together with other special characteristics of the inquiring process helps reduce the computational complexity of the dynamic programming problems.

The states or orders of the alternatives are completely specified by the initial state (no order) and the set of possible queries at each stage. The state space is then the set of all the possible orders that can be constructed from the set of alternatives. An alternate representation for the orders of the alternatives is the set of queries still available to assess in F_X. These two representations are equivalent in the sense that knowing the state of one representation we can obtain the corresponding state on the other representation. Still another representation for the order of the alternatives is the number of queries still available to assess in F_X. This last representation is sufficient for the purpose of the inquiring system. It also has the advantage that its value is scalar rather than the content of a set as in the case of the other two representations.

The purpose of the inquiring system is to control the process of selection of queries from the set of admissible queries at each stage so as to induce a linear order on the set of alternatives with minimum effort from the decision maker. To accomplish this, the system needs to determine which queries to select to minimize the total effort demanded by the decision maker. Hence, the measure of effectiveness being used is the total effort demanded by the decision maker which is a function of:

1. The total number of queries selected in order to reach a linear order
2. The complexity of such queries

We are assuming that time is an important factor in decision making and thus minimizing the number of queries will reduce the time spent in the process. Also, the cognitive complexity of the queries contributes to the total effort demanded because preference comparisons where a few attributes are held at different levels are easier to understand, and hence to respond to, than those with many attributes held at different levels.

Given the initial state R^0, the problem is to find an inquiry pattern $\{q_0, q_1, \ldots, q_N\}$, where N is the total number of stages that minimizes the total effort. To make this approach operational, we must define the scalar-valued function g_k considering that the objective of the inquiring system is to keep to a minimum the amount and complexity of the queries required to make a recommendation. The inquiring system will always select the best admissible query so as to induce a linear order on the set of feasible alternatives. Although the criterion of effectiveness is based on simplicity and relevancy to performing the task, it may be possible, depending on the decision situation, that the system would have to select queries not necessarily simple according to the decision maker's conception of simplicity.

Facilitation of the decision-making process in this and similar situations does not necessarily ease the task of the decision maker. On the other hand, it accomplishes the other part of our measure of effectiveness in focusing on those queries that are relevant to the objective of discovering a preference order. The effort function $g_k[R^k, w_k(q_k)]$ determines the behavior of the inquiring system. An effort function that penalizes complex queries will result in a different inquiry pattern than an effort function that does not.

We should stress that although other functional forms of g_k satisfy our requirements, a different form of g_k will only affect the behavior of the inquiring system and not its purpose of conducting inquiry in the most effective way.

The dynamic programming formulation for the search of a large knowledge base may result in a very complex and large set of dynamic programming problems. The set of final states grows very rapidly with increases in the number of alternatives (m) and the number of attributes (n). There are $2^{nm}!$ possible linear orders. This fact may render this approach inefficient due to the computational complexity required each time a response to a query is made. Obviously, this complexity is reduced when some information is learned. Thus, it is possible to start the process with heuristically generated queries and redirect our search each time we learn about the true order of the set of alternatives. This combination of heuristic and dynamic programming improves the efficiency of the search for inquiry patterns.

As mentioned earlier, solving the dynamic programming formulation of the inquiring process may be computationally impractical or infeasible due to the dimensionality of the problem. A good part of this difficulty is removed by the fact that exact knowledge of the state of the system at each stage is obtained during the inquiring process. In other words, the decision maker articulates a preference among alternatives when given a query. A valid assumption in facilitating the computation is to let the inquiry space be state dependent. Every time the decision maker responds to a query, the cardinality of the set of admissible queries is reduced. These conditions of the inquiring process give the system the characteristic of being adaptive. It uses knowledge of the state at the current stage to readjust the search toward the final stage in the most effective way.

V. POTENTIAL EXTENSIONS

There are several significant future contributions which we believe could result from extensions of this effort. A very basic and first part of such efforts should concern the use of very general "value of information" techniques to direct and control the assessment of information, including imperfect information, that is required for judgment and choice. In this, the concern should be developing an inquiry process whereby the queries that best accomplish the objective of inducing a linear order on the set of feasible alternatives will be identified. The knowledge support system to be developed, based on the research accomplished here, would have two major components: a model for the analysis of alternatives and a metamodel for inquiry in charge of directing and controlling the interactive nature of a system that allows adaptation and learning. In effect these will allow for higher level metaknowledge, perhaps represented as formal reasoning-based knowledge, to direct and control the use of skill and rule-based knowledge which will typically be schema-like. Thus, it may be possible to resolve the problem of lower level knowledge explosion which often occurs in an expert system as the accumulation of knowledge, some of which may be inconsistent, leading to an unmanageably large number of rules.

The research concerning knowledge imperfection provides an approach by which defects in the knowledge base can be identified, avoided, and resolved, but only when it is important that this be done. This may involve or even require a formal scheme to explain to the decision maker, in a suitable way, the nature of information imperfections and the effects that may result from using different techniques to resolve them. This and the results of the portions of the study dealing with knowledge acquisition and representation could lead to the development of information presentation aids for purposes such as training, skill building, and decision making in operational contexts. The operational evaluation study should result in useful information concerning the ubiquity of the knowledge support system concepts being investigated and, hopefully, useful information concerning evaluation procedures for knowledge support systems in general.

VI. POTENTIAL APPLICATIONS

The development of computerized systems that aid human decision makers in stressful and potentially crisis situations, such as command and control tasks, is a primary need at this time.[16] It is important that decision process modeling be accomplished but that the decision process itself be output oriented, as opposed to process oriented, so that it is driven by the objective of higher quality outputs and not just a better process. Available information, especially in distributed decision situations is imperfect. The concepts described here and more fully developed in the cited references should be applicable to the development of support systems having these characteristics and capabilities. The approach to more effective model base management and knowledge support discussed here should also have ultimate applicability as a training and tutorial aid that provides understanding of the value of information in decision situations.

ACKNOWLEDGMENT

This research was sponsored by the U.S. Army Research Institute under research Contract MDA 903-82-C-0124.

REFERENCES

1. **Stephanou, H. and Sage, A. P.,** Perspectives on imperfect information processing, *IEEE Trans. Syst., Man, Cybern.,* SMC 17 (3), 1987.
2. **Sage, A. P., Ed.,** *System Design for Human Interaction,* IEEE Press, New York, 1987.
3. **Sage, A. P.,** *Methodology for Large Scale Systems,* McGraw-Hill, New York, 1977.
4. **Sage, A. P. and White, E. B.,** Decision and information structures in regret models of judgment and choice, *IEEE Trans. Syst., Man, Cybern.,* 13(2), 136, 1983.
5. **Sage, A. P. and White, C. C.,** ARIADNE: a knowledge based interactive system for planning and decision support, *IEEE Trans. Syst., Man, Cybern.,* 14(1), 35, 1984.
6. **Sage, A. P.,** On Human Information Processing and Inference Analysis as Large Scale Systems Problems, *Proc. 1982 Am. Autom. Control Conf.,* Washington, D.C., June 1982, 693.
7. **Sage, A. P. and Botta, R. F.,** On human information processing and its enhancement using knowledge-based systems *Large Scale Syst.,* 5(1), 35, 1983.
8. **Sage, A. P. and Lagomasino, A.,** Knowledge representation and man machine dialogue, in *Advances in Man Machine Systems Research,* Rouse, W. B., Ed., JAI Press, Greenwich, CT, 1984, 223.
9. **Lagomasino, A. and Sage, A. P.,** Representation and interpretation of information for decision support with imperfect knowledge, *Large Scale Syst.,* 9, 169, 1985.
10. **Lagomasino, A. and Sage, A. P.,** Imprecise knowledge representation in inferential activities, in *Approximate Reasoning in Expert Systems,* Gupta, M., Kandel, A., Bandler, W., and Kiszka, J. B., Eds., Elsevier/North-Holland, Amsterdam, 1986, 473.
11. **Lagomasino, A. and Sage, A. P.,** An interactive inquiry system, *Large Scale Syst.,* 9, 231, 1985.
12. **Sage, A. P. and Lagomasino, A.,** Computer based intelligence support: an integrated expert system and decision support systems approach, in *Expert Systems for Managers,* Silverman, B. G., Ed., TIMS Series in Artificial Intelligence and Management Science, Addison-Wesley, Reading, MA, 1986.
13. **Sage, A. P.,** On the management of information imperfection in knowledge based systems, in *Information Processing and Management of Uncertainty,* Bouchon, B. and Yager, R., Eds., Springer-Verlag, New York, 1987.

14. **Sage, A. P.,** A Taxonomy of Models of Cognitive Efforts in Human Problem Solving, *Proc. 6th Triennial World Congr. Int. Fed. Autom. Control,* Budapest, Hungary, July 1984.

15. **Sage, A. P.,** Behavioral and organizational models for human decisionmaking, *Policy Anal. Inf. Sci.,* 7(2), 1, 1985.

16. **Sage, A. P. and Rouse, W. B.,** Aiding the human decision maker through the knowledge based sciences, *IEEE Trans. Syst., Man, Cybern.,* 16(2), 1986.

17. **Sage, A. P.,** Knowledge, skills and information requirements for systems design, in *System Design: Behavioral Perspectives on Designers, Tools, and Organizations,* Rouse, W. B. and Boff, K. R., Eds. Elsevier/North-Holland, Amsterdam, in press, 1987.

18. **Sage, A. P.,** Behavioral and organizational considerations in the design of information systems and processes for planning and decision support, *IEEE Trans. Syst. Man, Cybern.,* SMC 11(9), 640, 1981.

19. **Lagomasino, A. and Sage, A. P.,** A learning approach for incorporation of imperfect knowledge in decision support system design, in Lecture Notes in *Control in Information Science: Real Time Control of Large Scale Systems,* Schmidt, G., Singh, M., Titli, A., and Tzafestas, S. B., Eds., Springer-Verlag, 1984, 70.

20. **Sage, A. P., Galing, B., and Lagomasino, A.,** Methodologies for determination of information requirements for decision support systems, *Large Scale Syst.,* 5(2), 131, 1983.

21. **Sage, A. P.,** Knowledge support needs in C^{3}I systems, in *Artificial Intelligence and National Defense: Applications to C^{3}I and Beyond,* Andriole, S. J., Ed., Armed Forces Communications and Electronics Association Press, Washington, DC, 1986.

22. **Sage, A. P.,** The role of knowledge support in C^{3}I systems, in *Expert Systems in Government,* IEEE Computer Society Press, Washington, D.C., October, 1986.

23. **Sage, A. P.,** Information technology and organizational judgment and choice foundations for command and control, in *Principles of Command and Control,* Andriole, S. J., and Boyes, J. L., Eds., Armed Forces Communications and Electronics Association Press, Washington, DC, in press, 1987.

24. **Keeney, R. and Raiffa, H.,** *Decisions with Multiple Objectives,* John Wiley & Sons, New York, 1976.

25. **Kahneman, D., Slovic, P., and Tversky, A., Eds.,** *Judgment Under Uncertainty: Heuristics and Biases,* Cambridge University Press, New York, 1982.

26. **Fishburn, P. C.,** Analysis of decisions with incomplete knowledge of probabilities, *Oper. Res.,* 13(2), 217, 1965.

27. **Fishburn, P. C.,** *Utility Theory for Decision Making,* John Wiley & Sons, New York, 1970.

28. **Sarin, R. K.,** Screening of multiattribute alternatives, *Omega,* Westport, CT, 5(4), 481, 1977.

29. **von Neumann, J. and Morgenstern, O.,** *Theory of Games and Economic Behavior,* 3rd ed., John Wiley & Sons, New York, 1964.

30. **von Winterfeldt, D. and Edwards, W.,** *Decision Analysis and Behavioral Research,* Cambridge University Press, London, 1986.

Chapter 8

FUZZY LINGUISTIC INFERENCE NETWORK GENERATOR

Thomas Whalen and Brian Schott

TABLE OF CONTENTS

I. NETWORK FAULT DIAGNOSIS

A large number of practical problem situations involving economic, electronic, biomedical, and other systems involve complex collections of interacting components which can be modeled as a directed graph or network. Typically, most of the components or "nodes" of the system are difficult or costly to assess directly; a few nodes, designated as "outputs", "symptoms", "indicators", etc., are exceptions to this general rule.

The first step in solving a problem (or recognizing an opportunity for improvement) in such a system is to use information about the indicator nodes to determine which of the other nodes have values which are "too high" or "too low". This question is a fundamentally qualitative one, even though it will usually lead to a subsequent quantitative decision to apply a measured stimulus to raise or lower the offending nodes. In fact, some systems, such as financial accounting systems, abound in variables which can be measured very precisely in a numeric sense but which are very difficult to assess as high, low, or medium. These assessments, which are necessary for management, must be inferred indirectly.

Fault diagnosis is the process of isolating the defective node, which is causing the unacceptable behavior at the indicator nodes, by deducing possible candidates given the collective behavior of the outputs and the interconnections in the network. The structure of the fault diagnosis problem and the way human beings go about solving it have been extensively studied by the Rouses and their associates, both as an abstract problem[9,10] and in real or realistic applied contexts.[1] This work has all been done in a "crisp" context in which each node is either "on" (functioning optimally) or "off" (not functioning at all), with individual nodes modeled as either **and** gates or as **or** gates. A more recent paper[6] uses fuzzy logic to model human performance in this nonfuzzy task.

In a fuzzy network, on the other hand, nodes are not restricted to just the two values, "on" and "off", but may also function at various intermediate levels of partial degradation. The output or indicator nodes are distinguished by relatively easy access to objective measurements as they are in the crisp formulation and, more importantly, by the fact that the user can classify these measurements using the fuzzy sets of acceptable and unacceptable values.

For the majority of nodes, either the objective measurements are costly to observe or their qualitative assessment is much more difficult than for the indicator nodes. One important source of difficulty in assessing nonindicator nodes occurs when their fuzzy sets of acceptable values are context-sensitive, depending strongly on the values of other nodes in the network. Also, despite clear goals about system outputs, we may not know what numeric values of nonindicator nodes are most compatible with these goals.

When poor performance is detected at the indicator nodes of a fuzzy network, we need to use qualitative assessments of the indicator nodes and the pattern of interconnections in the network to deduce assessments of the nonindicator nodes. In the present research this is done using fuzzy logic and linguistic variables combined into a fuzzy production system[14,15] to propagate assessments backward in the network from indicator nodes to nonindicator nodes.

II. FAULT: A FUZZY PRODUCTION SYSTEM FOR FINANCIAL NETWORK DIAGNOSIS

A production system is a body of problem-specific knowledge represented in a computer by a collection of "**IF** condition **THEN** conclusion" rules.[2] In a fuzzy production system, the actions and conditions are represented in terms of fuzzy sets and the **IF-THEN** relationship is represented by one of several systems of fuzzy logic.[16] In FAULT,[18] the propositions in the **IF** clause and the **THEN** clause refer to financial accounting ratios and line items. These propositions are expressed using linguistic fuzzy variables;[12,19] the inputs to FAULT are assessments of the ratios and its outputs are assessments of the line items, also expressed using fuzzy linguistic variables. The fuzzy logic used is based on Brouwerian many valued logic.[11,16,17]

A large part of the financial accounting profession's body of knowledge consists of ways to reduce numerous monetary amounts, by a division and subtraction, to a smaller number of more easily interpreted summary figures.[3,5,13] The relationships among these quantities can be easily expressed in network form. In this context, the main benefit of a fuzzy production system is the ability to make formal deductions based on the vague linguistic assessments most appropriate to the analysis of overall financial conditions of ongoing business enterprise or by extension of any phenomenon being modeled by a fuzzy network. Also, unlike standard logic, the concept of ''close' is readily defined in fuzzy logic: If the data come close to matching the **if** clause of a particular rule, without actually matching it completely, then the outcome of the rule is somewhat less definite than it would have been if the match had been closer.

Table 1 shows an example of the type of fuzzy reasoning used in FAULT. The rule is used repeatedly in the system to infer the qualitative value of a line item which appears in the numerator of two different ratios. Since linguistic assessments, like ''rather low'' or ''upper medium'', refer to how quantities relate to a specific system rather than how they are represented notationally, the assessments are invariant to a logarithmic transformation. This invariance allows the rule in Table 1 to also be used to deduce an assessment for a line item which is the minuend in two subtractions whose results have been already assessed, or even to assess a line item which is the minuend in a subtraction and the numerator in a ratio. In each case, the two known results are bound to RATIO1 and RATIO2 and the deduction is returned as NUM2. Other rule structures used in FAULT are similarly versatile. This kind of multipurpose rule structure is intermediate between true frame-based processing and the table-driven rules of Pasik and Schor.[8]

Figure 1 shows the small network of financial ratios and line items used to validate the FAULT concept. The user examines the numeric values of the ratios (P:E, ROE, EP, OPM, and GPM) and judgmentally assigns them linguistic assessments. FAULT then uses these linguistic assessments to deduce linguistic assessments of the remaining line items.

III. FLING: A FUZZY LINGUISTIC INFERENCE NETWORK GENERATOR

FAULT is a single purpose program designed to analyze the particular network, given assessments of the five financial ratios (Figure 1). FLING (Fuzzy Linguistic Inference Network Generator) automates the coding and much of the analysis which was done manually in developing FAULT. Rather than being limited to a single view of a single small network of financial variables and ratios, FLING is a flexible tool for general systems modeling.

The representation of the external system which is assumed by the current version of FLING is a directed graph or network in which each node expresses a qualitative relationship among three variables: one effect variable and two cause variables — major cause variable and minor cause variable. The node is labeled with the name of the effect variable, and has exactly two incoming arcs corresponding to the two cause variables: They are either effect variables of other nodes or fundamental variables, represented by nodes with no incoming arcs. Each effect variable corresponds to a quantity or quality of the external system which depends, in a positive sense, on the major input variable and, in either a positive or negative sense, on the minor input variable. A node can have any number of outgoing arcs, depending on the number of other variables in the external system which depend on the effect variable of the node in question.

TABLE 1
Sample Fuzzy Production Rule
If Ratio 1 Is Close To Ratio 2, Then Num 2 is Between Ratio 1 & Ratio 2

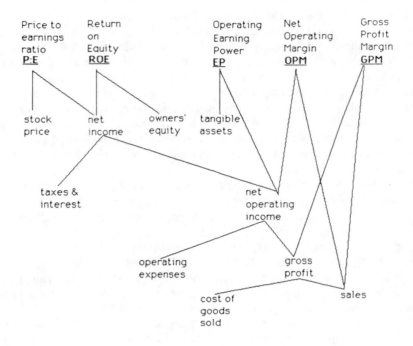

FIGURE 1.   Example network of financial ratios and line-items.

Division and subtraction, the only relationships recognized in FAULT, are represented to FLING in a straightforward manner. The ratio, or difference, becomes the effect variable; the numerator, or minuend, becomes the major cause variable; and the denominator, or subtrahend, becomes the minor cause variable. Because of the stated invariance of linguistic evaluations under logarithmic transformations, differences and ratios are not distinguished.

Addition and multiplication present more of a problem due to the FLING inference frames assuming a balanced relation of both positive and negative elements. This problem can be eliminated by "inverting" the relation to solve for the major cause variable, which is positively related to the former effect variable and negatively related to the minor cause variable. That is, its partial correlation is negative, with the minor cause variable controlling for the original effect variable.

However, FLING is not limited to the purely arithmetic relations of addition, subtraction, multiplication, and division. Any dependence relation of one variable on two others can be modeled, and any dependence relation of one variable on more than two others can be disaggregated into the requisite binary form. As a result, the FLING model has considerable potential generality beyond the original finance application.

FLING can generate a rule-based program comparable to FAULT to deduce which (if any) of the variables in an arbitrary external system have values that are too high or too low based on similar assessments of a subset of these nodes. There are two inputs to FLING. The first is a description of the network to be analyzed. This description is a list of triples each of which describes one node in terms of the effect variable, the major cause variable, and the minor cause variable. This list is divided into two sublists. The first sublist contains descriptions of nodes in which the relation between the effect variable and the minor cause variable is negative. The second sublist contains descriptions of nodes in which this relation is positive. FLING converts the nodes on the second sublist into the "inverted" internal representation discussed previously and adds these nodes to the first sublist, so that FLING works internally only with nodes in which the relation is balanced, like division or subtraction, or qualitative generalizations of these.

The second input to FLING is a list of variables in the network; these are the indicator nodes whose linguistic assessments will be the inputs to the rule-based system written by FLING. Using these two input lists, FLING writes a LISP program, which contains all necessary initialization and output codes, plus a collection of fuzzy production rules similar to the one in Table 1. These rules work together in a data-driven fashion to draw all possible conclusions regarding the network from the information about the input nodes which will be given the FLING-generated program by the user. Once generated by FLING, the resultant program can be saved and used whenever the user desires to evaluate the same network with new data values, e.g., at the end of every monthly business accounting period.

Note that the nodes whose assessments will form the *inputs* to the program written by FLING are the indicator nodes (outputs) of the external system being modeled and analyzed. The chain of events in a financial application is as follows:

1. The real firm generates the dollar amounts.
2. The summary figures are numerically calculated.
3. The user qualitatively assesses the summary figures.
4. The program written by FLING qualitatively assesses the remainder of the nodes in the system.

The qualitative assessments of the monetary amounts are not to be viewed as a substitute for or approximation of the actual dollar amounts, but rather as a necessary aid to making sense of the dollars. It is even more important for managerial decision making that a senior manager know whether a particular cost item is "too high" than it is to report the precise dollars-and-cents cost.

The advantages of an automatic programming approach such as FLING go beyond just reducing the programming labor involved in implementing a decision support system for financial analysts. In the first place, when FLING was validated, using the same network and input set used in FAULT, it detected and exploited a redundancy in the network which was not used in FAULT itself. The FLING-created program could detect when the set of assessments was internally inconsistent in some cases and flag this situation by adding the fuzzy modifier "possibly" to deductions based on conflicting data.

Another advantage is that it is possible to quickly change the list of inputs. When one of the regular inputs is unavailable, FLING can generate a set of rules designed to work around the gap. On the other hand, if extra information becomes available, FLING can write a program to optimally exploit it.

This ability enables a program to work in three modes depending on the list of indicator nodes. When the indicator nodes are all effects, then FLING writes a system which operates in a pure diagnosis mode. When the indicator nodes are all causes, the resulting program is a pure simulation. A mixture of causes and effects, among the designated indicators, results in a mixed simulation/diagnosis system. An example of such a hybrid system is discussed in Section V., "Common Size Analysis".

Finally, research in the selection and interpretation of financial accounting ratios can be facilitated by trying many different networks in conjunction with real or simulated business data.

IV. DEMONSTRATION OF NETWORK INFERENCE USING FAULT

The original FAULT model was exercised and validated using the facts from ECOFURN, a published case. The ECOFURN financial data were generated using PAVE,[4] a business simulation program developed to support **P**lanning **A**nd **V**aluation of actual business Enterprises. PAVE inputs are a set of detailed assumptions about the firm and its environment. The outputs are projected financial statements of the firm for current and future periods.

Three benchmark data sets were produced via simulation so that FAULT deductions could be compared with a known correct diagnosis. The first stage in generating the FAULT benchmark data was to modify the ECOFURN cast to generate three scenarios. In the first scenario the inputs give the simulated firm operating conditions which are quite favorable (but not absolutely ideal). In the second scenario, the cost of goods sold was higher than in the first and all other inputs were the same. In the third, operating expense was higher and all other inputs were as in the first. PAVE processed each set of inputs, generating three sets of financial statements which differed greatly from each other due to the direct and indirect effects of the different inputs.

In the second stage of the process, the five financial ratios used in FAULT were computed for each of the three scenarios and presented to an expert financial analyst. The expert's task was to provide verbal judgments of the five ratios in each set, using terms from Wenstop's[12] quasi-natural language, such as "high", "very low", or "upper medium".

The upper part of Table 2 shows the numeric values and the expert's linguistic interpretations of the ratios for the first of the three scenarios "near-normal operation". The lower part of Table 2 shows the results of FAULT deductions from these five linguistic inputs. As expected, nine of the ten line item assessments lie in the range between "upper medium" and "lower medium", reflecting the smooth operation of this simulated firm. The only exception is a linguistic value of "low" for stock price, which may reflect either an undervalued stock (and, therefore, a very good investment opportunity) or knowledge by investors of some problem outside the scope of this limited financial analysis.

Tables 3 and 2 differ most in regard to the cost of goods sold. The data in Table 3 were generated by increasing the cost of goods sold above the corresponding cost in the first scenario, and FAULT was able to detect this fact by deduction from the expert's linguistic evaluations of the five ratios. Cost of goods sold is assessed by FAULT as "high", which explains the assessments of "very low" for gross profit and operating income, and "rather low" for net income. Taxes and interest are removed from consideration as a cause by the assessment "not high". Since the expert's input gives no clear indications about operating expenses or tangible assets, these items are assessed as "unknown". A stock price assessment of "upper upper medium" means the poor performance of the firm has led to an overvalued stock whose price is very likely to fall soon unless management takes corrective action.

The assessment of owner's equity as "high" and sales as "very high" in Table 3 are somewhat paradoxical. In this context, they do not indicate good news about a firm with booming sales and high retained earnings, but rather bad news about a firm whose meager profits are not living up to what would be expected, based on the level of sales activity and the amount of assets entrusted to it by the owners. (See Section V. for an alternative way of handling these items.)

In Table 4 we see the results of the third scenario; the cost of goods sold was the same as in the first scenario, but operating expenses were higher. Again, FAULT detected this, assessing operating expenses as "high" and, thus, explaining the "low" operating income. Taxes and interest are eliminated again as a plausible cause by the assessment "not (more or less high)". Only the slightest shadow of suspicion is cast on cost of goods sold which, although numerically the same as in the first scenario, is assessed here in a different context

TABLE 2
Results for "Near-Normal" Scenario

Ratios	Numeric value (%)	Assessment
P:E	6.7	Low
ROE	11.2	Lower medium
EP	21.9	Medium
OpM	9.9	Lower medium
GPM	50.5	Upper medium

Line items	Numeric value ($)	Assessment
Owner's equity	225,106	Upper medium
Stock price	169,058	Low
Tangible assets	239,448	Medium
Sales	526,714	Medium
Gross profit	266,099	Medium
Operating income	52,355	Medium to lower medium
Net income	25,107	Upper medium
Cost of goods sold	260,615	Medium
Operating expenses	213,744	Medium to upper medium
Taxes and interest	27,248	Lower medium

TABLE 3
Results for "High Cost of Goods Sold" Scenario

Ratios	Numeric value (%)	Assessment
P:E	19.7	Upper upper medium
ROE	3.6	Very Low
EP	9.4	Very low
OpM	4.0	Very low
GPM	44.5	Low

Line items	Numeric value ($)	Assessment
Owner's equity	207,530	High
Stock price	148,186	Upper upper medium
Tangible assets	222,697	Unknown
Sales	526,714	Very high
Gross profit	234,604	Very low
Operating income	20,860	Very low
Net income	7,531	Rather low
Cost of goods sold	292,110	High
Operating expenses	213,744	Unknown
Taxes and interest	13,329	Not high

TABLE 4
Results For "High Operating Expenses" Scenario

Ratios	Numeric Value (%)	Assessment
P:E	12.4	Medium
ROE	5.9	Low
EP	13.1	Low
OpM	5.6	Low
GPM	50.5	Upper medium

Line Items	Numeric value ($)	Assessment
Owner's equity	212,642	High
Stock price	156,593	Medium
Tangible assets	222,089	Unknown
Sales	526,714	Upper medium
Gross profit	266,099	Lower medium
Operating income	29,760	Low
Net income	12,643	(Below medium) but not low
Cost of goods sold	260,615	Upper medium
Operating expenses	236,339	High
Taxes and interest	17,117	Not (more or less high)

as "upper medium". This is perhaps a reasonable assessment if lowering the cost of goods sold is viewed as a strategy to partly compensate for a perhaps unavoidable increase in operating expenses. As in Table 3, "high" assessment of sales and the "upper medium" assessment of owner's equity are to be interpreted as bad news rather than good news.

V. COMMON SIZE ANALYSIS

As mentioned in Section IV two of the ten line-item assessments deduced by FAULT, sales and owner's equity, are misleading unless very carefully interpreted. Simply ignoring these deduced assessments, however, is not advisable because the items are so closely interrelated. Instead, the FLING program generator was used to produce a new version of the program by treating sales and owner's equity as inputs whose value is always defined to be "medium". This approach is similar in philosophy to the "common size analysis" technique in the literature of financial analysis, in which all income statement items are expressed as a percentage of sales (making sales thus equal 100%) and all balance sheet items are expressed as a percentage of assets or equity.[7] However, expressing the line items as ratios themselves would greatly complicate the more sophisticated forms of financial indicator ratio analysis. Instead, the approach taken here is to reassess the linguistic evaluations of all line items, given the input assessments of indicator ratios, and also given the constraint that any departures from the nominal value of "medium" must be allocated other than to sales or owner's equity.

Using Brouwerian logic in each case, Tables 5, 6, and 7 compare the linguistic outputs of this program with those of the original FAULT system. The assessments of "medium" for sales and for owner's equity, in the column for the common-size model, are starred to indicate that they are inputs rather than outputs.

Not surprisingly, in Table 5 the results of the two systems are only trivially different from one another. Since the original FAULT system deduced "medium" and "upper medium", respectively, for sales and owner's equity, constraining them to be "medium" in the new system had little effect.

TABLE 5
Common-size Analysis, "Near-Normal" Scenario

Line Items	Original model	Common-size model
Owner's equity	Upper medium	Medium[a]
Stock price	Low	Low
Tangible assets	Medium	Medium
Sales	Medium	Medium[a]
Gross profit	Medium	Upper medium
Operating income	Medium to lower medium	Lower medium
Net income	Upper medium	Possibly (medium to upper medium)
Cost of goods sold	Medium	Lower medium
Operating expenses	Medium to upper medium	Upper medium
Taxes and interest	Lower medium	Possibly (medium to lower medium)

[a]   Inputs rather than outputs.

TABLE 6
Common-Size Analysis, "High Cost of Goods Sold" Scenario

Line items	Original model	Common-size model
Owner's equity	High	Medium[a]
Stock price	Upper upper medium	Upper upper medium
Tangible assets	Unknown	Unknown
Sales	Very high	Medium[a]
Gross profit	Very low	Low
Operating income	Very low	Very low
Net income	Rather low	Possibly (rather low)
Cost of goods sold	High	High
Operating expenses	Unknown	Unknown
Taxes and interest	Not high	(Very low) or possibly (not high)

[a]   Inputs rather than outputs.

TABLE 7
Common-size Analysis, "High Operating Expenses" Scenario

Line items	Original model	Common-size model
Owner's equity	High	Medium[a]
Stock price	Medium	Medium
Tangible assets	Unknown	Unknown
Sales	Upper medium	Medium[a]
Gross profit	Lower medium	Lower medium
Operating income	Low	Low
Net income	(Below medium) but not low	Possibly lower lower medium
Cost of goods sold	Upper medium	Lower medium
Operating expenses	High	High
Taxes and interest	Not (more or less high)	Low

[a]   Inputs rather than outputs.

In Table 6, defining sales and owner's equity to be "medium" slightly increases our confidence that the trouble in the firm is not due to excessive taxes and interest, since the mildly ambiguous "not high" changes from being the entire assessment of this line item to just a hedge, modifying the primary diagnosis of "very low". Another effect is a reduction in our confidence in the assessment of net income from "rather low" to "possibly rather low". This reflects a mild internal inconsistency among the seven inputs to the new system. (It is not reasonable to expect human judgment to achieve perfect internal consistency, especially in larger systems than the present pilot models. The ability to use inconsistent judgments and to flag the effects of inconsistency for possible reassessment is a major strength of the linguistic variables/approximate reasoning approach.)

Common size analysis is shown to its greatest advantage in Table 7, where the diagnosis of excessive operating expenses is thrown into sharper focus by the changes in the assessments of the other two expense items. In this case, cost of goods sold moves from "upper medium" to "lower medium" and taxes and interest moves from "not (more or less high)" to "low". Interestingly, the assessment for net income becomes simultaneously more precise and less definite as a result of having more information (seven inputs rather than five) brought to bear on it; the assessment is sharpened from "(below medium) but not low" to "lower lower medium" by the larger quantity of information but, at the same time, the mild internal inconsistency of the information requires that the hedge of "possibly" be prefixed to the latter assessment.

VI. CONCLUSION

FLING transforms a description of a general system, expressed as a directed fuzzy network with a subset of nodes designated as observable "indicator" nodes, into an executable LISP program. The resultant program performs diagnosis or simulation, or a mixture of both diagnosis and simulation, in order to propagate knowledge in the form of linguistic assessments, from the indicator nodes through the fuzzy network, to the remaining nonindicator nodes. FLING constructs all possible connections of the nodes and is capable of working around unknown facts, when necessary, and of constructing redundant knowledge links, when possible.

A prototyping strategy to knowledge base development is supported by FLING. New facts, relations, and assumptions are automatically and consistently accommodated.

Application contexts include network representations of financial ratios, social interaction networks, and causal modeling. The ability to handle cycles in networks is under study, especially in order to extend the model to dynamic systems.

REFERENCES

1. **van Eekhout, J. M. and Rouse, W. B.,** Human errors in detection, diagnosis, and compensation for failures in the engine control room of a supertanker, *IEEE Trans. Syst., Man, Cyber.* SMC-11, 813, 1981.
2. **Farley, A. M.,** Issues in knowledge-based problem solving *IEEE Trans. Syst. Man, Cybern.,* SMC 10, 446, 1980.
3. **Flores, I.,** *Data Structure and Management,* 2nd Ed., Prentice-Hall, Englewood Cliffs, NJ, 1977.
4. **Grawoig, D. and Hubbard, C.,** *Strategic Financial Planning with Simulation,* Petrocelli Books, New York, 1981.
5. **Lev, B.,** *Financial Statement Analysis: A New Approach,* Prentice-Hall, Englewood Cliffs, NJ, 1974.
6. **Hunt, R. M. and Rouse, W. B.,** A Fuzzy rule-based model of human problem solving, *IEEE Trans. Syst., Man, Cybern.,* SMC 14, 112, 1984.
7. **Lambrix, R. J. and Singhvi, S. S.,** How to set volume sensitive ROI targets, in *Using Logical Techniques for Making Better Decisions,* Dickson, D., Ed., John Wiley & Sons, New York 1983.

8. **Pasik, A. and Schor, M.,** Table-driven rules in expert systems, *ACM SIGART Newsl.*, No. 87, 31, 1984.
9. **Rouse, W. B. and Rouse, S. H.,** Measures of complexity of fault diagnosis tasks, *IEEE Trans. Syst. Man, Cybern.*, SMC 9, 720, 1979.
10. **Rouse, W. B., Rouse, S. H., and Pellegrino, S. J.,** A Rule-based model of human problem solving performance in fault diagnosis tasks, *IEEE Trans. Syst. Man & Cybern.*, SMC 10, 366, 1980.
11. **Sanchez, E.,** Solutions in composite fuzzy relation equations: Application to medical diagnosis in Brouwerian logic, in Gupta, M., et al., Eds., *Fuzzy Automata And Decision Processes,* Elsevier/North-Holland, New York, 1977.
12. **Wenstop, F.,** Quantitative analysis with linguistic variables, *Fuzzy Sets Syst.,* 4, 99, 1980.
13. **Weston, J. F. and Brigham, E. F.,** *Managerial Finance,* 6th ed., Dryden Press, Hinsdale, IL, 1978.
14. **Whalen, T. and Schott, B.,** Decision support with fuzzy production systems, in Wang, P., Ed., *Advances in Fuzzy Set Theory And Applications,* Plenum Press, New York, 1983.
15. **Whalen, T. and Schott, B.,** Issues in fuzzy production systems, *Int. J. Man-Mach. Stud.,* 19, 57, 1983.
16. **Whalen, T. and Schott, B.,** Alternative logics for approximate reasoning in expert systems. *Int. J. Man-Mach. Stud.,* 1985.
17. **Whalen, T. and Schott, B.,** Generalized network modeling and diagnosis using financial ratios, *Inf. Sci.,* 36, 179, 1985.
18. **Whalen, T., Schott, B., and Ganoe, F.,** Fault diagnosis in a fuzzy network, *Proc. 1982 Int. Conf. Cybern. Soc.,* 1982.
19. **Zadeh, L. A.,** The concept of a linguistic variable and its application to approximate reasoning, Part I: *Inf. Sci.,* 8, 199. Part II: Inf. Sci., 8, 301; Part III: Inf. Sci., 9, 43, 1975.

Chapter 9

ADVANCES IN AUTOMATED REASONING USING POSSIBILISTIC LOGIC

Didier Dubois, Jérôme Lang, and Henri Prade

TABLE OF CONTENTS

I. INTRODUCTION

Pieces of information stored in a knowledge base are not always regarded as equally certain. The lack of certainty is sometimes due to incomplete and/or conflicting evidence. Possibilistic logic, through the use of so-called necessity measures, offers a way of grading certainty on a numerical scale. As shown by Dubois,[1] these necessity measures are the unique numerical counterpart to a kind of qualitative ordering relations which model "is at least as certain as". These qualitative foundations of necessity measures suggest that when assessing degrees of certainty, the ordering of these numbers is more important than their absolute values. This notion of certainty is not probabilistic because its only ambition is to model the fact that, in a knowledge base, some sentences are more disputable than others, due to incomplete information, so that the body of knowledge can be stratified in layers of various strength levels. This view of uncertainty contrasts with the Bayesian paradigm where uncertainty derives from reasoning with exceptions; degrees of uncertainty are meant to summarize these exceptions (for instance, counting them as a surrogate of enumerating them). Of course, there is a debate regarding the Bayesian approach whose mathematics are often claimed to be tailored to the modeling of degrees of belief. They are also adapted to handle frequencies, be they subjectively estimated. This chapter does not mean to get into this controversy. It puts forward a more qualitative view of degrees of belief that relies on ordering. It is purposely weaker than a pure probabilistic approach.

Possibilistic logic is in accordance with Zadeh's possibility theory.[11] Possibility measures can also be viewed as consonant belief functions.[3] However, possibilistic logic is not a truth-functional, many-valued logic. Also it is not a logic of vagueness (as is fuzzy logic) because it pertains to nonfuzzy propositions, the truth of which is partially unknown due to incomplete information. It is, rather, a logic of partial ignorance. See Dubois and Prade[4] for the distinction between logic of vagueness and logic of uncertainty. Possibilistic logic allows for refutation techniques that stem from extensions of the resolution principle;[2,4] such an automated reasoning method has been implemented.[2] In this paper completeness properties of the resolution method are established. A nice feature of this logic is pointed out; namely, it can deal with partially inconsistent sets of uncertain sentences without resorting to extraneous control procedures.

II. BACKGROUND

Possibilistic logic uses propositional calculus formulas or first-order logic closed formulas, to which a possibility degree or a necessity degree, between 0 and 1, is attached. Let $\Pi(p)$ (respectively, $N(p)$) be the possibility (respectively, necessity) degree of p. We adopt the following conventions:

- $N(p) = 1$ means that, from the available knowledge, p is certainly true; $\Pi(p) = 0$, that it is impossible for p to be true.
- $\Pi(;\emptyset) = N(;\emptyset) = 0$; $\Pi(\mathbb{1}) = N(\mathbb{1}) = 1$ where $;\emptyset$ and $\mathbb{1}$ denote the contradiction and the tautology, respectively.
- $\forall p, \Pi(p) = 1 - N(\neg p)$, i.e., to say that p is impossible is equivalent to saying that $\neg p$ is certainly true.
- $\Pi(p) = \Pi(\neg p) = 1$ (equivalent to $N(p) = N(\neg p) = 0$) expresses that, from the available knowledge, nothing can disprove or confirm p (this is the case of total ignorance).
- $\forall p, \forall q, \Pi(p \vee q) = \max(\Pi(p),\Pi(q))$; this is the basic axiom of possibility measures,[3,11] which supposes that the imprecise of vague knowledge upon which the

attribution of possibility or necessity degrees is based, is coherent (i.e., can be described in terms of fuzzy sets or equivalently, in terms of a nested family of ordinary subsets); see Dubois Prade.[3] This is equivalent to $N(p \wedge q) = \min(N(p),N(q))$. However, we only have $\Pi(p \wedge q) \leq \min(\Pi(p),\Pi(q))$ (no equality in the general case) and, similarly, $N(p \vee q) \geq \max(N(p),N(q))$.

- Besides, we have $\max(\Pi(p),\Pi(\neg p)) = 1$ for any classical formula p, i.e., which does not involve any vague predicate or vague quantifier, indeed $p \vee \neg p = 1$ in this case. This implies that $N(p) > 0 \rightarrow \Pi(p) = 1$ (i.e., $N(\neg p) = 0$), which means that a formula is completely possible before being somewhat certain.

- This contrasts with fuzzy logic where the degree of truth v satisfies both $v(\tilde{p} \wedge \tilde{q}) = \min(v(\tilde{p}),v(\tilde{q}))$ and $v(\tilde{p} \vee \tilde{q}) = \max(v(\tilde{p}),v(\tilde{q}))$, as well as $v(\tilde{p}) = 1 - v(\neg \tilde{p})$ for vague propositions $\tilde{p}$ and $\tilde{q}$. But these propositions no longer obey excluded-middle or contradiction laws..

III. THE RESOLUTION RULES

Lee[9] proposed a deduction method in fuzzy logic based on mechanical theorem proving; it is a generalization of the resolution principle. This approach has been pursued by several authors, especially Mukaidono et al.[10]

The resolution rules we present here are not in the framework of fuzzy logic, but in possibilistic logic. The latter, closer to classical logic than fuzzy logic, enables us to represent knowledge whose truth or falsity is uncertain, but whose content is not vague. The classical rule for propositional clauses is generalized by:

$$\frac{\begin{array}{c} N(p \vee q) \geq \alpha \\ N(\neg p \vee r) \geq \beta \end{array}}{N(q \vee r) \geq \min(\alpha,\beta)} \tag{1}$$

in case of lower bounds on necessity measures[4] and the particularization rule by:

$$\frac{N(\forall x p(x)) \geq \alpha}{N(p(a)) \geq \alpha} \tag{2}$$

and, more generally, for any substitution σ, and any clause C:

$$\frac{N(C) \geq \alpha}{N(C\sigma) \geq \alpha} \tag{3}$$

Besides the following rule:[5]

$$\frac{\begin{array}{c} N(p \vee q) \geq \alpha \\ \Pi(\neg p \vee r) \geq \beta \end{array}}{\Pi(q \vee r) \geq \alpha \, l_m \beta} \tag{4}$$

where

$$\alpha \, l_m \, \beta = \begin{cases} \beta & \text{if } \alpha + \beta > 1 \\ 0 & \text{if } \alpha + \beta \leq 1 \end{cases}$$

holds when one of the lower bounds qualifies a possibility measure, as well as the counterpart of Equation 2:

$$\frac{\Pi(\forall x p(x)) \geq \propto}{\Pi(p(a)) \geq \propto} \tag{5}$$

An uncertain clause will be a first-order logic clause C to which a valuation is attached; it is a lower bound of its necessity or possibility measure. Thus, in the following we shall write (C (N $\propto$)) (respectively, (C Π $\propto$))) as soon as the inequality N(C) $\geq$ $\propto$ (respectively, Π(C) $\geq$ $\propto$) is known. Since N(p) > 0 implies Π(p) = 1, it is sufficient to consider clauses which are weighted either in terms of necessity or in terms of possibility.

We may notice the analogy of Equations 1 and 4 with the modal resolution rules (see Farinãs del Cerro[7]:

$$\frac{\square\ E \\ \square\ F}{\square\ \mathscr{R}(E,F)} \tag{6}$$

and

$$\frac{\square\ E \\ \Diamond\ F}{\Diamond\ \mathscr{R}(E,F)} \tag{7}$$

where $\mathscr{R}$ (E,F) is the resolvant of E and F. However, whereas E and F in Equations 6 and 7 may contain modal operators, possibilistic resolution is not allowed to handle knowledge such as N[Π(p) $\geq$ $\propto$] $\geq$ β yet (such an expression can be formally manipulated, but its semantics remain partially unclear for the moment). See Dubois, Prade, and Testemale[6] for a preliminary discussion of the links between possibilistic logic and modal logic. A stronger form of Equation 7 is valid where $\Diamond$($\mathscr{R}$ (E,F) $\wedge$ F) is derived; analogously, we can infer in Equation 4 the stronger conclusion Π(($\neg$p $\wedge$ q) $\vee$ r) $\geq$ $\propto$ 1_m β.

IV. REFUTATION IN POSSIBILISTIC LOGIC

The refutation method is generalized to possibilistic logic (Dubois, Prade.[4,5] Indeed, if we are interested in proving that p is true, necessarily or possibly to some degree, we add to the knowledge base $\mathscr{H}$ the assumption:

$$N(\neg\ p) = 1$$

i.e., that p is false (with total certainty). Let $\mathscr{H}'$ be the new knowledge base. Then it is proved that any valuation $\propto$ attached to the empty clause produced by the extended resolution from $\mathscr{H}'$ is a lower bound of the necessity (respectively, possibility) measure of the conclusion p, if its form is (N $\propto$) (respectively, (Π $\propto$)). It entails the existence of "optimal refutations," i.e., derivations of an empty clause with a maximal valuation, the valuations being ordered by:

$$(N\ \propto) \leq (N\ \beta) \text{ if and only if } \propto\ \leq\ \beta$$
$$(\Pi\ \propto) \leq (\Pi\ \beta) \text{ if and only if } \propto\ \leq\ \beta$$
$$(\Pi\ \propto) \leq (N\ \beta) \text{ for any } (\propto,\beta) \in [0,1] \times [0,1]$$

V. EXAMPLE

Let $\mathcal{H}$ be the following knowledge base:

C1   If Bob attends a meeting, then Mary does not.
C2   Bob comes to the meeting tomorrow.
C3   If Betty attends a meeting, then it is likely that the meeting will not be quiet.
C4   It is highly possible (but not certain at all) that Betty comes to the meeting tomorrow.
C5   If Albert comes tomorrow and Mary does not, then it is almost certain that the meeting will not be quiet.
C6   It is likely that Mary or John will come tomorrow.
C7   If John comes tomorrow, it is rather likely that Albert will come.
C8   If John does not come tomorrow, it is almost certain that the meeting will be quiet.

This can be represented by the following weighted clauses:

C1	$\neg$Bob(x) $\vee$ $\neg$Mary(x)	(N 1)
C2	Bob(m)	(N 1)
C3	$\neg$Betty(x) $\vee$ $\neg$quiet(x)	(N 0.7)
C4	Betty (m)	(Π 0.8)
C5	Mary(m) $\vee$ $\neg$Albert (m) $\vee$ $\neg$ quiet(m)	(N 0.8)
C6	John(m) $\vee$ Mary(m)	(N 0.7)
C7	$\neg$John(m) $\vee$ Albert (m)	(N 0.6)
C8	John(m) $\vee$ quiet (m)	(N 0.8)

If we want to try to prove that the meeting tomorrow will not be quiet, we add the clause C0: quiet(m) (N 1). Then it can be checked that there exist two possible refutations:

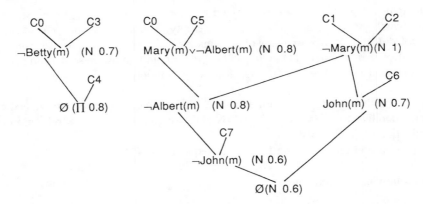

The second refutation is the optimal one. We proved that $N(\neg quiet(m)) \geq 0.6$, i.e., it is rather likely that the meeting tomorrow will not be quiet.

In order to find an optimal refutation from a set a weighted clauses $\mathcal{H}$, we have to define resolution strategies like in classical logic. A resolution strategy for weighted clauses will be said to be directly complete if and only if its application to a set of clauses $\mathcal{H}$ gives an optimal refutation of $\mathcal{H}$ (if it exists), in a finite number of applications of the resolution principle. Directly complete linear resolution strategies have been defined by adapting the ordered search algorithm A*, as follows. To each clause produced by the linear strategy we associate a state S, and for every state S we compute the functions g(S) and h(S), where g(S) is the cost of the path from the initial state to S and h(S) is the evaluation of the cost of the optimal path from S to a goal state. The involved costs are pairs of the form (N $\propto$) or (Π $\propto$), being ordered as stated in Section III. Then we compute:

$$f(s) = g(s)*h(s) \text{ where } (N \propto) * (N \beta) = (N \min (\propto,\beta))$$
$$(N \propto) * (\Pi \beta) = (\Pi (\propto 1_m \beta))$$
$$(\Pi \propto) * (\Pi \beta) = (\Pi 0)$$

and each time we expand the nondeveloped state maximizing f. If the heuristic function h verifies $\forall S$, $h(S) \geq \hat{h}(S)$, where $\hat{h}(S)$ is the real optimal cost from S to a goal state, then the strategy is directly complete. See Dubois, Lang, and Prade[2] for the case of (N $\propto$)-valued clauses.

VI. SEMANTIC ASPECTS

A semantic has been defined for clauses weighted by lower bounds of a necessity measure. If p is a closed formula, M(p) the set of the models of p, then the models of (p (N $\propto$)) will be defined by a fuzzy set M(p (N $\propto$)) with a membership function:

$$\mu_{M(p \ (N\propto))} (1) = 1 \text{ if } 1 \in M(p)$$
$$= 1 - \propto \text{ if } 1 \in M(\neg p)$$

Then the fuzzy set of the models of a knowledge base $\mathcal{H} = \{C_1, C_2,...,C_n\}$, where C_i is a closed formula with its weight, will be the intersection of the fuzzy sets $M(C_i)$, i.e.:

$$\mu_{M(\mathcal{H})}(1) = \min \mu_{M \ (Ci)} (1)$$
$$i = 1,...,n$$

The *consistency* degree of $\mathcal{H}$ will be defined by $c(\mathcal{H}) = \max_1 \mu_{M(\mathcal{H})}(1)$; it estimates the degree to which the set of models of $\mathcal{H}$ is not empty. The quantity $\text{Inc}(\mathcal{H}) = 1 - c(\mathcal{H})$ will be called degree of *inconsistency* of $\mathcal{H}$. Finally, we say that $\mathcal{F}$ is a *logical consequence* of $\mathcal{H}$ if and only if $\forall 1$, $\mu_{M(\mathcal{F})}(1) \geq \mu_{M(\mathcal{H})}(1)$. Let us note that all these definitions recover those of classical logic. We shall use the following notations:

- $\mathcal{H} \vdash C$ with $C = (C^* (N \propto))$ if and, only if, from the set of necessity-valued clauses corresponding to $\mathcal{H}' = \mathcal{H} \vee \{ \neg C (N ;1)\}$, we can produce an $\propto$-refutation (i.e., a deduction of $(0 (N \propto))$).
- $\mathcal{H} \models C$ if and only if C is a logical consequence of $\mathcal{H}$.

Then the following theorems hold:

THEOREM 1
For any necessity-valued clauses $C = (C^* (N \propto))$ and $C' = (C'^* (N \beta))$, $C,C' \vdash C''$ implies $C,C' \models C''$, where C'' is a weighted clause.

Proof
We have to prove that $\forall 1$ $\mu_{M(C'')} (1) \geq \mu_{M(C \wedge C')}(1)$.

- If 1 is a model of the classical formula $C^* \wedge C'^*$, then $\mu_{M(C \wedge C')}(1) = 1$ and the soundness of the classical resolution principle enables us to say that $1 \in M(C''^*)$, where C''^* is the resolvant of (C^*, C'^*). Thus we have $\mu_{M(C'')}(1) = 1 \geq \mu_{M(C \wedge C')}(1)$.

• If 1 is not a model of $C^* \wedge C'^*$, then $\mu_{M(C \wedge C')}(1) \in \{1 - \alpha, 1 - \alpha'\}$, and $\mu_{M(C \wedge C')}(1) \leq \max (1 - \alpha, 1 - \alpha') \leq 1 - \min (\alpha, \alpha')$. Besides, by definition we have $\mu_{M(C'')}(1) = 1 - \min (\alpha, \alpha')$ and then $\mu_{M(C'')}(1) \geq \mu_{M(C \wedge C')}(1)$.

Q.E.D.

Corollary

Let $\mathcal{H}$ be a set of necessity-valued clauses, then any necessity-valued clause C derived from $\mathcal{H}$ is a logical consequence of $\mathcal{H}$, i.e., the resolution principle for necessity-valued clauses is sound.

Proof

By induction on the refutation, using Theorem 1.

THEOREM 2

Let $\mathcal{H}$ be a set of necessity-valued clauses. The $\mathcal{H}$ is α-**inconsistent** ($\alpha > 0$) $\rightarrow$ **there is an α-refutation from** $\mathcal{H}$,, i.e., the resolution principle for necessity-valued clauses is complete for the refutation.

LEMMA 2.1: $\mathcal{H}$ is α-inconsistent
$\Leftrightarrow (;\emptyset (N \alpha))$ is a logical consequence of $\mathcal{H}$.
Proof
Let us note $V_\alpha = (;\emptyset (N \alpha))$.
First, let us notice that $\forall 1, \mu_{M(V \alpha)} (1) = 1 - \alpha$ (trivial). Then we have $\mathcal{H}$ is α-inconsistant $\Leftrightarrow \forall 1, \mu_{M(\mathcal{H})}(1) \leq 1 - \alpha \Rightarrow \forall 1, \mu_{M(V \alpha)} (1) = 1 - \alpha \geq \mu_{M(\mathcal{H})}(1) \Leftrightarrow V_\alpha$ is a logical consequence of $\mathcal{H}$.

LEMMA 2.2: V_α is a logical consequence of $\mathcal{H}$ ($\alpha > 0$) $\Rightarrow V_\alpha$ is a logical consequence of $\mathcal{H}_\alpha$, where $\mathcal{H}_\alpha$ is the subset of the clauses of $\mathcal{H}$ whose valuation is greater than or equal to $(N \alpha)$.
Proof
$\quad V_\alpha$ is a logical consequence of $\mathcal{H}$
$\Rightarrow \forall 1, \mu_{M(V \alpha)}(1 \geq \mu_{M(\mathcal{H})}(1))$
$\Rightarrow \forall 1, 1 - \alpha \geq \mu_{M(\mathcal{H})}(1)$
$\Rightarrow \forall 1, \min \{\mu_{M(Ci)}(1), 1 \leq i \leq n\} \leq 1 - \alpha$
$\Rightarrow \forall 1, \min (\min \{\mu_{M(Ci)}(1), Ci \in \mathcal{H}_\alpha\}, \min \{\mu_{M(Ci)}(1), Ci \in \mathcal{H}\alpha\}) \leq 1 - \alpha$ (A)
Besides we have $\forall Cj \in \mathcal{H} - \mathcal{H}_\alpha, Cj = (Cj^*, \alpha')$ with $\alpha' < \alpha$. Hence $\mu_{M(Cj)}(1) \geq 1 - \alpha' > 1 - \alpha$ and $\mu_{M(\mathcal{H} - \mathcal{H}\alpha)}(1) > 1 - \alpha$ and (A) $\Rightarrow \forall 1, \mu_{M(\mathcal{H}\alpha)}(1) \leq 1 - \alpha \Rightarrow \forall 1, \mu_{M(V\alpha)}(1) = 1 - \alpha \geq \mu_{M(\mathcal{H}\alpha)}(1) \Rightarrow V\alpha$ is a logical consequence of $\mathcal{H}_\alpha$

LEMMA 2.3: V_α is a logical consequence of $\mathcal{H}_\alpha \Rightarrow \mathcal{H}_\alpha^*$ is inconsistent in classical logic, where $\mathcal{H}_\alpha^*$ is the set of clauses of $\mathcal{H}_\alpha$ without their weights.
Proof
$\mathcal{H} = \{C_i 1 \leq i \leq n\} = \{(C_i^* (N \alpha_i)), 1 \leq i \leq n\}$
$(\emptyset (N \alpha))$ is a logical consequence of $\mathcal{H}_\alpha$
$\Rightarrow \mathcal{H}_\alpha$ is α-inconsistent (Lemma 2.1)
$\Rightarrow \forall 1, \mu_{M(\mathcal{H}\alpha)}(1) \leq 1 - \alpha < 1$
$\Rightarrow \forall 1, \exists i(1), \mu_{M(Ci(1))}(1) < 1$
$\Rightarrow \forall 1, \exists i(1)$ such that 1 is not a model of $C_{i(1)}^*$ in classical logic (else we would have $\mu_{M(Ci(1))}(1) = 1$)
$\Rightarrow \mathcal{H}_\alpha^*$ is inconsistent in classical logic

LEMMA 2.4: $(\emptyset(N \propto))$ is a logical consequence of $\mathcal{H}_\propto \Rightarrow$ there is a $(N \propto)$-refutation from $\mathcal{H}$.

Proof

$(\emptyset (N \propto))$ is a logical consequence of $\mathcal{H}_\propto \Rightarrow \mathcal{H}_\propto^*$ is inconsistent in classical logic (Lemma 2.3)$\Rightarrow$ there is a refutation by resolution form $\mathcal{H}_\propto^*$ (completeness of the resolution principle in classical first-order logic), and the refutation of $\mathcal{H}_\propto$ using the same clauses (with the associated valuations) has a valuation greater or equal than $(N \propto)$ (since each clause of $\mathcal{H}_\propto$ has a valuation greater than or equal than $(N \propto)$). The soundness of the resolution principle enables us to say that this refutation cannot have a valuation strictly greater than $\propto$; besides, a refutation of $\mathcal{H}_\propto$ is also a refutation of $\mathcal{H}$ so we conclude that there is a $(N \propto)$-refutation from $\mathcal{H}$.

Proof of theorem 2:

   $\mathcal{H}$ is $\propto$-inconsistent

$\Leftrightarrow (\emptyset (N \propto))$ is a logical consequence of $\mathcal{H}$ (Lemma 2.1.)

$\Leftrightarrow (\emptyset (N \propto))$ is a logical consequence of $\mathcal{H}_\propto$ (Lemma 2.2.)

$\Leftrightarrow$ there is a $(N \propto)$-refutation from $\mathcal{H}$ (Lemma 2.4.)

Q.E.D.

Completeness when possibility-valued clauses are allowed remains a topic for further research. However, the model $M(p (\Pi \propto))$ of a possibility-valued clause $(p (\Pi \propto))$ should be a fuzzy set with a membership function partially defined by the constraint:

$$\mu_{M(p (\Pi \alpha))} (1) \geq \alpha \text{ if } 1 \in M(p)$$
$$\geq 0 \text{ otherwise}$$

Note that the clauses weighted in terms of possibility have no influence on $c(\mathcal{H})$.

VII. NONMONOTONICITY

   It is possible to work with a partially inconsistent knowledge base $\mathcal{H}$. Let $\propto$ be the inconsistency degree of $\mathcal{H}$. From $\mathcal{H}$ we want to prove C with some necessity degree. Let β be the necessity degree of the optimal empty clause derived from $\mathcal{H}' = \mathcal{H} \cup \{(\neg C (N 1))\}$. Then, if $\beta > \propto$, † there exists a consistent sub-base $\mathcal{S}$ of $\mathcal{H}$ from which we can infer $(C (N \beta))$ by resolution, and then we will consider the proof of $(C (N \beta))$ as valid. Indeed, the β-refutation uses only clauses with a necessity degree greater or equal to β, i.e., strictly greater than $\propto$; $\mathcal{S}$ is the subset of $\mathcal{H}$ containing the clauses of $\mathcal{H}$ used in this β-refutation, and we cannot produce any refutation from $\mathcal{S}$ only, since this refutation would have a valuation strictly greater to $(N \propto)$, and then $\mathcal{H}$ would have had an inconsistency degree greater than $\propto$.

   Moreover, as pointed out in Reference 8, adding to a consistent knowledge base $\mathcal{H}$, a clause $(C (N\propto))$ that makes it partially inconsistent, produces a nonmonotonic behavior. Namely, if from $\mathcal{H}$ a conclusion $(p (N \beta))$ can be obtained by refutation, it may happen that from $\mathcal{H}' = \mathcal{H} \cup \{(C (N \propto))\}$, an opposite conclusion $(\neg p (N \gamma))$ with $\gamma > 1 - c(\mathcal{H}') \geq \beta$ can be derived.

EXAMPLE: We again work with the knowledge base of Section IV. Suppose we add to $\mathcal{H}$ the clause $(\neg John(m) (N1))$, i.e., $\propto = 1$, expressing that we are now certain that John

---

† We always have $\beta \geq \propto$. More precisely, we have $\min(\beta,\beta') = \propto$ where β' is the degree attached to the optimal empty clause derived from $\mathcal{H} \cup \{C (N 1))\}$.

will not come tomorrow. Let $\mathcal{H}'$ be the new knowledge base. The inconsistency degree of $\mathcal{H}'$ is 0.7, i.e., $1 - c(\mathcal{H}') = 0.7$:

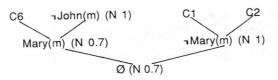

Now the proof of ($\neg$quiet(m) (N 0.6)) (it corresponds to $\beta = 0.6$) is no longer valid; but we prove (quiet(m) (N 0.8)), i.e., $\gamma = 0.8$:

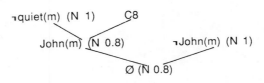

using only a consistent part of $\mathcal{H} \cup \{(\neg\text{John}(m)\ (N\ 1))\}$. Thus, a nonmonotonic behavior can be captured in this framework.

VIII. CONCLUDING REMARKS

The resolution patterns and the refutation strategy presented in this chapter are implemented in the system POSLOG (for POSsibilistic LOGic) on a microcomputer.

Besides, extensions of the resolution principles Equations 1 and 4 presented here to clauses with **vague** predicates have been proposed in Dubois and Prade.[5]

ACKNOWLEDGMENT

A previous version of the ideas discussed in this paper were the topic of oral communication at JILIA-88, European Workshop on Logical Methods Artificial Intelligence, Roscoff, France, June 27—30, 1988.

REFERENCES

1. **Dubois, D.,** Possibility theory: towards normative foundations, in *Decision, Risk, and Rationality,* Munier, B., Ed., D. Reidel, Dordrecht, 1988, 601.
2. **Dubois, D., Lang, J., and Prade, H.,** *Theorem proving under uncertainty — A possibility theory-based approach,* Proc. 10th Int. Joint Conf. Artif. Intell., Milano, August 1987, 984.
3. **Dubois, D. and Prade, H., (with the collaboration of Farreny, H., Martin-Clouaire, R., and Testemale, C.,)** Théorie des Possibilités, Applications à la Représentation des Connaissances en Informatique, Masson, Paris, (2nd rev. ext. ed. 1987), English version: *Possibility Theory, An Approach to Computerized Processing of Uncertainty,* Plenum Press, New York, 1988.
4. **Dubois, D. and Prade, H.,** Necessity measures and the resolution principle, *IEEE Trans. Syst., Man Cybern.,* 17, 474, 1987.
5. **Dubois, D. and Prade, H.,** Resolution principles in possibilistic logic, *Int. J. Approx. Reason.,* 4, 1, 1990.
6. **Dubois, D., Prade, H., and Testemale, C.,** In search of a modal system for possibility theory, *Proc. 8th Europ. Conf. Artif. Intell.,* Pitman Publishing, London, 1988, 501.
7. **Fariñas del Cerro, L.,** Resolution modal logic, *Logique et Analyse,* 110—111, 153, 1985.

8. **Besnard, P., Cordier, M. O., Dubois, D., Fariñas del Cerro, L., Froidevaux, C., Moinard, Y., Prade, H., Schwind, C., and Siegel, P.** (Groupe Léa Sombé) Raisonnements sur des informations incomplètes en intelligence artificielle, Teknea, Toulouse, *Int. J. of Intell. Syst.,* John Wiley & Sons, New York., 1989 (English trans.).
9. **Lee, R. C. T.,** Fuzzy logic and the resolution principle, *J. Assoc. Comput. Mach.,* 19, 109, 1972.
10. **Mukaidono, M., Shen, Z. L., and Ding. L.,** Fundamentals of fuzzy Prolog, *Int. J. Approx. Reason.,* 3, 179, 1989.
11. **Zadeh, L. A.,** Fuzzy sets as a basis for a theory of possibility, *Fuzzy Sets Syst.,* 1, 3, 1978.

Chapter 10

FUZZY ASSOCIATIVE MEMORY SYSTEMS

Bart Kosko

TABLE OF CONTENTS

I. INTRODUCTION: ASSOCIATION GENERALIZES CONDITIONING

Association generalizes conditioning. When B is conditioned on A, B is associated with A. The association pair (A,B) then *recalls* B when presented with A:

$$A \rightarrow (A,B) \rightarrow B$$

Associative memories map data to data. Random access memories (RAMs) map addresses to data. Content addressable memories (CAMs) map data to addresses. Parallel distributed associative memories superimpose associated data on the same memory medium. RAMs are not parallel distributed; they are serial and local. Neural-network CAMs are parallel distributed.[46] Autoassociative[32,33] associative memories store patterns from the same vector space. A piece of a stored pattern recalls the entire pattern. More generally, heteroassociative memories store pairs of unrelated data: $(A_1, B_1), \ldots, (A_m, B_m)$, where A_i is an n-vector and B_i is a p-vector. The stored pair (A_i, B_i) behaves as a logical biconditional, or metarule: If A_i, Then B_i; If B_i, Then A_i.

Association strength is a matter of degree, just as conditioning is a matter of degree. *Fuzzy association* is storing and recalling uncertain associations. The associations can be uncertain, or the associated patterns A and B can be uncertain, or both. The storage media are *fuzzy associative memories* (FAMs). Following we discuss several easy-to-construct FAMs, how they behave, and how the associative recall process in neural networks/associative memories often amounts to minimizing a fuzzy entropy.

Modus ponens is a paradigm of associative memory. In propositional logic, modus ponens is the theorem that if A is true and the conditional $A \rightarrow B$ is true, then B is true. The pair (A,B) is stored by the conditional $A \rightarrow B$, which serves as a type of memory. When an input or search key C is presented to the memory, B is recalled if, and only if, C $=$ A. When the fuzzy truth values $t(A)$, $t(B)$, $t(A \rightarrow B) \in [0,1]$, the conditional is a skeletal FAM. When the stored pair (A, B) is presented with the approximate datum $A' \approx A$, then $B' \approx B$ is recalled. For example, if $p = t(A \rightarrow B) = \min(1, 1 - t(A) + t(B)) < 1$, then $t(B) = p + t(A) - 1$. Hence $t(B') \approx t(B)$ if $t(A') \approx t(A)$.

How are the associations (A_i, B_i) obtained? In expert-systems theory the corresponding conditionals are obtained by the *knowledge engineering process* by asking for them. In FAM theory, the pattern associations are obtained by the *association engineering process* by asking for them, by abstracting situation-response prototypes from historical data, or by autonomously growing them in adaptive learning networks.

II. FUZZY THEORY REVIEW AND NEW FOUNDATIONS

We present a geometric interpretation of sets and fuzzy sets as points on or in unit hypercubes. A set A is a binary n-vector — a *bit* vector. A $= (1\ 0\ 1\ 1)$ is a set, or message. A set A is a point in the Boolean n-cube $B^n = \{0,1\}^n$, the set of all subsets of the space $X = \{x_1, \ldots, x_n\}$. X has 2^n subsets. A is equivalent to one of the 2^n-many mappings $m_A: X \rightarrow \{0,1\}$; $m_A(x_i)$ is the *degree* of membership of the ith element x_i in A. The numeral 0 indicates no membership; 1 indicates total membership. The space X is the unit vector $(1, \ldots, 1)$. The empty set $\emptyset$ is the null vector $(0, \ldots, 0)$. The *power set* of X, denoted 2^X, is the set of all subsets of X: $2^X = \{A \in B^n: A \subset X\}$. Hence, $2^X = B^n$. In general, the power set of B, 2^B, is the set B subsets. A is a *subset* of B if $m_A(x_i) \leq m_B(x_i)$ for all i; equivalently, if $x_i \in A$ implies $x_i \in B$. Hence $2^B \subset 2^X$, a subcollection of B^n. The intersection $A \cap B = \{x: x \in A \text{ and } x \in B\}$, the set of elements common to A and B; equivalently, $m_{A \cap B} = \min(m_A, m_B)$. Dually, the union $A \cup B$ is the set of elements in either A and B, or $m_{A \cup B} = \max(m_A, m_B)$. The complement $A^c = X/A = \{x: x \notin A\}$, all elements of X not in A; equivalently, $m_A c = 1 - m_A$. The *cardinality* or size of A is the number of elements in A. If $A = (1\ 0\ 1\ 1)$, the cardinality of A is 3. If A is not finite — if A can be put into a one-to-one mapping with a proper subset of itself — then its cardinality is ∞. The l^p *distance* between sets A and B is the pth root of the sum of the pth absolute differences:

$$l^p(A,B) = \left(\sum_i |m_A(x_i) - m_B(x_i)|^p \right)^{1/p}$$

$1^1(A,B) = H(A,B)$ is Hamming distance, the number of vector slots in which the binary patterns A and B differ. If $A = (1\ 0\ 1\ 1)$, then $H(A, X) = 1$.

A *fuzzy set* A is a *fit* vector. $A = (.2\ 0\ .5\ .7)$ is a fuzzy set. A is a point in the unit hypercube $I^n = [0,1]^n$. Every set is a fuzzy set since $B^n \subset I^n$. (An amazing fact is $B \infty = I \infty$, since $I \infty = c \infty = 2^{\infty \cdot \infty} = 2 \infty = B \infty$, where ∞ is countable infinity and c is the cardinality of the continuum.) A is a proper fuzzy set if $A \in I^n/B^n$, if some fit value of A is not 0 or 1. $F(2^X)$ is the fuzzy power set of X, the nonfuzzy set of all fuzzy subsets in X. A fuzzy set A is equivalent to some mapping m_A into fit values: $m_A: X \to [0,1]$. As stated, there are as many fuzzy subsets of any set as there are real numbers. Therefore, $F(2^X) = I^n$. If A is a finite discrete fuzzy set — i.e., m_A maps into a finite discretized set, $m_A: X \to \{d_1, \ldots, d_m\}$ — then there are m^n discrete fuzzy sets or messages. As such, fuzziness is characterized by the cardinality of range sets. The *fit value*, or fuzzy unit value, $m_A(x)$ indicates the degree to which x fits in the subset A, its elementhood or belongingness.

Fuzzy set intersection, union, complement, l^p distance, and cardinality have been defined previously for nonfuzzy sets. In particular, the cardinality[12] of A is called the *sigma-count*:[52-54] $\Sigma \text{Count}(A) = \Sigma m_A(x_i)$, the sum of fit values. If $A = (.2\ 0\ .5\ .7)$, $\Sigma \text{Count}(A) = 1.4$. Therefore, fuzzy cardinality is a real number, not necessarily an integer. The sigma-count is a positive measure[36] that generalizes classical counting measure in combinatorics. The *height* H(A) of A is the largest fit value in A: $H(A) = \max\{m_A(x_1), \ldots, m_A(x_n)\}$.

The fit vector A can also be interpreted as a *possibility* distribution.[51] I^n is the box of possibilities. In the special case when $\Sigma \text{count}(A) = 1$, A is a *probability* distribution. The set of probability distributions (convex coefficients) in I^n forms a simplex of dimension $n - 1$. In I^2 the probability simplex is a line. In I^3 it is a solid triangle. In I^4 it is a tetrahedron, and so on up. For example, the probability simplexes in the unit square and unit cube are illustrated here with appropriate shading. Note that the locus of equipossible distributions always intersects the probability simplex exactly at the point $(1/n, \ldots, 1/n)$, the uniform distribution with maximum probabilistic (Boltzmann/von Neumann/Shannon) entropy,[50] as discussed later (see Figure 1). Hence, our geometric interpretation of fuzzy sets puts probability theory in a box — and in a corner at that!

We now introduce some new fuzzy concepts.[39] How fuzzy is A? If A is nonfuzzy, a vertex of I^n, then A has 0 fuzziness, or 0 membership in the set of fuzzy or uncertain objects. If $A = (.5, \ldots, .5)$, then A has maximal fuzziness 1 since $A = A \cap A^c = A \cup A^c = A^c$! All other fuzziness lies between these extremes. Fuzziness is captured by the nonprobabilistic *entropy*[13] of A, denoted E(A). We can naturally define a fuzzy entropy from first-principles reasoning. Given the fuzzy power set I^n, we naturally have some distance measure available, i.e., a l^p distance, which essentially arises because [0,1] is linearly ordered (as in the theory of utility functions in microeconomics). As such, for any $A \in I^n$, there exists a *nearest nonfuzzy neighbor* $\overline{A}$ and a *farthest nonfuzzy neighbor* $\underline{A}$. $\overline{A}$ is obtained for any l^p metric by rounding the fit value to 0 or 1 and $\overline{A} = \underline{A}^c$. If $A = (.2\ 0\ .5\ .7)$, then $\overline{A} = (0\ 0\ 1\ 1)$ and $\underline{A} = (1\ 1\ 0\ 0)$. Let $a = l^p(A, \overline{A})$ and $b = l^p(A, \underline{A})$ (see Figure 2).

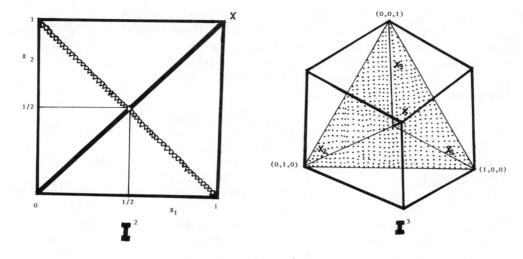

FIGURE 1.

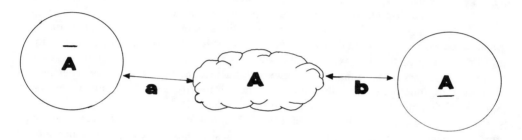

FIGURE 2.

Then $E(A) = a/b$. For if A is nonfuzzy, then $A = \overline{A}$ and, thus, $E(A) = 0$. If $A = (.5, \ldots ,.5)$, then A is equidistant to every nonfuzzy set (vertex), since it is the midpoint of the cube I^n. Thus $a = b$, and thus $E(A) = 1$. If $A = (.2\ 0\ .5\ .7)$, then $E(A) = a/b = 1/3$. These relationships ultimately follow from the metrical property:

$$0 \leq l^p(A,\overline{A}) \leq n^{1/p}/2 \leq l^p(A,\underline{A}) \leq n^{1/p} \tag{1}$$

The key theorem[39] on l^1 fuzzy entropy is

$$E(A) = \frac{\Sigma \text{Count}(A \cap A^c)}{\Sigma \text{Count}(A \cup A^c)} \tag{2}$$

the ratio of fuzzy overlap to *underlap*, with a similar result for general l^p metrics. Hence, the entropy measure a/b captures the characteristic aspects of fuzziness since A is properly fuzzy if $A \in I^n/B^n$ if $A \cap A^c \neq \emptyset$ if $A \cup A^c \neq X$. Phrased another way, if we do not know something A with certainty, then we do not know its opposite A^c with certainty either, and, therefore, there is overlap (and underlap) between the two concepts. Fuzzy theorists will also note that for t-norms[29,31,45] T and t-conorms S, that since $T \leq \min \leq \max \leq S$, replacing min and max in Equation 2 with any other T and S produces an E' such that $E' < E$; in particular, $E(A) = 1$ for $A = (.5, \ldots ,.5)$ in Equation 2 if $T = \min$ and $S = \max$. As such, in some sense we have derived the fundamental fuzzy operations of min, max, and negation from entropy first principles.

Another new concept is degree of *subsethood* S(A,B), the degree to which A is a subset of B. Analyses of subsethood are relatively recent. Bandler and Kohout[5] pointed out that, with nonfuzzy sets, A ⊂ B if A ∈ 2^B. They suggested the fuzzification strategy of interpreting S(A,B) as the degree of membership of A in B fuzzy power set F(2^B), i.e., $m_F(_2B)(A)$ = S(A,B). Kosko[39] showed how to operationally measure S(A,B) by counting *violations* of the dominated membership function relationship $m_A ≤ m_B$.

Suppose m_A is dominated by m_B for all elements x except the violation x_v: $m_A(x_v)$ > $m_B(x_v)$. Then should not A still be a subset of B to a high degree? Should not this degree depend on the magnitude and relative frequency of violations? Consider the fuzzy sets A = (.2 0 .5 .7) and B = (.3 1 .8 .6) and C = (1 .2 .8 .8). Then A ∉ F(2^B) and A ∈ F(2^C). The violation x_4, with magnitude .7 − .6 = .1, prevents A from being a proper fuzzy subset of B. We solve this problem by defining nonsubsethood in terms of normalized violations, then negate that measure. $\Sigma max(0,m_A(x_i) − m_B(x_i))$ counts violations of $m_A ≤ m_B$. Dividing by $\Sigma Count(A)$ adjusts for the *mass* of A and forces the nonsubsethood measure to vary between 0 and 1.

Negating this measure defines fuzzy subsethood:

$$m_{F(2B)}A = S(A,B) = 1 - \frac{\Sigma max(0,m_A(x_i) - m_B(x_i))}{\Sigma Count(A)} \tag{3}$$

Therefore, S(A,B) = 1 if $m_A(x_i) ≤ m_B(x_i)$ for all i. If A = (.2 0 .5 .7) and B = (.3 1 .8 .6), then S(A,B) = 1 − .1/1.4 = 13/14. Fuzzy theorists will note that 1 − max(0, $m_A − m_B$) = min(1 − m_A + m_B) = $t_L(A → B)$, the Lucasiewicz implication operator for fuzzy logic. The key theorem[39] is that the local definition of S(A,B) given by Equation 3 in terms of pairwise violations is equivalent to a global ratio of cardinalities:

$$S(A,B) = \frac{\Sigma Count(A ∩ B)}{\Sigma Count(A)} \tag{4}$$

Hence, for nonfuzzy A and B, S(A,B) is a rational number.

Of course, Equation 4 has the look of a conditional probability P(B | A) = P(A ∩ B)/P(A). Zadeh[52-54] even defines his *relative sigma-count* $\Sigma count(B/A)$ according to the right-hand side of Equation 4, though he notes S(A,B) + S(A,B^c) ≥ 1. Who can *derive* a conditional probability from first principles? The Radon-Nikodym theorem in measure theory allows the derivation of the conditional expectation E(m_A | m_B) = P(B | A), if A and B are nonfuzzy, but this is hardly a first-principles derivation! Fuzzy theorists will also note that if the intersection A ∩ B in Equation 4 is defined with any other t-norm T than min, and if max in Equation 3 is or is not replaced with the dual (or any) other t-conorm S, then, in general, S(A,B) is strictly greater than the ratio of sigma-counts since T ≤ min and max(0, x) = S(0, x) for all S.

The following theorem is the key result of the new fuzzy concepts, which follows trivially from theorems 2 and 4, and provides a definitive answer to the often heard objection to fuzzy theory, "What can you get with fuzzy theory that you cannot get with probability theory?":

$$E(A) = S(A ∪ A^c, A ∩ A^c) \tag{5}$$

Suppose a dogmatic probabilist claims some probability measure P measures the uncertainty of the situation A and, thus, that fuzzy entropy is but a disguised notational variant of classical measure theory. Then P cannot be identically 0, since $E(A) > 0$ if $A \in I^n/B^n$. Nevertheless, $P(A) > 0$ implies that $A = X = \emptyset$, since in a probability space $A \cap A^c = \emptyset$ and $A \cup A^c = X$. Then the sure event X is impossible: $P(X) = 0$, not 1. Contradiction!

We also note that, using a traditional logarithm-of-probability measure of entropy, the maximum entropy[27,50] set is the uniform distribution $U = (1/n, \ldots, 1/n)$. However, $E(U) = 1/n - 1$ by Equation 2, revealing how fast U approaches $\emptyset$ with increasing dimensionality n. Of course, $E(\emptyset) = 0$. This raises the question: which distribution best characterizes maximum uncertainty, the midpoint $M = (1/2, \ldots, 1/2)$ or the uniform distribution $U = (1/n, \ldots, 1/n)$? It seems odd that $m_U(x)$ depends on n but $m_M(x)$ does not. An appeal to *relative* constancy is unpersuasive because the entropy of the constant set $C = (c, \ldots, c)$ is $E(C) = c/1 - c$ if $c \leq 1/2$, $1 - c/c$ if $c \geq 1/2$, which again is maximized if $c = 1/2$. For every uniform distribution U there is a constant distribution C so that $U = C$, the point of intersection of the probability simplex with the locus of equipossible events. (There are uncountably many irrational-coordinate $C \neq U$.) More generally we can show that *every* probability distribution P is such that $E(P) \leq E(U) = 1/n - 1$, and, consequently, as dimensionality increases, probability distributions approach entropic degeneracy! For the metric associative memories discussed later that store patterns at I^n vertices, M not U corresponds to the most ambiguous memory cue.

Another remark is in order about fuzzy vs. probabilistic foundations. There is the Cox theorem[11] that Bayesians cite to show that non-Bayesian statisticians and probabilists, in particular frequentists, are ultimately in error. Thus, Jaynes[28] proclaims, "Cox proved that any method of inference, in which we represent degrees of plausibility by real numbers, is necessarily either equivalent to Laplace's, or inconsistent." Cox used *bivalent* logic (Boolean algebra) to show that the "conditions of consistency can be stated in the form of functional equations,"[28] namely, the probabilistic product and sum rules:

$$P(A \cap B|C) = P(A|B \cap C) \, P(B|C)$$
$$P(B|A) + P(B^c|A) = 1$$

A quick check with Equation 4 shows that $S(C, A \cap B) = S(B \cap C, A) \, S(C, B)$ and that, using $\min(x,y) \geq x\,y$ for x and y in $[0,1]$, $S(A,B) + S(A,B^c) \geq 1$, as stated before. Cox's argument is limited by its forthright bivalent-logic framework and its assumption that propositional combination operators are twice differentiable! Max and min, like most t-norms and t-conorms, are not differentiable. Indeed, the only popular differentiable t-norm is the probabilistic or product (sum) t-norm (t-conorm) $x\,y$ $(x + y - xy)$. Outside the Bayesian/ frequentist debate in probability theory (in the $n - 1$ simplex in I^n), it is hard to imagine citing Cox's theorem as a constraint on anything, and yet the citation is common in AI polemics against fuzzy theory.[8]

We note another property of subsethood that follows from the geometric interpretation of fuzzy power sets. The geometric interpretation of B fuzzy power set $F(2^B) = \{A \in I^n: m_A(x) \leq m_B(x) \text{ for all s}\}$ is the *hyper-rectangle* H_B whose sides are defined by the coordinate bounds $0 \leq x_i \leq m_B(x_i)$. Thus, the point B in I^n is the vertex of H_B connected by a long diagonal of H_B to the origin. We cannot count H_B in any interesting way because every nondegenerate hyper-rectangle in I^n contains the same uncountable number of points. We must appeal, instead, to Lebesgue measure, which on I^n is simply the *volume* of $F(2^B)$, which we denote in slightly misleading form as $v(B)$. Analogous to the simple sigma-count sum of fit values, $v(B)$ is simply the product of fit values: $v(B) = m_B(x_1) \, m_B(x_2) \ldots m_B(x_n)$. For example, the rectangular fuzzy power sets of $A = (1/3, 3/4)$, with $v(A) = 3/12$, and $B = (3/4, 1/4)$, with $v(B) = 3/16$, are indicated with appropriate diagonal shading (see Figure 3).

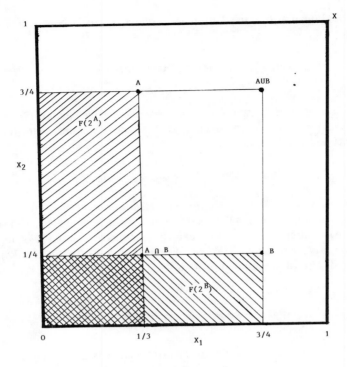

FIGURE 3.

This example illustrates that fuzzy power sets can be represented in a sort of hyper-rectangular Venn diagram. It also is of interest that $H_{A \cap B}$ and $H_{A \cup B}$ correspond to rectangular combinations of H_A and H_B only if intersection and union are defined with min and max. This geometry further suggests an alternative definition of subsethood, the *volume subsethood* $V(A,B) = v(A \cap B)/v(A)$ provided A is in the interior of I^n, with corresponding *volume fuzzy entropy* $E_V(A) = v(A \cap A^c)/v(A \cup A^c)$. (Observe that $V(A,B)$ would not depend on A if intersection were defined with product.) In the example, $V(A,B) = 1/3$ and $E_V(A) = 1/6$. However, $S(A,B) = 7/13$ and $E_1(A) = 7/12$, each of which is strictly larger than its volume counterpart. In general, we have the theorem $V(A,B) \le S(A,B)$ with equality if, and only if, $A = B$, as a straightforward derivation shows. This theorem is further evidence for identifying subsethood with $S(A,B)$. Indeed it implies that $E_V(A) \le E(A)$, which reminds us that $E_V(A) = 0$, if, and only if, A is on the boundary of I^n, *not* necessarily a vertex. We might call E_V a *pseudo-entropy measure* since it otherwise satisfies the De Luca Termini axioms.[13] The reason for these strict inequalities is, of course, that we are multiplying unit-interval valued quantities. Such products automatically approach 0 as the number of factors increases, regardless of their individual magnitudes. However, since the Σ-count generalizes counting measure, it is not fundamentally affected by large sums. Indeed, it is ultimately defined on countably infinite sets.

Returning to standard fuzzy concepts, a two-place fuzzy *relation* R is a point in $[0,1]^n \times [0,1]^p$, i.e., a fuzzy subset of the product space $X \times Y$, i.e., an n-by-p matrix of elements in $[0,1]$. Similarly defined are n-place fuzzy relations/arrays. If R is an n-by-p relation and S is a p-by-q relation, the max-min n-by-q *composition* or product R o S is formed analogously to the matrix product R S by replacing pairwise products with pairwise minima and replacing row/column sums with row/column maxima. Then:

$$m_{R \circ S}(x_i, z_k) = \max_j \min(m_R(x_i, y_j), m_S(y_j, z_k)) \tag{6}$$

A special case of Equation 6 is the 1-by-p *set product* A o R, where A is a 1-by-n fuzzy set:

$$m_{A \circ R}(y_j) = \max_i \ \min(m_A(x_i), m_R(x_i, y_j)) \tag{7}$$

Another special case of Equation 6 is the n-by-p *set outer composition* A o B, which we shall denote as a vector outer product A^T o B, where A^T is the vector transpose of A, and A is an n-set and B is a p-set:

$$m_{A^T \circ B}(x_i, y_i) = \min(m_A(x_i), m_B(y_j)) \tag{8}$$

Incidentally, A^T o B = A × B, the *fuzzy cartesian product* of A and B. We note that the previously stated concepts of intersection, union, complementation, l^p distance, cardinality, subsethood, and entropy pass over exactly to n-place relations. For instance, the l^1 entropy of the n-by-p relation R is $E(R) = \Sigma Count(R \cap R^c)/\Sigma Count(R \cup R^c)$; if S is also n-by-p, then the degree of subsethood of R in S is

$$S(R,S) = 1 - \frac{\displaystyle\sum_i \sum_j \max(0, m_R(x_i, y_j) - m_S(x_i, y_j))}{\Sigma Count(R)} \tag{9}$$

Finally, if A is an n-set and R is a n-by-p relation, then A is a *fuzzy eigenset* of R if A o R = A.

Zadeh[51] has also shown that if we interpret the n-set A as a possibility distribution, we can interpret the n-by-p relation R as a *conditional possibility distribution:* A o R = B; or, in Zadeh's notation, P_Z o $P_{Y|X}$ = P_Y, where the values of the variables X, Y, and Z have been restricted[51] to the fuzzy sets, C, B, and A, and C is also an n-set. Zadeh defines the matrix elements of P(Y|X) as literal fuzzy logical (Lucasiewicz) implications: $p_{ij} = \min(1, 1 - m_C(x_i) + m_B(y_j))$. This leads to the *compositional rule of inference:*[53] P(X') o P(Y|X) $\approx$ P(Y) if X' $\approx$ X. Accordingly, the compositional rule of inference is a set-level generalization of modus ponens. It is also a prototypical FAM.

III. FUZZY EIGENSET SUBSET FAMS

The compositional rule of inference is a compositional rule of fuzzy association. The set pair (A,B) is stored in parallel distributed fashion in the n-by-p relation R. Zadeh's suggests defining R pointwise with the fuzzy logical implication values $r_{ij} = \min(1, 1 - m_A(x_i) + m_B(y_j))$ so that, in effect, A will push B out of R in a set-level modus ponens. The FAM interpretation of the compositional rule of inference, however, does not dictate a particular selection of R. For instance, if we insist that A compositionally recall B if (A,B) is stored in R, then R cannot be defined by Lucasiewicz implication. Other operators must be sought.

A *fuzzy eigenset FAM* is an n-by-n relation M that stores an n-set A as an eigenset: A o M = A. This is an autoassociative FAM. An heteroassociative *eigenset* FAM stores the set pair (A,B), where A is an n-set and B is a p-set, as left/right eigensets of some n-by-p relation M: A o M = B and B o M^T = A. A *subset* FAM maps arbitrary sets A' $\approx$ A into subsets of A or B: A' o M $\subset$ A or A' o M^T $\subset$ B. Intuitively a subset FAM maps the unknown into the known (stored). The less similar A' is to A, the closer A' o M should be to the empty set Ø.

Let us examine the behavior of the Lucasiewicz L used by Zadeh:[51] $l_{ij} = \min(1, 1 - m_A(x_i) + m_B(y_j)) = 1$ if $m_A(x_i) \leq m_B(y_j)$. Therefore L tends to be widely populated with 1 values and values near 1. Consequently, $m_B,(y_j) = m_{AoR}(y_j) \approx \max\{m_A(x_1), \ldots, m_A(x_n)\} = H(A)$ tends to hold. Suppose we store $A = (.2\ 0\ .5\ .7)$ and $B = (1\ .5\ .6)$ in L:

$$
L = \begin{pmatrix} 1 & 1 & 1 \\ 1 & 1 & 1 \\ 1 & 1 & 1 \\ 1 & .8 & .9 \end{pmatrix}
$$

Then $A \circ L = (.7\ .7\ .7)$ and $B \circ L^T = (1\ 1\ 1)$ as expected, with $H(A) = .7$ and $H(B) = 1$. Therefore, L is not an eigenset FAM or a subset FAM. Indeed this behavior suggests abandoning the selection $R = L$ in the compositional rule of inference and, in general, in conditional possibility theory. Theorem 2 following suggests an alternative.

We propose the $M = A^T \circ B$ as a fuzzy eigenset subset FAM. The connection strength $m_{ij} = \min(m_A(x_i), m_B(y_j))$ is a *fuzzy Hebb law*[23] since it encodes correlation or conjunctive learning. A fuzzy logical implication operator is m_{ij} since it satisfies the only one invariant axiom[14] of implication — that $t(A \rightarrow B) = 0$ if $t(A) = 1$ and $t(B) = 0$, or "Truth cannot imply falsehood" as Aristotelian philosophers say.

First, we examine the autoassociative case $M = A^T \circ A$. The fit vector $A = (.2\ 0\ .5\ .7)$ is then stored in the symmetrical relation

$$
M = \begin{pmatrix} .2 & 0 & .2 & .2 \\ 0 & 0 & 0 & 0 \\ .2 & 0 & .5 & .5 \\ .2 & 0 & .5 & .7 \end{pmatrix}
$$

Then $A \circ M = A$. Consider $A' = (.3\ .2\ 1\ .6)$: $A' \circ M = (.2\ 0\ .5\ .6)$ A. These properties, in fact, hold in general and follow from two facts: Diagonal$(M) = A$ and M is *diagonal-dominant*, $m_{ij} \leq m_{jj}$ for all i. This observation allows us to prove that there exist families of fuzzy eigenset subset FAMs.

Theorem 1. If Diagonal$(M) = A$ and M is diagonal dominant, then M is a fuzzy eigenset subset FAM.

Proof. $m_{AoM}(x_j) = \max_i \{\min(a_i, m_{ij})\} = \min(a_j, m_{jj}) \vee \max_{ij} \{\min(ai, mij)\} = a_j \vee \max_{ij} \{\min(a_i, m_{ij})\} = a_j$, where "v" stands for max, since $a_j = m_{jj} \geq m_{ij} \geq \min(a_i, m_{ij})$ for all i. Thus, $A \circ M = A$.

Next suppose $A' \sim A$. Then $m_{A'oM}(x_j) = \max_i \{\min(a'_i, m_{ij})\} \leq \max_i \{m_{ij}\} = m_{jj} = a_j$. Thus, $A' \circ M \subset A$.

Q. E. D.

Accordingly, the matrix Diagonal(A) is always a fuzzy eigenset subset FAM, as well as $A^T \circ A$. The latter FAM, of course, has greater recall or pattern-completion power.

The fuzzy Hebbian FAM $M = A^T \circ A$ admits no feedback error-correction and does not allow several patterns $A_1, \ldots, A_m$ to be superimposed in one relation. Fuzzy eigenset FAMs are projection operators: $A \circ M^2 = (A \circ M) \circ M = A \circ M = A$. Consequently, nothing is gained by feeding the output $A \circ M$ back to M. Fuzzy eigenset FAMs are "one-shot'' associative memories. A natural way to store the n-sets $A_1, \ldots, A_m$ would be to memorize the patterns in eigenset FAMs $A_i^T \circ A_i$ and then superimpose them by taking pointwise maxima: $M = \max\{A_1^T \circ A_1, \ldots, A_m^T \circ A_m\}$. We then require at a minimum that every stored pattern A_i recall itself: $A_i \circ M = A_i$. Unfortunately $M = A_{max}^T \circ A_{max}$, where A_{max} is the superimposed n-set $\max\{A_1, \ldots, A_n\}$. Accordingly, M maps all patterns either into A_{max} or some proper subset of A_{max} — in accordance with Theorem 1! A similar consequence befalls many other pointwise superimposition operators, including minimum.

We propose $M = A^T \circ B$ as an heteroassociative fuzzy eigenset subset FAM. Suppose we wish to store $A = (.2\ 0\ .5\ .7)$ and $B = (1\ .5\ .6)$ in the 4-by-3 M:

$$M = \begin{pmatrix} .2 & .2 & .2 \\ 0 & 0 & 0 \\ .5 & .5 & .5 \\ .7 & .5 & .6 \end{pmatrix}$$

Then $A \circ M = (.7\ .5\ .6) = B' \ne B$ but $B \circ M^T = A$. The recall disparity arises in the forward direction $(A \rightarrow B)$ because $H(A) < H(B)$, as is made clear in the proof of the following theorem.

Theorem 2. If $M = A^T \circ B$ and $H(A) = H(B)$, then M is a fuzzy left/right eigenset subset FAM.

Proof. $m_{A \circ M}(y_j) = \overset{max}{i}\{\min(a_i, m_{ij})\} = \overset{max}{i}\{\min\{a_i, a_i, b_j\}\} = \min(H(A), b_j) = b_j$, provided $H(A) \ge b_j$ for each j, i.e., provided $H(A) \ge H(B)$. Similarly $m_{B \circ M}T(x_i) = a_i$ provided $H(B) \ge H(A)$. Hence $H(A) = H(B)$ guarantees perfect forward and backward recall of (A,B). If $A' \sim A$, then $m_{A \circ M}(y_j) = \min(H(A \cap A'), b_j)$, and thus $A' \circ M \subset B$ whatever the height of B. Similarly, M^T is a subset FAM. Q. E. D. Of course, the autoassociative FAM $A^T \circ A$ always satisfies $H(A) = H(A)$. For binary sets A and B, the height can only be violated if A or B is the empty set, which is always a subset.

Theorem 2 recommends $A^T \circ B$ over L as the choice for R in the compositional rule of inference[52,53] and for $P(Y|X)$ in the theory of conditional possibility distributions.[51] Nevertheless, $M = A^T \circ B$ suffers from the same problems that the autoassociator $A^T \circ A$ suffers from. M is not a feedback *bidirectional* associative memory[37] and the pairs $(A_1, B_1), \ldots, (A_m, B_m)$ cannot be reliably stored by superimposing them by pointwise maxima (minima, etc.) in $M = \max\{A_1^T \circ B_1, \ldots, A_m^T \circ B_m\} = A_{max}^T \circ B_{max}$.

Theorem 2 also underlies the algorithm in the first fuzzy logic VLSI chip developed at AT&T Bell Labs by Togai and Watanabe (see Chapter 18).[48,49] The "inference engine on a chip" — that performs 80,000 FLIPS (fuzzy logical implications per second) — is, in fact, the world's first VLSI heteroassociative eigenset subset FAM. The chip instantiates the compositional rule of inference and uses the logical operation min(x,y) instead of min(1,1 − x + y)! Togai[55] chose this operator on the basis of extensive simulation runs because it worked and alternative operators did not.

Another recent VHSIC chip is the TRW FAM chip[24] based on the TRW WAM (window addressable memory) VHSIC chip. The TRW FAM, however, is not strictly speaking fuzzy or an associative memory. It is, like the TRW CAM VHSIC chip, a *metric* CAM in which data vectors are addressed in parallel. Data vectors (192-bit or 96-bit words) are not distributed in the storage medium. The TRW FAM, however, allows a discrete window of partial matches of input vectors to stored vectors; thus, the "fuzzy" in FAM. The TRW FAM allows fuzzy membership functions to be approximated with hyperrectangles and trades parallel distributed storage and access for raw VHSIC processing power — up to 7.68 billion comparisons per second.

IV. METRICAL FAMS THAT MINIMIZE FUZZY ENTROPY

Many popular neural-net associative memories minimize fuzzy entropy as they recall stored patterns. Binary patterns are stored at some of the 2^n vertices of I^n. An input partial pattern $A' \sim A$ enters the memory as some point inside I^n, as a fuzzy set. The networks distributed "computation" proceeds so as to drive A' to its nearest stored neighbor A. A' quickly travels from inside the n-cube to the nearest vertex. Consequently, such an associative memory minimizes fuzzy entropy. It *disambiguates* input fuzzy patterns by defuzzifying them pointwise in the direction of their nearest nonfuzzy patterns. The FAM is *metric* because similarity is measured by distance, usually l^1 or sometimes l^2. Metric FAMs are less general than subset FAMs because pattern distance in effect measures pattern subsethood and supersethood. For example, l^p metrics consist of the absolute sums:

$$|m_A - m_B| = \max(m_A - m_B, m_B - m_A)$$
$$= \max(0, m_A - m_B) + \max(0, m_B - m_A)$$

which measure pointwise violations of subsethood $m_A \leq m_B$ and supersethood $m_B \geq m_A$, as discussed in Section III.

We shall briefly illustrate fuzzy-entropy minimization by metric FAMs that are designed to minimize a Lyapunov or "energy" function. The neural network literature is too vast to survey here and the workings of all but the simplest nets are too lengthy to detail. The interested reader should consult the monumental works of Grossberg[20-22] and Kohonen[32,33] for the serious study of neural networks, brains, and minds as *programmable dynamic systems*.

The easiest hetero-/autoassociative metric FAM to study is Kohonen's optimal linear associative memory (OLAM).[33] The OLAM is not a dynamic associative memory. All recall takes place in one synchronous step. Therefore, the OLAM is not a fuzzy entropy minimizer. It is, however, optimal in the least-squares (l^2) sense, and simple OLAMs can be constructed with pen and paper. We present it because it illustrates the properties of most metric associative memories and because it is the metric analogue of Theorems 1 and 2.

The OLAM stores m-many associations $(A_1, B_1), \ldots, (A_m, B_m)$ in an n-by-p matrix M. A_i can be any real n-vector and B_i any real p-vector, not necessarily points in I^n and I^p. B is recalled given A if $B = AM$. Then recall (decoding) is a linear procedure. We demand two things from M. First, $A_i M = B_i$ must hold for all i; A_i must perfectly recall its associated output B_i. Second, if some input A is closer to A_i than to all other A_j, then the output $B = AM$ must be closer to B_i than to all other B_j. We seek a parallel distributed construction (encoding) procedure for such an optimal M.

To simplify notation, we rewrite the recall equation in matrix notation as $B = AM$. Here $A^T = [A_1^T | \ldots | A_m^T]$ and $B^T = [B_1^T | \ldots | B_m^T]$ are rectangular matrices. A_i is the ith row of the m-by-n matrix A. B_i is the ith row of the m-by-p matrix B. Norm(M) is the matrix norm of M compatible with the vector l^2 (Euclidean) norm, the square root of Trace($M^T M$). We search the space of n-by-p real matrices to find that matrix M that minimizes Norm(B $-$ AM). In the special case where $n = m = p$ and A^{-1}, the matrix inverse of A, exists, the optimal M is simply $A^{-1}B$. In general, the solution is $M = A*B$, where $A*$ is the Moore-Penrose pseudo-inverse of A. The bidirectional[37] OLAM for recalling A given B is $M* = B*A$. The OLAM autoassociation matrix is $M = A*A$. $A*$ uniquely exists for every matrix and a variety of recursive procedures exist to calculate it. The popular Greville's algorithm for computing $A*$ enjoys an equivalence to the time-independent Kalman filter,[30] which underlies the practical power of the OLAM.

The *memory capacity* of the OLAM M is $m < n$. No more associations can be stored in M and reliably recalled than the dimensionality of the pattern space — a result true for most associative memories.[1] Kohonen[33] has shown that the OLAM attenuates/amplifies circularly symmetric noise according to $\sqrt{m}/n$, again reflecting the capacity bound. The autoassociative OLAM A*A is an optimal projection operator. The OLAM uniquely decomposes any pattern P into signal and noise $P_s + P_n$ in a Hilbert-space version of Pythagoras's theorem. P is the hypotenuse, P_s and P_n are the orthogonal legs. A*A projects P onto the subspace spanned by the patterns in A. The projection $P_s = P A*A$ is the best signal estimate (best "prediction") of P given A. $P_n = P - P_s$ is the orthogonal *novelty* ("error") in P. $I - A*A$ is a *novelty filter,* where I is the n-by-n identity matrix. It optimally measures what is new or unexplained in P with respect to what is known (stored in A). These properties rest on the properties of Fourier coefficients in Hilbert space and are thus structurally analogous to linear regression properties of uniquely decomposing a data vector into a best prediction vector and residual or error vector. Some neural net critics have even claimed that associative memory theory is disguised regression. The fundamental difference is that regression models assume a functional *dependency* between A and B, namely, B = cA + e, and attribute all disparities between B and cA to some stochastic error e. On the contrary, in associative memory theory we assume no dependencies between the arbitrary data vectors or sets A_i and B_i. We instead seek encoding structures, often matrices, that form an association between them in some appropriate way.

We now describe the convergence, coding, and decoding properties of a binary heteroassociative CAM whose continuous extension minimizes fuzzy entropy. The memory is a *bidirectional associative memory* (BAM).[37] BAMs generalize the unidirectional autoassociative CAMs popularized by Hopfield.[25,26] A BAM is represented by an n-by-p real matrix M. Two-way associative search of stored heteroassociative patterns (A_i, B_i) occurs by iteratively passing information in the forward direction through M and in the backward direction through M^T: $A \to M \to B \to M^T \to A' \to M \to B' \to \ldots \to A_i \to M \to B_i \to M^T \to A_i \to \ldots$ The interpretation of this schema is that the BAM stabilizes or resonants on the nearest stored pair (A_i, B_i). If M is symmetric ($M = M^T$), unidirectional and bidirectional behavior coincides. We prove that *every* n-by-p matrix M is a discrete and continuous bidirectionally stable heteroassociative CAM. Accordingly, every matrix can be decoded.

The m pairs (A_i, B_i) are encoded in a BAM by summing bipolar correlation matrices. This sculpts a scalar energy surface that governs the BAM nonalgorithmic computation, in principle producing O(1) search — classification speed is *independent* of the number of patterns stored! The BAM storage-capacity upper bound is m < min(n,p).

The neural-net interpretation of a BAM is a two-layer hierarchy of symmetrically connected neurons with fixed connection weights. There are n neurons in the A field F_A and p neurons in the B field F_B (see Figure 4).

When an arbitrary pattern A, B, or (A,B) is presented to the BAM, the respective neurons in the F_A or F_B fields are activated. The network quickly equilibrates to a state of two-pattern reverberation, or pseudo-*adaptive resonance.*[6,7,10,17-21] The stable resonance corresponds to a system energy local minimum. Geometrically, an input pattern is placed on the BAM energy surface as a ball-bearing in the binary product space $\{0,1\}^n \times \{0,1\}^p$. The correlation encoding scheme sculpts the energy surface so that the patterns (A_i, B_i) are stored as local energy minima. The input ball-bearing rolls down into the nearest basin of attraction, dissipating energy as it rolls. Frictional damping brings it to rest at the bottom of the energy well, and the pattern is classified accordingly. We remark that temporal patterns or limit cycles can similarly be stored at local energy minima in large product spaces.

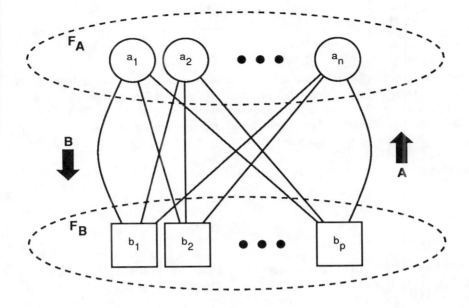

FIGURE 4.

The binary pair $(A,B) \in \{0,1\}^n \times \{0,1\}^p$ defines a *state* of the BAM M. The *energy* $E(A,B)$ of the state is defined by the quadratic form:

$$E(A,B) = -1/2\ AMB^T - 1/2\ BM^TA^T$$
$$= -AMB^T$$
$$= -\sum_i \sum_j a_i b_j m_{ij} \tag{10}$$

The bivalent neurons turn on or off according to a McCulloch-Pitts[42] sign law:

$$a_i = \begin{cases} 1 \text{ if } BM_i^T > 0 \\ 0 \text{ if } BM_i^T < 0 \end{cases} \tag{11}$$

$$b_j = \begin{cases} 1 \text{ if } AM^j > 0 \\ 0 \text{ if } AM^j < 0 \end{cases} \tag{12}$$

where M_i is the ith row (column) of M (M^T) and M^j is the jth column (row) of M (M^T). If a neuron equals 0, it maintains its current state. If the default 0 thresholds in the hyperplane, threshold laws in Equations 11 and 12 are respectively replaced with $T = (t_1, \ldots, t_n)$ and $S = (s_1, \ldots, s_p)$, then the BAM energy is generalized by replacing Equation 10 with Equation 13:

$$E(A,B) = -AMB^T + AT^T + BS^T \tag{13}$$

We shall, however, assume 0 thresholds.

BAM convergence is proved by showing that synchronous or asynchronous state changes decrease the energy and that the energy is bounded below, so the BAM gravitates to fixed points. Boundedness follows from:

$$E(A,B) \geq - \sum_i \sum_j |m_{ij}|$$ (14)

for all A and B.

Synchronous behavior occurs when all or some neurons in a field change their state at the same clock cycle. Asynchronous behavior occurs in the special case where only one neuron changes state per cycle and, thus, can be interpreted as each neuron in the field randomly and independently changing state. Field F_A change is denoted by $\Delta A = A_2 - A_1 = (\Delta a_1, \ldots, \Delta a_n)$ and energy change by $\Delta E = E_2 - E_1$. Thus, $\Delta a_i = -1, 0,$ or $+1$. Then

$$\Delta E = - \Delta A M B^T$$

$$= - \sum_i \Delta a_i \sum_j bj \, m_{ij}$$

$$= - \sum_i \Delta a_i \, B \, M_i^T$$ (15)

We only care about nonzero state changes. But, if $\Delta a_i > 0$, then Equation 11 implies $BM_i^T > 0$; if $\Delta a_i < 0$, Equation 11 implies $BM_i^T < 0$. Hence state change and input sum agree in sign and their product is positive: $\Delta a_i \, BM_i^T > 0$. Hence $\Delta E < 0$. (Note that every energy change exceeds some fixed positive quantity and, thus, patterns do not slide negligibly down the energy surface.) Similarly Equation 12 implies $\Delta E = - A M \Delta B^T < 0$. Since M was arbitrary, this proves that *every matrix is bidirectionally stable* (and that every symmetric matrix is unidirectionally stable).

The m binary pairs (A_i, B_i) are encoded by transforming them to bipolar pairs $(X_i, Y_i) \in \{-1, 1\}^n \times \{-1, 1\}^p$ (where $X_i = 2 A_i - I$ and $Y_i = 2 B_i - I$ and I is a vector of n-many or p-many 1 values and then binary 0 values are replaced with -1 values), interpreting association as logical equivalence and thus mathematical correlation $X_i^T Y_i$ as in OLAM encoding, then summed (superimposed) pointwise to yield $M = X_1^T Y_1 + \ldots + X_m^T Y_m$. (Accordingly, to encode the binary vectors $A_1, \ldots, A_m$ in an autoassociative unidirectional CAM, bipolar autocorrelation matrices are summed: $M = X_1^T X_1 + \ldots + X_m^T X_m$.) Bipolar correlation matrices naturally represent excitatory ($m_{ij} > 0$) and inhibitory ($m_{ij} < 0$) synaptic connections. Summed bipolar matrices naturally represent the combined strength of excitatory or inhibitory synapses. The summing process represents Hebbian[23] learning or correlation learning. A pair (A_i, B_i) is unlearned or forgot by summing $- X_i^T Y_i$ or, equivalently, by encoding (A_i^c, B_i) since $X_i^c = - X_i$. It turns out[37] that recall accuracy is improved on average if bipolar state vectors are used instead of binary state vectors. Then, for instance,

$$X_i M = (X_i, X_i^T) Y_i + \sum_{j \neq i} (X_i X_j^T) Y_j$$

$$= n Y_i + \sum_{j \neq i} (X_i X_j^T) Y_j$$

$$\sim c Y_i, \, c > 0$$ (16)

which tends to threshold to Y_i (B_i) if $m < n$. Similarly $Y_j M^T$ tends to threshold to X_i (A_i) if $m < p$. Therefore, $m < \min(n,p)$ is a rough estimate of the BAM storage capacity. BAM recall reliability is further improved if $l_1(A_i, A_j) \sim l_1(B_i, B_j)$, where l_1 is here simply Hamming distance.

For concreteness we will store the four nonorthogonal binary associations:

A_1 = (1 0 1 0 1 0 1 0 1 0 1 0 1 0 1) B_1 = (1 1 1 1 0 0 0 0 1 1)
A_2 = (1 1 0 0 1 1 0 0 1 1 0 0 1 1 0) B_2 = (1 1 1 0 0 0 1 1 1 0)
A_3 = (1 1 1 0 0 0 1 1 1 0 0 0 1 1 1) B_3 = (1 1 0 0 1 1 0 0 1 1)
A_4 = (1 1 1 1 0 0 0 0 1 1 1 1 0 0 0) B_4 = (1 0 1 0 1 0 1 0 1 0)

where m = 4 < min(n,p) = min(15,10) = 10. These binary associations are transformed to bipolar associations:

X_1 = (1 −1 1 −1 1 −1 1 −1 1 −1 1 −1 1 −1 1)
X_2 = (1 1 −1 −1 1 1 −1 −1 1 1 −1 −1 1 1 −1)
X_3 = (1 1 1 −1 −1 −1 1 1 1 −1 −1 −1 1 1 1)
X_4 = (1 1 1 1 −1 −1 −1 −1 1 1 1 1 −1 −1 −1)

Y_1 = (1 1 1 1 −1 −1 −1 −1 1 1)
Y_2 = (1 1 1 −1 −1 −1 1 1 1 −1)
Y_3 = (1 1 −1 −1 1 1 −1 −1 1 1)
Y_4 = (1 −1 1 −1 1 −1 1 −1 1 −1)

Then $M = X_1^T Y_1 + X_2^T Y_2 + X_3^T Y_3 + X_4^T Y_4$:

$$
\begin{pmatrix}
4 & 2 & 2 & -2 & 0 & -2 & 0 & -2 & 4 & 0 \\
2 & 0 & 0 & -4 & 2 & 0 & 2 & 0 & 2 & -2 \\
2 & 0 & 0 & 0 & 2 & 0 & -2 & -4 & 2 & 2 \\
-2 & -4 & 0 & 0 & 2 & 0 & 2 & 0 & -2 & -2 \\
0 & 2 & 2 & 2 & -4 & -2 & 0 & 2 & 0 & 0 \\
-2 & 0 & 0 & 0 & -2 & 0 & 2 & 4 & -2 & -2 \\
0 & 2 & -2 & 2 & 0 & 2 & -4 & -2 & 0 & 4 \\
-2 & 0 & -4 & 0 & 2 & 4 & -2 & 0 & -2 & 2 \\
4 & 2 & 2 & -2 & 0 & -2 & 0 & -2 & 4 & 0 \\
0 & -2 & 2 & -2 & 0 & -2 & 4 & 2 & 0 & -4 \\
0 & -2 & 2 & 2 & 0 & -2 & 0 & -2 & 0 & 0 \\
-2 & -4 & 0 & 0 & 2 & 0 & 2 & 0 & -2 & -2 \\
2 & 4 & 0 & 0 & -2 & 0 & -2 & 0 & 2 & 2 \\
0 & 0 & -2 & -2 & 0 & 2 & 0 & 2 & 0 & 0 \\
0 & 0 & -2 & 2 & 0 & 2 & -4 & -2 & 0 & 4
\end{pmatrix}
$$

which stores (A_1,B_1), . . . , (A_4,B_4) as stable points in $\{0,1\}^{15} \times \{0,1\}^{10}$ with respective energies -56, -48, -60, and -40. So (A_3,B_3) is the dominant attractor. For example, if B = (1 1 0 0 1 0 0 0 0 0) ~ B_3, with $l^1(B,B_3)$ = 3, then (A_3,B_3) is recalled in one iteration. The blended pairs (A_1,B_4), (A_2,B_3), (A_3,B_2), and (A_4,B_1) each respectively recall the stored pairs (A_1,B_1), (A_2,B_2), (A_3,B_3), and (A_4,B_4) as expected, since the A_i matches correspond to the correct specification of 15 variables, while the B_i matches only correspond to the correct specification of 10 variables.

We now turn to continuous BAMs[37] described by the dynamic equations:

$$\dot{a}_i = \sum_j S(b_j)m_{ji} - a_i \tag{17}$$

$$\dot{b}_j = \sum_i S(a_i)m_{ij} - b_j \tag{18}$$

$S(a_i(t))$ and $S(b_j(t))$ take values in [0,1], and then $A(t) = (S(a_1(t)), \ldots, S(a_n(t)))$ and $B(t)$ $= (S(b_1(t)), \ldots, S(b_p(t)))$ define fuzzy sets in I^n and I^p at any time t. In other words, Equations 17 and 18 equate analogue neurons to fuzzy sets! Ultimately this is perhaps the deepest connection between neural networks and fuzzy theory.

The *signal function* S in the input sums to a_i and b_j denotes an arbitrary monotone nondecreasing function and, thus, its derivative S' with respect to the argument a_i or b_j is nonnegative, $S' \geq 0$. In particular, S is a *sigmoid* or S-shaped curve such as $S(x) = (1 + e^{-bx})^{-1}$ for positive b. It is well known that neuron firing frequency is sigmoidal. Indeed, the fact that neuron input-output curves are S-shaped has the status of a neurobiological constant. Nature selected sigmoidal behavior for some good reason. Grossberg[15-18] has proven the remarkable fact that, roughly speaking, a sigmoid signal function is optimal because it computes a *quenching threshold* below which activity is suppressed as noise and above which activity is contrast enhanced (then stored in short-term memory). Unfortunately linear signal functions ($S(a_i) = a_i$) amplify noise as faithfully as they amplify signals. At first sight, then, the hyperplane threshold laws in Equations 11 and 12 appear inadmissible. However, threshold linear (step function) behavior is the limiting case of a steep sigmoid curve. The first terms on the right-hand side of Equations 17 and 18 are again the input sums to a_i and b_j. The second terms are passive decay terms. If no activation flows into them, a_i and b_j exponentially decay to zero. We note that constant input I_i (J_j) or constant thresholds $-A_i$ ($-B_j$) can be added to the right-hand sides of Equations 17 and 18. We shall omit these terms for notational convenience in the dynamic analysis following.

The global stability of the BAM system defined in Equations 17 and 18 can be analyzed by applying the Lyapunov techniques of the Cohen-Grossberg theorem[9,24] (which was proved for general unidirectional autoassociators). Lyapunov techniques offer a shortcut to global analysis of a dynamic system without solving the underlying differential equations. For very steep S, we can prove BAM convergence for Equations 17 and 18 by taking the time derivative of the bounded energy function:

$$E = -\sum_i \sum_j S(a_i)S(b_j)m_{ij} \qquad (19)$$

The steep S assumption essentially allows us to ignore the passive decay terms $-a_i$ and $-b_j$ in Equations 17 and 18. Then the chain and product rules of differentiation (and relabeling of m_{ij} as needed) give:

$$\dot{E} = -\sum_i S'\dot{a}_i \sum_j S(b_j)m_{ji} - \sum_j S'\dot{b}_j \sum_i S(a_i)m_{ij}$$

$$= -\sum_i S'\dot{a}_i^2 - \sum_j S'\dot{b}_j^2$$

$$\leq 0 \qquad (20)$$

upon substitution of Equations 17 and 18 (with zeroed decay terms) for input sums. If passive decay terms are used, appropriate sums of integrals must be added to Equation 19 to allow the substitution trick of Equation 20 to work. Similarly, if inputs and thresholds are used in Equations 17 and 18, sums of the terms $S(a_i) I_i$ and $-S(a_i) A_i$ must be added. Since the constant connection matrix M used was an arbitrary n-by-p real matrix, this proves that *every matrix is continuously bidirectionally stable.*

We next examine Hopfield's[26] neural-circuit interpretation of Cohen-Grossberg unidirectional autoassociators. Hopfield proposes a resistance-capacitance charging equation of the form:

$$C_i \dot{u}_i \; = \; \sum_j V_j m_{ji} \; - \; u_i/R_i \qquad\qquad (21)$$

for symmetrical M, where C_i is the input capacitance of the ith cell membrane, R_i is the transmembrane resistance, and $1/m_{ij}$ is the finite impedance between the output voltage V_j and the cell body of cell i. The voltage vector $V(t) = (V_1(t), \ldots , V_n(t))$ defines a time-varying fuzzy subset in I^n. The voltage signal is a sigmoid function of its input $V_i = g(u_i)$, where $g(u_i)$ describes the input-output behavior of a nonlinear amplifier with negligible response time and which is one-to-one, $u_i = g^{-1}(V_i)$. Therefore, as a function of V_i, the integral of $g^{-1}(x)\,dx$ on $[0,V_i]$ describes a U-shaped curve, and m_{ij} is a resistor whose magnitude is $1/R_{ij}$. Therefore,

$$1/R_i \; = \; 1/p_i \; + \; \sum_j 1/R_{ji} \qquad\qquad (22)$$

where p_i is the input resistance of the ith amplifier. Then the u_i input sum $V\,M_i^T$ is the electrical current input to the ith neuron.

As per the Cohen-Grossberg theorem, Hopfield's symmetrical electrical network admits the global Lyapunov function:

$$E \; = \; -1/2 \sum_i \sum_j V_i V_j m_{ij} \; + \; \sum_i 1/R_i \int_0^{V_i} g^{-1}(V)\,dV \qquad\qquad (23)$$

since E is bounded below and $\dot{E} \leq 0$, which follows upon differentiation, substitution of Equation 21, use of symmetrical M, and noting $g^{-b}(V_i) \geq 0$ since g^{-1} is a monotone increasing (sigmoid) function of V_i.

Hopfield uses a clever argument to show that the integral term on the right-hand side of Equation 23 is zero or nearly zero and, thus, that stable points corresponds to sets on B^n. Writing $V_i = g_i(s\,u_i)$ allows the voltage gain g_i to be scaled without altering the sigmoidal asymptotes. Then $u_i = 1/s\,g_i^{-1}(V_i)$. Therefore we can factor $1/s$ out of the integral term on the right-hand side of Equation 23. Accordingly, this term approaches zero in the high-gain case as s approaches infinity, and the sigmoid becomes a threshold-linear operation, as in the bivalent case. Then the minima of $E = -VMV^T$ are the same as if V is a binary vector, namely, (usually) the vertices of I^n. As s decreases, energy minima move to the interior of I_n, creating a more proper FAM, though more difficult to program. The less steep the sigmoid, the fewer energy minima occur at vertices of I^n. In sum, the trajectory of an initial input pattern is from somewhere inside I^n to the nearest vertex. This *disambiguation process* is exactly the minimization of fuzzy entropy.

Another popular associative memory that turns out to minimize fuzzy entropy is Anderson's[2-4] brain state in a box (BSB), called this because, once more, a brain state or neuron activation vector is represented as a point in I^n, as a fuzzy set. BSB is an example of *supervised learning* using the celebrated Widrow-Hoff rule of feedback error-correction. A_i and B_i are vectors of length n. The connection matrix is formed by summing the correlation matrices $A_i^T B_i$. To teach the network B_i, the difference is computed between the desired signal B_i and the (unthresholded) actual output $A_i M$ to produce the error $B_i - A_i M$. M is then incremented additively with the error signal:

$$\Delta M \; = \; r\,(B_i - A_i M)^T A_i \qquad\qquad (24)$$

where r is a learning rate parameter. Now suppose $A_i = B_i$. BSB dynamics are described by $A(t + 1) = A(t) + A(t)\,M = (I + M)\,A(t)$. M has the form of a sample covariance matrix and, thus, admits standard principal component or factor analysis in terms of dominant eigenvectors/eigenvalues, which contain most of the variance of the system. Equation 24 produces differential weighting of eigenvectors. Eigenvector enhancement drives brain states to corners of the I^n box. After learning, Anderson describes the qualitative nonlinear dynamics of the BSB as follows: "If we start with an activity pattern inside the box, it receives positive feedback on certain components which have the effect of forcing it outward. When its elements start to limit (i.e., when it hits the walls of the box), it moves into a corner of the box where it remains for eternity." In other words, the BSB dynamics produce a state trajectory so that, with respect to the nearest (stored) vertex, each change in fit value is monotonically toward 0 or 1 depending whether the fit value was initially below or above $^1/_2$. Therefore, the BSB model is a fuzzy entropy minimizer.

A final example of neural network minimization of fuzzy entropy occurs in some sense in Grossberg competitive learning.[16-18] These networks are *lateral inhibition* or recurrent *on-center off-surround* networks. Much biological evidence suggests they naturally occur in visual information processing. Here M has positive diagonal and negative off-diagonal elements. This every-neuron-for-himself connection topology forces every neuron to positively reinforce itself and negatively reinforce its neighbors. The neurons, or feature detectors, compete for activation.[46] The winner(s) of the competition classify the input pattern according to their feature characteristic. Activation thresholds regulate competition. The larger the threshold, the more sum positive activation a neuron must receive to fire and, thus, the fewer feature-detecting winners. As mentioned before, Grossberg[15-18] has shown that sigmoid signal functions compute a quenching threshold for competitive networks that decides which competitors will be contrast enhanced or suppressed as noise. If the input patterns are fuzzy sets, if competitive contrast enhancement and noise suppression produce bit vectors from fit vectors, and if the quenching threshold is taken to be $^1/_2$, then competitive learning also minimizes fuzzy entropy.

Two comments are in order about these examples of fuzzy entropy minimization occurring in popular associative-memory/learning models. First, fuzzy entropy minimization is a candidate design principle or architecture constraint for associative memories that disambiguate fit vectors into stored bit vectors. Grossberg competitive learning illustrates that fuzzy entropy minimization can occur in the absence of a minimized Lyapunov or energy function. We can require that proposed associative memories on I^n minimize fuzzy entropy, at least on average. More reasonably, we ought always examine such associators to see whether they minimize fuzzy entropy and why. Conversely, we can construct FAMs that change state if, and only if, they minimize fuzzy entropy. For instance, if $I_i(t)$ is some suitably preprocessed input to fit element or neuron $a_i(t)$, then an interesting FAM is defined by the rule: $a_i(t + 1) = I_i(t)$ if $I_i(t) \leq a_i(t) \leq ^1/_2$ or $I_i(t) \geq a_i(t) \geq ^1/_2$; otherwise, $a_i(t + 1) = a_i(t)$.

Second, probability theory did not play a fundamental role in any of the examples cited. This is ultimately for two reasons. Convergence to vertices means possible convergence to nonprobability distributions, namely, the $2^n - n$ such bit vectors (possibility distributions). The other reason is that FAM disambiguation involves ambiguity, and this is elementhood in something and its opposite, i.e., nondegenerate overlap $A \cap A^c$ and underlap $A \cup A^c$, which measure fuzziness and are inherent to I^n state-space representations.

V. COMBINING FUZZY COGNITIVE MAP FAMS

We briefly describe a new technique[40] for combining *fuzzy cognitive maps* (FCMs).[34,38] A FCM is a fuzzy signed digraph with feedback. Nodes represent fuzzy sets or variable concepts like *social stability*. Fuzzy directed edges represented partial causality. Positive ("excitatory") edges represent causal increase. Negative ("inhibitory") edges represent

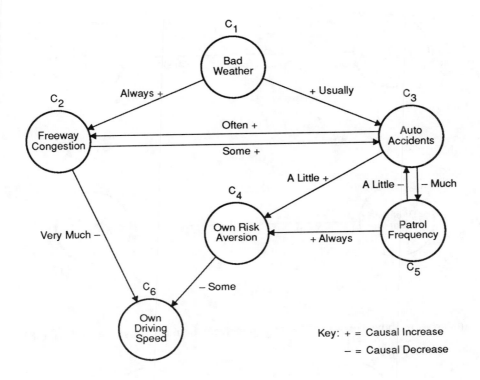

C_1

Bad Weather

C_2

Freeway Congestion

C_3

Auto Accidents

C_4

Own Risk Aversion

Patrol Frequency

C_5

C_6

Own Driving Speed

Always +

+ Usually

Often +

Some +

A Little +

A Little −

− Much

+ Always

Very Much −

− Some

Key: + = Causal Increase
− = Causal Decrease

FIGURE 5.

causal decrease. More generally, both nodes and edges are functions of time. What a FCM fragment of how fast one drives on a freeway might be, can be seen in Figure 5.

Here the causal weights are linguistic fuzzy quantifiers from some partially ordered set. One causal path on this FCM from bad weather to own driving speed is "bad weather *usually* increases auto accidents, which *much* decreases highway patrol frequency (though this decreases auto accidents *a little*), which in turn *always* increases own risk aversion, which *some*what decreases own driving speed."

Causality can be represented as a fuzzy relation on the product space $I^n \times I^n$, where I^n is the *concept space* of the m-many concepts (fuzzy sets) $C_1, \ldots, C_m$. Accordingly, $I^n = F(2^X)$ for some space X of primitives or *cognitons*. A *concept* is then any fuzzy subset of X, which implies that any finite set of primitives (e.g., sensibilia) gives rise to an infinitude of concepts or thoughts. (Thus, we view a thought as a fuzzy set, not as a *language string*.) It is natural to identify the degree to which C_i causes C_j, the edge function $e(C_i,C_j)$, with subsethood: $e(C_i,C_j) = S(C_i,C_j)$. The fuzzy causal interval, however, is the bipolar interval $[-1,1]$ not $[0,1]$ since we allow negative as well as positive causality. Fortunately we can transform all negative causality to positive causality (and vice versa) by introducing m-many dis-concepts[34] $\sim C_i$ and by replacing $e(C_i,C_j) < 0$ with $e(C_i,\sim C_j) > 0$. Since the concept space $F(2^X)$ is I^n, fuzzy dis-concepts reside in $F(2^X)$ as well. For example, we can eliminate the grim sociological causal relationship as seen in Figure 6.

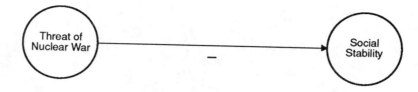

FIGURE 6.

with

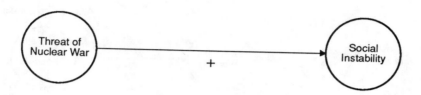

FIGURE 7.

Similarly the above freeway FCM in Figure 7 can be recast as shown in Figure 8.

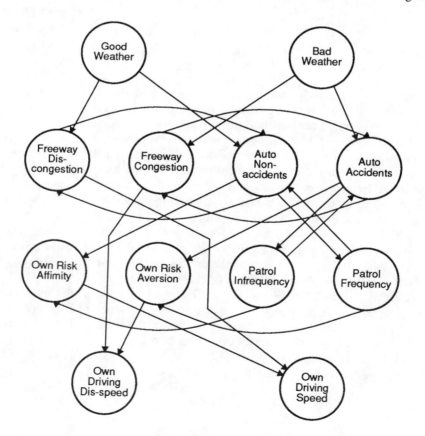

FIGURE 8.

where it is understood that $e(C_i,C_j) \approx e(\sim C_i,\sim C_j)$ if $e(C_i,C_j) > 0$, and $e(C_i,C_j) \approx e(C_i,\sim C_j)$ if $e(C_i,C_j) < 0$, i.e., that original and transformed causal strengths are similar if not equivalent.

FCMs have two properties that distinguish them from other knowledge representation schemes in artificial intelligence. First, they can be quickly and painlessly acquired from knowledge sources. The knowledge source or expert can use his own partial order to describe the partial causal relationships or, as we shall assume, he either speaks in unit-interval quantities or we preprocess his input into unit-interval quantities. Second, *graph search on a FCM is nearly impossible*. FCMs are full of feedback, of logical cycles. Standard graph-search logical inference techniques can only get caught in thickets of infinite loops, and the N-P complexity of checking graphs for cycles is well known. AI search trees, such as frame structures or Markov or Bayesian trees,[43,44,47] are *trees*. They are acyclic by design. That is why they are locally tractable. The price paid for these computations, however, is the inability to accurately represent causally connected phenomena, fuzzy or not.

Ironically, FCM graph-search intractability is solely responsible for FCM FAM inferencing. Graphs and trees are represented by connection matrices which, in turn, house dynamical systems. Feedback or cyclicness produces interesting dynamic systems. Trees, in particular AI decision trees (expert systems), are pure feed forward structures. *They have no interesting dynamic behavior*. They map all input into the null set.

A FCM m-by-m connection matrix F is a FAM. Its *complexity* is ultimately bounded by m^2. The ith row of F, F_i, lists the excitatory and inhibitory causal connections from C_i to the other causal concepts C_j. The ith column, F^i, lists the causal connections directed to C_i. We assume no concept causes itself; so F is zero-diagonal. Here we shall assume the simple binary-state model as used in the BAM which, nevertheless, is a useful practical tool and complex theoretical structure. We assume concepts are either on (1) or off (0), knowing that we can generalize this with a differential model. FCM weights $F_{ij} = e(C_i,C_j)$ are causally excitatory or inhibitory weights in the fuzzy causal interval $[-1,1]$, knowing that dis-concepts can be used to maintain unit-interval causality. An *event* or *state* of the FCM is the bit (fit) vector of causal concept activations $c(t) = (C_1(t), \ldots ,C_m(t)) \in B^m$ at time t in the dynamic process. We assume the causal concepts in C are *synchronously* updated knowing that, as in the BAM case, asynchronous update schedules (causal time constants) can be used as well. The input to $C(t + 1)$ is some nonlinear function n of the input-sum vector $C(t)$ M that keeps $C(t + 1)$ binary. We assume the simplest nonconstant function n, the threshold-linear function. Then:

$$C_i(t + 1) = \begin{cases} 1 \text{ if } C(t) \, M^i > 0 \\ \\ 0 \text{ if } C(t) \, M^i \le 0 \end{cases} \tag{25}$$

Equation 25 differs from the BAM threshold-linear law, as seen in Equation 12, when the causal input sum equals the threshold (here the default threshold 0). In Equation 12 the neuron maintains its current state, either on or off, when this occurs. In Equation 25 we encode the causal intuition that something happens if, and only if, something makes it happen. Put another way, we assume passive decay of causal activation.

A FCM inference is a *resonant* state or pattern. The FCM is *turned on* by an initial or continuous input pattern that quickly evolves to some steady resonant state of FCM activity. FCMs only make predictions or forward-chaining inferences because of the underlying nonlinear (irreversible) dynamics of the state-transition model. A resonant state is a limit cycle, sequence of states, or a fixed point (one-step limit cycle). The causal interpretation of a resonant limit cycle is a predicted *sequence of events,* or spatiotemporal pattern. Some complex differential FCMs with time-varying edge functions can, in principle, resonate on chaotic attractors.

Binary FCMs governed by Equation 25 always converge to a stable limit cycle in no more than 2^m synchronous update iterations. The first state of the resonant limit cycle is the first state that is causally recalled twice. Energy minimization arguments[37] can also be used to show rapid convergence to stable limit cycles. If l_i is the length of the ith limit cycle and there are r distinct limit cycles, then $l_1 + \ldots + l_r \leq 2^m$ since a distinct binary state can occur in, at most, one limit cycle and need not occur in any. If A_r is the arithmetic-mean length of distinct limit cycles, then $r \leq 2^m/A_r$ holds with strict inequality in most cases. In general there are few limit cycles relative to the total number 2^m of possible input states.

FCM resonant states can be interpreted as *hidden patterns* in the F connections that summarize an expert's or experts' knowledge. Experts tend to map all possible what-if questions (2^m binary states) into relatively few responses. We identify these characteristic responses with FCM limit cycles; so, once again, $r/2^m$ tends to be very small. The limit-cycle patterns are *hidden* in the FCM connection topology. As shown by Equation 25, different fuzzy weights with the same sign can produce different hidden patterns. Yet also by Equation 25, FCMs tend to be *robust* when edges are perturbed. This follows since typically in large FCM matrices only large changes can change the sign of the input sum, and even then only one or two bit values may change and the trajectory may still converge to the original limit cycle or some modification of it.

Suppose we have somehow combined the knowledge of several experts in the FCM F. We can intuitively interpret a resident hidden pattern in F as the consensus eventually reached by a round-table discussion among the experts. A topic or question (causal state description) is proposed; then fairly soon a rough agreement or fixed pattern of responses is reached — no matter how many experts or how rich the causal environment. In rare cases a unanimous opinion is reached, which corresponds to quick convergence to a fixed point. In general, the stable responses will be some causal sequence or set of conditions ("Yes, but only if this and that . . . "), which corresponds to a limit cycle, or a clear-cut two-sided disagreement, which corresponds to a two-step limit cycle or BAM fixed point. The evolved pattern of responses will typically differ from the complete position of each expert. Indeed, even in casual conversation, like questions tend to map into like response configurations, and perturbations of the causal environment or problem domain usually do not affect the evolved outcome.

We now summarize the FCM knowledge combination scheme in Kosko.[40] A more general, but somewhat less practical, scheme for combining arbitrary partially ordered quantities can be found in Kosko.[38] The present scheme combines arbitrary FCMs from arbitrarily many knowledge sources with arbitrary credibility weights w_i in [0,1]. Accordingly, unlike traditional AI knowledge engineering, where there is only one or two *domain experts,* here there can be as many experts as pleased, with as diverse knowledge as pleased, and with different levels (degrees) of expertise. The author has used this technique to systematically combine over 30 experimental sociopolitical FCMs in a matter of minutes.

A FCM connection matrix stands behind every FCM graph. These matrices are causal FAMs and can be easily manipulated. However, fuzzy knowledge is much easier to acquire FCM graphic form. This suggests a mathematical transform strategy: *transform FCM graphs to FCM matrices, combine the matrices, then inverse transform back to a combined FCM graph.* The key insight is that a FCM is a causal FAM. Then, the natural way to combine or superimpose the FCM matrices is simply to add them pointwise. Moreover, we can gate[16-19] the ith FCM matrix F_i with the weight w_i simply by multiplying the matrix by w_i, yielding w_iF_i. (Here we assume $w_i = 0$ indicates no credibility and $w_i = 1$ indicates maximum credibility.) Alternatively we can gate F_i with w_i by taking the pairwise minimum (t-norm) of w_i with each element of F_i. For simplicity we assume multiplicative gating.

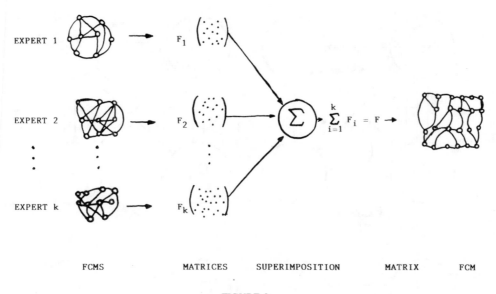

FCMS   MATRICES SUPERIMPOSITION MATRIX FCM

FIGURE 9.

FCMs that differ in number or kind of concept nodes can still be combined by introducing *augmented connection matrices*. Intuitively, if only the ith expert discusses concept C, then C is a phantom node in the other FCMs, where it has no causal connectivity. Thus, it *is* represented in the other FCM matrices, except the C row and columns contain only values of 0. Specifically, suppose the total number (union) of nodes discussed by all the experts is n, which may be quite large with respect to the dimensions of a particular FCM matrix. Then we augment each FCM matrix F_i to be an n-by-n matrix, perhaps quite sparse. We permute row and columns as necessary to bring them into mutual coincidence, labeling the row/column concepts 1 to n. Then the values of augmented F_i can be combined by adding pointwise as shown in Figure 9.

The combined FCM F naturally weights the knowledge of the experts. If only one expert out of k asserts a particular causal connection, then that connection can have a maximum magnitude of 1/k. If 50 experts out of 100, all equally weighted, assert that a causal connection between C_i and C_j has weight $+1$ and the other 50 assert that it has weight -1, then no causal connection occurs between C_i and C_j — $f_{ij} = e_F(C_i, C_j) = 0$ — in the combined FCM F, and so on. We also note that if simple 0-thresholding is used in Equation 25, we can normalize F by either the combined weight with the maximum magnitude or by k to keep all f_{ij} in $[-1, 1]$, though this is not necessary for FAM or for causal-interpretation purposes. In general, instead of normalizing by k, when credibility weights $w_1, \ldots, w_k$ are used, the *weight sum* $W = w_1 + \ldots w_k$ should be used for normalization.

Let us examine a simple example: four unweighted experts provide the FCMs as shown in Figure 10.

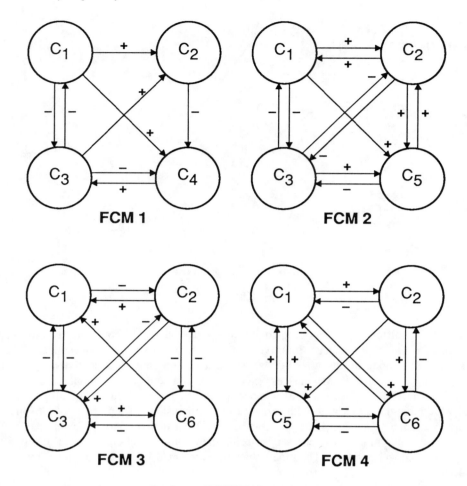

FIGURE 10.

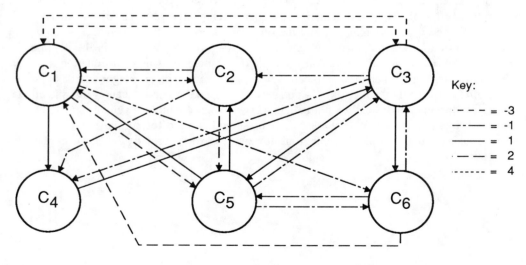

FIGURE 11.

For simplicity each edge is either -1, 0, or $+1$. There are six distinct concept nodes. Therefore, we can represent these FCMs in four 6-by-6 augmented connection matrices:

$$F_1 = \begin{pmatrix} 0 & 1 & -1 & 1 & 0 & 0 \\ 0 & 0 & 0 & -1 & 0 & 0 \\ -1 & 1 & 0 & -1 & 0 & 0 \\ 0 & 0 & 1 & 0 & 0 & 0 \\ 0 & 0 & 0 & 0 & 0 & 0 \\ 0 & 0 & 0 & 0 & 0 & 0 \end{pmatrix} \quad F_2 = \begin{pmatrix} 0 & 1 & -1 & 0 & 1 & 0 \\ 1 & 0 & -1 & 0 & 1 & 0 \\ -1 & -1 & 0 & 0 & 1 & 0 \\ 0 & 0 & 0 & 0 & 0 & 0 \\ 0 & 1 & -1 & 0 & 0 & 0 \\ 0 & 0 & 0 & 0 & 0 & 0 \end{pmatrix}$$

$$F_3 = \begin{pmatrix} 0 & 1 & -1 & 0 & 0 & 0 \\ -1 & 0 & 1 & 0 & 0 & -1 \\ -1 & -1 & 0 & 0 & 0 & 1 \\ 0 & 0 & 0 & 0 & 0 & 0 \\ 0 & 0 & 0 & 0 & 0 & 0 \\ 1 & -1 & -1 & 0 & 0 & 0 \end{pmatrix} \quad F_3 = \begin{pmatrix} 0 & 1 & 0 & 0 & 1 & -1 \\ -1 & 0 & 0 & 0 & 1 & 1 \\ 0 & 0 & 0 & 0 & 0 & 0 \\ 0 & 0 & 0 & 0 & 0 & 0 \\ 1 & 0 & 0 & 0 & 0 & -1 \\ 1 & -1 & 0 & 0 & -1 & 0 \end{pmatrix}$$

which combine to yield:

$$F = \begin{pmatrix} 0 & 4 & -3 & 1 & 2 & -1 \\ -1 & 0 & 0 & -1 & 2 & 0 \\ -3 & -1 & 0 & -1 & 1 & 1 \\ 0 & 0 & 1 & 0 & 0 & 0 \\ 1 & 1 & -1 & 0 & 0 & -1 \\ 2 & -2 & -1 & 0 & -1 & 0 \end{pmatrix}$$

which *inverse transforms* to the FCM shown in Figure 11.

Suppose we input the state vector $S = (1\ 1\ 0\ 0\ 0\ 0)$ into F. Then
$S\ F = (-1\ 4\ -3\ 0\ 4\ -1) \rightarrow (0\ 1\ 0\ 0\ 1\ 0) = S'$. Then $S'\ F = (0\ 1\ -1\ -1\ 2\ -1) \rightarrow (0\ 1\ 0\ 0\ 1\ 0) = S'$. So input S stabilizes or resonates on S'. Similarly, the input patterns $(1\ 1\ 1\ 1\ 1\ 1)$ and $(0\ 0\ 0\ 0\ 1\ 1)$ evoke the resonant *hidden pattern* S', which forms a dominant basin of attraction in the state space $\{0,1\}^6$ with energy $E(S') = -S'\ FS^T = -3$. It is interesting to note how S' behaves in the original matrices F_i. On F_1, S' maps into the empty set $(0\ 0\ 0\ 0\ 0\ 0)$. On F_2, S' maps into $(1\ 1\ 0\ 0\ 1\ 0)$, which has Hamming distance 1 from S'. On F_3, S' maps into the limit cycle $(0\ 0\ 1\ 0\ 0\ 0) \rightarrow (0\ 0\ 0\ 0\ 0\ 1) \rightarrow (1\ 0\ 0\ 0\ 0\ 0) \rightarrow (0\ 1\ 0\ 0\ 0\ 0) \rightarrow (0\ 0\ 1\ 0\ 0\ 0)$. On F_4, S' maps into the limit cycle $S' \rightarrow (0\ 0\ 0\ 0\ 1\ 0) \rightarrow (1\ 0\ 0\ 0\ 0\ 0) \rightarrow S'$.

We briefly mention extensions of the FCM framework to learning networks. Here both edges and nodes are time-varying real functions. Node functions essentially sum sigmoid signal functions and combine passive decay, external input, and variable thresholds in forms similar to Equations 17 or 21. Learning occurs by modifying the causal edges in response to evidential changes. A combined FCM merely initiates a learning FCM, or periodically updates it. Two generic learning laws are the Hebbian[20,23,33] and *differential* Hebbian[35,37,41] learning laws:

$$\dot{e}_{ij} = C_i C_j \tag{26}$$

$$\dot{e}_{ij} = \dot{C}_i \dot{C}_j \tag{27}$$

both of which usually contain many other similar terms. The classical Hebbian law, Equation 26, captures simple correlation or conjunctive learning. Though it abounds in neural models, it is unacceptable in causal models. For Equation 26 grows causal connections on the basis of *concomitant activation*. Consequently, if any two nodes are active in the network, Equation 26 inductively infers a causal connection between them. These causal connections are almost always spurious. Moreover, since the node functions C_i and C_j are nonnegative, Equation 26 implies that e_{ij} exponentially increases to its maximum value, after which no learning occurs. The differential Hebbian law, Equation 27 was motivated by the causal descriptions of the empiricist philosophers[35,41] David Hume ("constant conjunction of events") and John Stuart Mill ("concomitant variation").

Concomitant variation, not activation, drives learning (inductive causal inference) in Equation 27. This produces an intuitive sign law. For although C_i and C_j cannot be negative, their derivatives can be negative and are negative if, and only if, C_i and C_j move in opposite directions. If they move in the same directions — they both increase or both decrease — then the derivative product is positive. In this sense the differential Hebbian more aptly captures the correlation learning that formed the basis for Hebbian learning. It is an open, but intriguing, question whether Equation 27 has any validity *in vivo*. Yet there is some reason to suspect it might, which follows from the natural relation between Equations 26 and 27. Any dynamic node equation for C_i contains a passive decay term $-d\,C_i$, where d is a positive decay constant (or function), and other terms, which we lump in the parameter 0_i:

$$\dot{C}_i = -dC_i + O_i \tag{28}$$

Therefore, substituting Equation 28 in Equation 27 for $\dot{C}_i$ and $\dot{C}_j$ gives:

$$\dot{e}_{ij} = \dot{C}_i \dot{C}_j$$
$$= d^2\, C_i C_j + O_{ij} \tag{29}$$

where again O_{ij} is a catchall symbol representing all other terms. Thus, Equation 29 shows that differential Hebbian learning implies Hebbian learning. It further shows how this subsumption comes about — by interaction of passive-decay, or resting-potential, phenomena. When Equation 27 is used to autonomously grow fuzzy logic (causal) paths on a FCM from raw data, it allows one, in some sense, to grow his own expert system.

In summary, the FCM FAM framework provides an alternative to traditional AI expert-system knowledge representation and inferencing. FCMs are comparatively easy to gather from experts, many experts, allowing rich and uncertain causal problem domains to be systematically represented and combined. Unlike expert-system logic trees, FCM FAMs do not directly allow a logical audit trail to be exhibited by which the system *explains itself*. Also the state-transition nonlinearities do not permit backward chaining (although more complicated knowledge networks can incorporate goal information in update thresholds and feedback loops). Instead emphasis is on *decoding* and *accurate prediction*. FCM node size does not, in principle, affect FCM convergence speed. Unambiguous predictions are speedily made given any input state or sustained input. These predictions can be compared with the available evidence; discrepancies can suggest changes in causal connections. Indeed, in sociopolitical problem domains, different FCMs can represent conflicting ideologically motivated world views. An evidential procedure can then adjudicate FCM modifications and, perhaps, world views. In other words, instantiated FCM FAMs are *testable* causal theories.

ACKNOWLEDGMENTS

This research was supported by the Air Force Office of Scientific Research (AFOSR F49620-86-C-0070) and the Advanced Research Projects Agency of the Department of Defense under ARPA Order No. 5794.

REFERENCES

1. **Abu-Mostafa, Y. S. and St. Jacques, J.,** Information capacity of the Hopfield model, *IEEE Trans. Inf. Theor.,* IT 31(4), 461, July 1985.
2. **Anderson, J. A.,** Cognitive and psychological computation with neural models, *IEEE Trans. Syst. Man, Cybern.,* SMC 13(5), September/October 1983.
3. **Anderson, J. A. and Mozer, M.,** Categorization and selective neurons, in *Parallel Models of Associative Memory,* Hinton, G. and Anderson, J. A., Eds., Erlbaum, Hillsdale, NJ, 1981.
4. **Anderson, J. A., Silverstein, J. W., Ritz, S. A., and Jones, R. S.,** Distinctive features, categorical perception, and probability learning: some applications of a neural model, *Psych. Rev.,* 84, 413, 1977.
5. **Bandler, W. and Kohout, L.,** Fuzzy power sets and fuzzy implication operators, *Fuzzy Sets Syst.,* 4, 13, 1980.
6. **Carpenter, G. A. and Grossberg, S.,** A massively parallel architecture for self-organizing neural pattern recognition machine, in *Comput. Vision, Graphics, Image Process.,* 1986.
7. **Carpenter, G. A. and Grossberg, S.,** Associative learning, adaptive pattern recognition, and cooperative-competitive decision making by neural networks, in *Proc. SPIE: Hybrid and Optical Systems,* Szu, H., Ed., 1986.
8. **Cheeseman, P.,** In defense of probability, *Proc. of the IJCAI-85,* August 1985, 1002.
9. **Cohen, M. A. and Grossberg, S.,** Absolute stability of global pattern formation and parallel memory storage by competitive neural networks, *IEEE Trans. Syst. Man, Cybern.,* SMC 13, 815, September/October 1983.
10. **Cohen, M. A. and Grossberg, S.,** Masking fields: a massively parallel neural architecture for learning, recognizing, and predicting multiple groupings of patterned data, *Appl. Optics,* 1986.
11. **Cox, R. T.,** Probability, frequency and reasonable expectation, *Am. J. Phys.,* 14(1), 1, January/February 1946.
12. **Dubois, D. and Prade, H.,** Fuzzy cardinality and the modeling of imprecise quantification, *Fuzzy Sets Syst.,* 16(3), 199, August 1985.
13. **De Luca, A. and Termini, S.,** A definition of a nonprobabilistic entropy in the setting of fuzzy sets theory, *Inf. Control,* 20, 301, 1972.
14. **Gaines, B. R.,** Foundations of fuzzy reasoning, *Int. J. of Man-Mach. Stud.,* 8, 623, 1976.
15. **Grossberg, S.,** Contour enhancement, short term memory, and constancies in reverberating neural networks, *Stud. Appl. Math.,* 52, 217, 1973.
16. **Grossberg, S.,** Adaptive pattern classification and universal recoding, I. parallel development and coding of neural feature detectors, *Biol. Cybern.,* 23, 121, 1976.
17. **Grossberg, S.,** A theory of human memory: self-organization and performance of sensory-motor codes, maps, and plans, in *Progress in Theoretical Biology,* Vol. 5, Rosen, R. and Snell, F., Eds., Academic Press, New York, 1978.
18. **Grossberg, S.,** How does a brain build a cognitive code? *Psych. Rev.,* 1, 1, 1980.
19. **Grossberg, S.,** Adaptive resonance in development, perception, and cognition, in *Mathematical Psychology and Psychopshysiology,* Grossberg, S., Ed., American Math Society, Providence, RI, 1981.
20. **Grossberg, S.,** *Studies of Mind and Brain: Neural Principles of Learning, Perception, Development, Cognition, and Motor Control,* Reidel Press, Boston, 1982.
21. **Grossberg, S.,** *The Adaptive Brain, I & II,* North-Holland, Amsterdam, 1986.
22. **Grossberg, S. and Kuperstein, M.,** *Neural Dynamics of Adaptive Sensory-Motor Control: Ballistic Eye Movements,* North-Holland, Amsterdam, 1986.
23. **Hebb, D.,** *The Organization of Behavior,* John Wiley & Sons, New York, 1949.
24. **Hecht-Nielsen, R.,** Performance limits of optical, electro-optical, and electronic artificial neural system processors, *Proc. SPIE: Hybrid and Optical Systems,* Szu, H., Ed., 1986.
25. **Hopfield, J. J.,** Neural networks and physical systems with emergent collective computational abilities, *Proc. Natl. Acad. Sci. U.S.A.,* 79, 2554, 1982.

26. **Hopfield, J. J.,** Neurons with graded response have collective computational properties like those of two-state neurons, *Proc. Natl. Acad. Sci. U.S.A.,* 81, 3088, 1984.

27. **Jaynes, E. T.,** *E. T. Jaynes: Papers on Probability, Statistics, and Statistical Physics,* Rosenkrantz, R. D., Ed., Reidel Press, Boston, 1983.

28. **Jaynes, E. T.,** Where do we stand on maximum entropy?, in *The Maximum Entropy Formalism,* Levine, R. and Tribus, A., Eds., MIT Press, Cambridge, MA, 1979.

29. **Kandel, A.,** *Fuzzy Mathematical Techniques with Applications,* Addison-Wesley, Reading, MA, 1986.

30. **Kishi, F. H.,** On line computer control techniques, in *Advances in Control Systems: Theory and Applications, Vol. 1,* Leondes, C. T., Ed., Academic Press, New York, 1964.

31. **Klement, E. P.,** Operations on fuzzy sets and fuzzy numbers related to triangular norms, in *Proc. 11th Int. Symp. Multi-Valued Logic,* 218, 1981.

32. **Kohonen, T.,** *Content-Addressable Memories,* Springer-Verlag, New York, 1980.

33. **Kohonen, T.,** *Self-Organization and Associative Memory,* Springer-Verlag, New York, 1984.

34. **Kosko, B.,** Fuzzy cognitive maps, *Int. J. Man-Mach. Stud.,* 24, 65, January 1986.

35. **Kosko, B.,** Differential Hebbian Learning, in *Am. Inst. Physics Conf. Proc.: Neural Networks for Computing,* Denker, J. S., Ed., Snowbird, Utah, April 1986, 277.

36. **Kosko, B.,** Counting with fuzzy sets, *IEEE Trans. Pattern Anal. Mach. Intell.,* PAMI 8(4), 556, July 1986.

37. **Kosko, B.,** Bidirectional Associative Memories, September 1986.

38. **Kosko, B.,** Fuzzy knowledge combination, *Int. J. Intell. Syst.,* 1, 1986.

39. **Kosko, B.,** Fuzzy entropy and conditioning, *Inf. Sci.,* 1986.

40. **Kosko, B.,** Hidden patterns in combined knowledge networks, 1986.

41. **Kosko, B.,** Vision as causal activation and association, in *Proc. SPIE: Intell. Robots Computer Vision,* Vol. 579, Casasent, D., Ed., 104, September 1985.

42. **McCulloch, W. S. and Pitts, W.,** A logical calculus of the ideas immanent in nervous activity, *Bull. Math. Biophys.,* 5, 115, 1943.

43. **Pearl, J.,** On evidential reasoning in a hierarchy of hypotheses, *Artif. Intell.,* 28, 9, 1986.

44. **Pearl, J.,** Fusion, Propagation, and Structuring in Bayesian Networks, *Artif. Intell.,* 29, 241, 1986.

45. **Prade, H.,** A computational approach to approximate and plausible reasoning with applications to expert systems, *IEEE Trans. Pattern Anal. Mach. Intell.,* 7(3), 260, 1985.

46. **Rumelhart, D. E. and McClelland, E., Eds.,** *Parallel Distributed Processing, Vols. 1 and 2,* MIT Press, Cambridge, MA, 1986.

47. **Shenoy, P. and Shafer, G.,** Propagating belief functions with local computations, *IEEE Expert,* 1(3), 43, 1986.

48. **Togai, M. and Watanabe, H.,** A VLSI Implementation of Fuzzy Inference Engine: Toward an Expert System on a Chip, in *Proc. 2nd IEEE CAIA, Miami Beach, FL,* December 1985.

49. **Togai, M. and Watanabe, H.,** Expert system on a chip: an engine for realtime approximate reasoning, *IEEE Expert,* 1(3), 55, 1986.

50. **Van Campenhout, J. M. and Cover, T. M.,** Maximum entropy and conditional probability, *IEEE Trans. Inf. Theory,* IT-27(4), July 1981.

51. **Zadeh, L. A.,** Fuzzy sets as a basis for a theory of possibility, *Fuzzy Sets Syst.,* 1, 3, 1978.

52. **Zadeh, L. A.,** A computational approach to fuzzy quantifiers in natural languages, *Comput. Math. Appl.,* 9(1), 149, 1983.

53. **Zadeh, L. A.,** The role of fuzzy logic in the management of uncertainty in expert systems, *Fuzzy Sets Syst.,* 11, 199, 1983.

54. **Zadeh, L. A.,** Syllogistic reasoning in fuzzy logic and its application to usuality and reasoning with dispositions, *IEEE Trans. Syst. Man, Cybern.,* SMC 15(6), 754, November/December 1985.

55. **Togai, M.,** personal communication.

Applications of Fuzzy Expert Systems

Chapter 11

THE ROLE OF APPROXIMATE REASONING IN A MEDICAL EXPERT SYSTEM

D. L. Hudson and M. E. Cohen

TABLE OF CONTENTS

I. INTRODUCTION

The first medical expert system, MYCIN,[1] was developed in 1975, shortly after the initial introduction of the expert system approach to decision making.[2,3] Since that time, expert systems have been a popular approach to medical problem solving. These systems include broad range projects such as INTERNIST,[4,5] which encompasses the entire field of internal medicine, as well as systems dealing with more specific medical problem domains.[6-11] A number of survey papers describing expert systems in medicine have also appeared in the literature.[12-15] The need for dealing with uncertain information was recognized even in the first of these systems.[16] A number of expert systems have developed ad hoc techniques of dealing with uncertainty.[1,17,18]

In the same general time frame, the theoretical development of fuzzy logic techniques was proceeding, fueled by a desire to better represent the real world. This field was initiated by Zadeh in his 1965 paper.[19] Since that time, theoretical development,[20-26] as well as practical application[27,28] of fuzzy logic and fuzzy set theory, in general, has progressed rapidly and, in particular, its application to expert systems.[29-34] Fuzzy techniques have also been applied in a number of medical applications.[35-37]

The development of EMERGE, a rule-based medical expert system for the analysis of chest pain in the emergency room environment, was begun in 1979. EMERGE was designed to address some of the difficulties encountered in earlier expert systems, including machine dependence, time-consuming consultations, and difficulty in replacement of the knowledge base.[38] Earlier systems were written in LISP and generally run on large-scale main frame computers possessed by a few medical institutions. In addition, rule searching was generally done by either forward or backward chaining, which resulted in very time-consuming consultations. Although the systems gave high level expert advice, they were not used in practice for these reasons. In order to counter the difficulties of machine dependence, EMERGE was written in standard Pascal in a memory-efficient manner to allow its use on microcomputers which, at that time, were just becoming popular.[39] A hierarchical approach to rule searching was designed in order to speed the consultation process, a necessity in emergency situations. The knowledge base for EMERGE was derived from existing medical logic outlines known as criteria maps,[40] a portion of which is shown in Figure 1. These maps had been developed over a number of years in consultation with medical experts and exist, not only for chest pain, but for approximately 20 medical applications.[41] Conversion from one rule base to another was simplified since the overall structure of the maps was the same for all applications.

The EMERGE system is arranged in a hierarchical structure, with questions contained in the highest level which pertain to symptoms with far-reaching consequences. It is denoted the level zero control flow. When a rule is substantiated in the level zero control flow, rapid focus of attention is achieved by transfer to a subcontrol flow which contains questions pertinent to that disorder. This structure permits rapid decision making and avoids asking unnecessary questions.

The original version of EMERGE used certainty factors as an indication of the degree of seriousness of an illness.[42] These were computed from scoring trees which were developed for each independent finding. They were then combined by ad hoc rules to give an overall certainty factor for each patient consultation.[43]

Current work on the EMERGE system involves inclusion of more sophisticated techniques of approximate reasoning.[44] Three major areas have been revised. First, user input has been modified from simple *y, n,* or *?* responses to permit input of a value indicating degree of presence of symptoms. Second, the original production rule format has been modified to permit weighting of antecedents. Finally, the traditional and/or binary logic has been generalized to permit inferences of rules in the new format. These three areas are the topic of this chapter and are discussed, in turn, in the following sections.

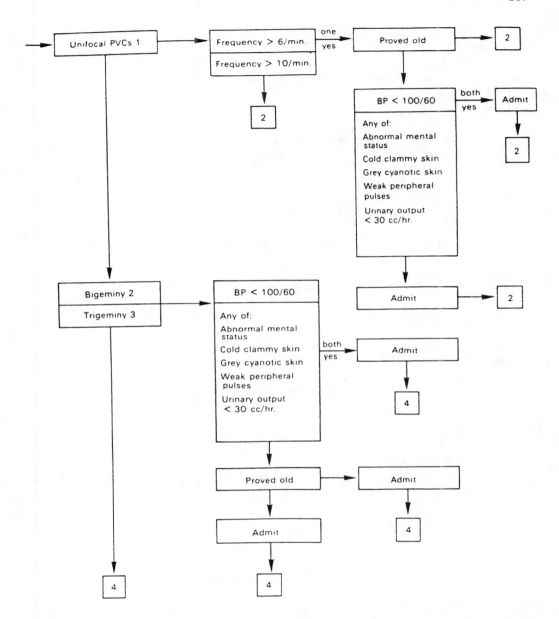

FIGURE 1.   Portion of chest pain criteria map.

II. PARTIAL PRESENCE OF SYMPTOMS

The EMERGE system was designed with two user interfaces: a question mode, in which the system answered questions with y, n, or ? and a data-driven mode, in which the user enters information. These two modes are illustrated in Figures 2 and 3. The data-driven mode incorporates an algorithm for inexact word and phrase matching to permit acceptance of altered word order and minor misspellings. Note the altered word order of ''Abnormal Chest X-ray'' in Figure 3 as compared to the question mode in Figure 2. As the system developed, a third menu-driven mode was developed. This interface was, in fact, developed during prospective analysis of the system at the request of the end users, who had become familiar with user-oriented microcomputer programs. Some sample menus are shown in

```
Enter name of patient:  WD
Age of patient:  58
Sex of patient:  m or f:  m
DO you wish to enter any clinical information on this patient?
Answer y or n:  n
Answer y if present, n if not, ? if no information

ECG ABNORMAL:  n
BP < 100/60:  n
BP > 150/90:  n
CHEST X-RAY ABNORMAL:  y
PULMONARY VENOUS HYPERTENSION:  n
PULMONARY EDEMA:  y
PVH OR EDEMA PROVED OLD:  n
DEPENDENT RALES:  n
RALES > 50% OF CHEST:  n
BRONCHOSPASM:  y
SHORTNESS OF BREATH:  y
NEW OR INCREASED BRONCHOSPASM:  y


Based on this information, the patient should be admitted to the
hospital.
```

FIGURE 2.   Question mode of user interface.

```
Enter name of patient :  WD
Age of patient:  58
Sex of patient, m or f:  m
Do you wish to enter any clinical information on this patient?
Answer y or n:  y

Type the clinical information one item per line.  When you have entered
all the information, type an asterisk (*) on a separate line.

ABNORMAL CHEST X-RAY
PULMONARY EDEMA
BRONCHOSPASM
*

PVH OR EDEMA PROVED OLD:  n
DEPENDENT RALES:  n
RALES > 50% OF CHEST:  n
SHORTNESS OF BREATH:  y
NEW OR INCREASED BRONCHOSPASM:  y


Based on this information, the patient should be admitted to the
hospital.
```

FIGURE 3.   Data-driven mode of user interface.

Figure 4. These two menus contain the information addressed in the level control flow. In all of the previous user interfaces, however, the symptoms were considered to be either present or absent. A third alternative was possible on the question mode. If the user had no information, a question mark could be entered. The question mark was treated as a "no" in the inference rules, but was treated differently in the certainty factor computation.

Difficulties arose in the preceding scheme in that the user had to make a yes/no decision about each test result, thus resulting in loss of information regarding severity of the symptoms. It was, in fact, discovered that the standard procedure for residents doing a physical examination was to enter a number between 0 and 10, inclusive, which indicated severity of the symptom. It was decided to adopt this procedure. A sample consultation using this new version is shown in Figure 5. The values entered are then divided by 10 to give a value between 0 and 1, inclusive. All values are given as positive symptoms. If a negation occurs in a production rule, the value is subtracted from 1. Similarly, in the data-driven mode, a numerical value can be entered after each clinical finding, as shown in Figure 6. If the

```
Enter name of patient:  WD
Age of patient:  58
Sex of patient, m or f:  m
Which interface do you wish to use?
Question (Q), Data-Driven (D), or Menu (m):  M
```

```
                    EMERGE CONSULTATION SOFTWARE

Enter the numbers of all positive findings, separated by commas.
Press enter when finished.

1  Abnormal ECG                    11 Pain Sharp
2  BP < 100/60                     12 ANY 2
3  Abnormal Chest X-ray                    Sweating
4  Syncope                                 Nausea
5  ANY OF                                  Dizziness
       Pain Excruciating          13 Dependent Rales
       Pain Unremitting           14 Rales > 50% of Chest
6  Pain Reproduced at One or More  15 ALL OF
       Loci in Chest                       Wheezing
7  Chest X-Ray Shows Broken Rib            Shortness of Breath
8  History of MI                   16 S3 Gallop
9  ANY OF                          17 ANY OF
       Pain Angina   Pain Squeezing        Pedal Edema
       Pain Heavy    Pain Aching           ^ JVP
       Pain Crushing Pain Dull             HJR
10 Pain Substernal or Left-Sided   18 Hemoptysis

  2, 15
```

```
                    EMERGE CONSULTATION SOFTWARE

Enter the numbers of all positive findings, separated by commas.
Press enter when finished.

19 Dyspnea                         24 ANY OF
20 Pleuritic Chest Pain                    Asymmetry of Carotid/Brachial/
21 Pain Relieved by Leaning Forward            Femoral Pulses
22 ANY OF                                  Absence of Lower Extrem. Pulses
       Childbirth                          Sternoclavicular Joint Pulsations
       Hospitalization             25 ANY OF
       Trauma                              > 50% Unilateral Pneumothorax
       Surgery                             Bilateral Pneumothorax Any Degree
       Prolonged Immobility        26 ANY OF
23 ANY OF                                  Palpable Cords in Calf
       P2 > A2                             Swollen Calf or Thigh
       Enlarged PA Segment                 Tenderness in Calf or Femoral
       Pulsation Prominent 2nd Left            Triangle
           Interspace                      Positive Homan's Sign

  0
```

FIGURE 4.   Menu mode of user interface.

number is omitted, a "yes," which is equivalent to a value of 10, is assumed. In the menu mode the number of the finding is entered, followed by the value. Separate findings are separated by commas. This is illustrated in Figure 7. In the question mode, the user is still given the option of responding with "y," "n," or "?" rather than with a numerical value.

III. UNEQUAL CONTRIBUTION OF ANTECEDENTS

In the original version of EMERGE a modified production rule format was used. All antecedents in the production rule were conjoined. However, an antecedent could take the form of standard check (SC). An SC could consist of a conjunction (AND), a disjunction (OR), or a COUNT. In a COUNT, n out of m premises were required to hold. These

```
Enter name of patient:  WD
Age of patient:  58
Sex of patient:  m or f:  m
Do you wish to enter any clinical information on this patient?
Answer y or n:  n

Answer with a value between 0 and 10, inclusive, with 0 indicating no
presence of the symptom.  If you wish, you may enter y (equivalent to
10) or n (equivalent to 0) instead.

ECG ABNORMAL:  0
BP < 100/60:  0
BP > 150/90:  0
CHEST X-RAY ABNORMAL:  10
PULMONARY VENOUS HYPERTENSION:  5
PULMONARY EDEMA:  9
PVH OR EDEMA PROVED OLD:  0
DEPENDENT RALES:  3
RALES > 50% OF CHEST:  3
BRONCHOSPASM:  8
SHORTNESS OF BREATH:  8
NEW OR INCREASED BRONCHOSPASM:  5


Based on this information, the patient should be admitted to the
hospital.
```

FIGURE 5.   Question mode with partial presence of symptoms.

```
Enter name of patient :  WD
Age of patient:  58
Sex of patient, m or f:  m
Do you wish to enter any clinical information on this patient?
Answer y or n:  y

Type the clinical information one item per line.  If you wish, a numeric
value between 0 and 10 indicating the degree of presence of a symptom
may be entered after the information in the following format:
ECG ABNORMAL:8
The value 0 indicates complete absence of the symptom.  If you do not
enter a number, 10 will be assumed.  When you have entered all the
information, type an asterisk (*) on a separate line.

ABNORMAL CHEST X-RAY:10
PULMONARY EDEMA:9
BRONCHOSPASM:8
*

PULMONARY VENOUS HYPERTENSION:  5
PVH OR EDEMA PROVED OLD:  0
DEPENDENT RALES:  3
RALES > 50% OF CHEST:  3
SHORTNESS OF BREATH:  8
NEW OR INCREASED BRONCHOSPASM:  5


Based on this information, the patient should be admitted to the
hospital.
```

FIGURE 6.   Data-driven mode with partial presence of symptoms.

```
┌─────────────────────────────────────────────────────────────────────┐
│                    EMERGE CONSULTATION SOFTWARE                       │
│                                                                       │
│ Enter the numbers of all positive findings, followed by a colon (:) and the │
│ degree of presence of that symptom.  Separate different findings with commas. │
│ Press enter when finished.                                            │
│                                                                       │
│ 1  Pulmonary Venous Hypertension      9 ^ JVR                         │
│ 2  Pulmonary Edema                   10 HJR                           │
│ 3  Dependent Rales                   11 Enlargement of Aortic Shadow  │
│ 4  Rales > 50% of Chest              12 Widening of Mediastinum       │
│ 5  Bronchospasm                      13 Bilateral Pleural Effusion    │
│ 6  Shortness of Breath               14 Wedge-Shaped Density          │
│ 7  S3 Gallop                         15 Large Pulmonary Artery        │
│ 8  Pedal Edema                                                        │
│                                                                       │
│  1:5,2:9,3:3,4:3,5:8,6:8                                              │
└─────────────────────────────────────────────────────────────────────┘
```

FIGURE 7.   Menu mode with partial presence of symptoms.

Number	Type	Threshold	Contents	a(i)
Rule 400		0.7	IF SC32	.3
			(Not) PVH or EDEMA PROVED OLD	.1
			(Not) DEPENDENT RALES	.1
			(Not) RALES > 50% OF CHEST	.1
			SC13	.2
			NEW OR INCREASED BRONCHOSPASM	.2
			THEN Admit	
SC12	AND	0.7	BRONCHOSPASM	.6
			SHORTNESS OF BREATH	.4
SC15	COUNT(2)	0.6	SWEATING	.4
			DIZZINESS	.3
			NAUSEA	.3
SC32	OR	0.5	PULMONARY VENOUS HYPERTENSION	.5
			PULMONARY EDEMA	.5

FIGURE 8.   Sample rule and standard checks (SC).

structures are illustrated in Figure 8. These SCs permitted representation of logical processes which were apparently common in medical reasoning, as noted in the criteria mapping procedure.

Upon further analysis it can be noted that, in fact, ANDs, ORs, and COUNTs are special cases of a general SC in that:

$$SC(m,1) \equiv OR$$

$$SC(m,m) \equiv AND$$

$$SC(m,n) \equiv COUNT\ (n),\ 1 < n < m \tag{1}$$

where m is the total number of antecedents in the SC.

As illustrated in Figure 8, for example, suppose that all antecedents are not of equal importance at arriving at a decision. We can then attach a weighting factor a(i) to the ith antecedent indicating its relative significance in the reasoning process. The requirement

$$\sum_{i=1}^{m} a(i) = 1$$

is added in order to achieve a normalizing process. As a practical consideration, values must be determined from experts in the application area as to the weighting of each antecedent. The default is to make all of equal weight, in which case all computations revert to previous values. However, upon questioning, it is often discovered that the expert does, in fact, consider some findings to be more significant than others. It appears that this information was not obtained previously simply because it could not be represented on traditional expert systems. The question then arises as to how to interpret these values in the context of the general SC described previously.

IV. GENERALIZATION OF BINARY INFERENCE

First, consider the SC structure. Initially, there were three rules for determining an SC value (true or false). It was an AND, OR, or COUNT. As described in the previous section, it is, in fact, unnecessary to make this distinction, as all can be treated as substantiation of a certain number in a list. In this structure, we are still dealing with crisp numbers. However, we can define the generalized SC in the crisp sense by:

$$SC = \sum_{i=1}^{m} \delta(i)$$

where

$$\delta(i) = 1 \text{ if ith antecedent is substantiated}$$

$$0 \text{ otherwise}$$

Then SC is substantiated if SC $\geq$ n, where n is the required number of premises $1 \leq n \leq m$. We, therefore, have essentially a threshold value for the substantiation, where the threshold is determined by the number of premises required to hold. The same system can be applied to rule substantiation but, in this case, since the antecedents are always conjoined, n = m.

If we include the generalization described in the last two sections, we now have two new values to deal with in the rule inference scheme. First, there is the a(i) value attached to the ith antecedent of every rule and SC. Second, there is the user response for each symptom, which we shall denote d(i). The following definition is made:

$$S = \sum_{i=1}^{m} a(i)d(i) \text{ where } a(i), d(i) \text{ defined on } [0,1]$$

$$\sum_{i=1}^{m} a(i) = 1$$

where a(i) is the weighting factor for the ith antecedent, d(i) is the degree of presence of the ith antecedent, and m is the total number of antecedents for the rule or SC in question.

The result of evaluation of an SC is thus a numerical value. The next question is how to interpret this numerical value.

We can still consider substantiation to be a thresholding problem, but the question is now how to determine the threshold in the fuzzy case. The determination of the threshold depends on the intent of a rule or SC, and must be arrived at in conjunction with experts. For example, consider the SC:

a(i)

.2 History of Chronic Atrial Flutter

.5 Atrial Flutter Resolved on 2nd ECG

.3 Heart Rate 120

which was an AND in the original system. One could clearly agree that all symptoms must be present in order to substantiate the SC. However, the second question is to what degree each symptom must be present. If one chooses a value of .7, then the threshold for the SC would be

$$S = (.7)(.33) + .7(.33) + .7(.33) = .7$$

if equal contribution of antecedents is assumed. However, since we have the requirement:

$$\sum_{i=1}^{m} a(i) = 1$$

the value will, in fact, be .7 regardless of the weighting factors on each antecedent. The threshold will change, however, if a different value for each symptom is required to confirm presence of a symptom.

If the value of s exceeds the threshold for an SC, the s value is used in the rule, otherwise 0 is used.

As an example, consider the run in Figure 5 and the weighting factors a(i) in Figure 8. The following computations would take place:

SC32

$$S = (.5)(.5) + (.5)(.9) = .25 + .45 = .70 > .5$$

SC13

$$S = (.6)(.8) + (.4)(.8) = .48 + .32 = .80 > .7$$

Rule 400

$$(.3)(.7) + (.1)(1) + (1)(.7) + (.1)(.7) + (.2)(.8) + (.2)(.5) = .71 > .7$$

Thus, the rule would be substantiated.

V. APPROXIMATE REASONING

In order to put the previous example on a sound theoretical basis, approximate reasoning techniques have been incorporated.[46,47]

As suggested by Zadeh,[20] expert system data can be expressed in the form:

$$V \text{ is } A$$

where V is a variable and A is its current value. An example from the EMERGE chest pain rule base would be "blood pressure is low". To give some more precise meaning to the word "low", fuzzy subsets can be used. Assume A is a fuzzy subset of a base set. For the preceding example, we consider the base set to be the interval of possible blood pressures [0,300]. Let X be the set of all values that V can assume. Then one can consider the possibility distribution $P_x(x)$ over X so that:

$$P_x(x) = A(x)$$

where $A(x)$ on the unit interval [0,1] is the grade of membership for each $x \in X$.

More commonly, EMERGE has statements of the form "sweating is present" as implied by the SC in Figure 8. The user response then indicates the degree of presence of the symptom.

Following the work by Yager,[32] for antecedents in conjunctive form:

$$V_1 \text{ is } A_1 \text{ AND } V_2 \text{ is } A_2 \text{ AND} \ldots V_n \text{ is } A_n$$

then

$$P_{V_1,\ldots,V_n}(x_1,\ldots,x_n) = \min_{i=1,\ldots,n} [A_i(x_i)]$$

where A_i is the membership function for the ith proposition. For disjunctive form:

$$V_1 \text{ is } A_1 \text{ OR } V_2 \text{ is } A_2 \text{ OR} \ldots V_n \text{ is } A_n$$

then

$$P_{V_1,\ldots,V_n}(x_1,\ldots,x_n) = \max_{i=1,\ldots,n} [A_i(x_i)]$$

Linguistic quantifiers are necessary to accommodate the COUNT form of the original SC. Two kinds of quantifiers are of interest here. A kind 1, or absolute quantifier, represents a specified amount, such as about three, at least four, or all. A kind 2, or relative quantifier, represents an approximate amount, such as most or some.[45] The COUNT can be represented through the use of a kind 1 or absolute quantifier of the form "at least n," as suggested by Zadeh,[45] where n is the number required by the COUNT. In order to accommodate quantifiers, $H(x_1, \ldots ,x_n)$ must be redefined:[40]

Let $D(x_1, \ldots ,x_n) = A_1(x_1), \ldots ,A_n(x_n)$

and let $D_i(x_1, \ldots ,x_n)$ be the ith largest element in the set $D(x_1, \ldots ,x_n)$; that is

$$D_1(x_1, \ldots ,x_n) = \max_{a \in D(x_1, \ldots ,x_n)} [a] = A_1(x_1) \vee A_2(x_2) \vee, \ldots ,\vee A_n(x_n)$$

and

$$D_2(x_1, \ldots, x_n) = \max_{a \in D(x_1, \ldots, x_n) - D_1(x_1, \ldots, x_n)} [a],$$

the second largest element in $D(x_1, \ldots, x_n)$,

.

.

.

$$D_n(x_1, \ldots, x_n) = \min_{a \in D(x_1, \ldots, x_n)} [a] = A_1(x_1) A_2(x_2), \ldots, A_n(x_n)$$

Then for any quantifier Q

$$H_1(x_1, \ldots, x_n) = \max_{i=1, \ldots, n} [Q_i \wedge D_i(x_1, \ldots, x_n)]$$

Consider the following case, which combines the above two instances:

IF (V_1 is R_1 AND V_2 is R_2) THEN L is C

where

V_1 is $R_1 \equiv$ (U_1 is S_1 OR U_2 is S_2 OR U_3 is S_3)

with the following conditions attached to each symbol:

	U_i	S_i
1	Sweating	Present
2	Nausea	Present
3	Dizziness	Present

V_2	R_2
Symptoms	Associated with onset

L	C
Patient	Admissible

In the original EMERGE formulation, this took the form of a rule containing an SC in the COUNT format:

IF SC15 AND (symptoms associated with onset)
THEN patient is admissible

where SC15 is shown in Figure 8.

In order to apply degrees of membership to this example, first we have:

$$P_{n1,n2,n3}(x_1, x, x_3) = \max_{i=1,2,3} [Q(i) \wedge D_i(x_1, x_2, x_3)]$$

where S_i is the membership function for the antecedent of the ith component of SC1, and

$$P_{m1,m2}(y_1y_2) = \min_{j=1,2} [R_i(y_i)]$$

$$= \min_{i=1,2,3}^{\max} [Q(i)\wedge D_i(x_1,x_2,x_3),R_2(y_2)]$$

where R_i is the membership function for the ith component of the rule. In order to accomplish the inference we use the form from Yager:[32]

V_1 is A_1 AND V_2 is A_2 AND V_n is A_n THEN U is B

then

$$P_{u/V_1,...,V_n}(x_1,...,x_n,y_1) = \min (1,1 - H(x_1,...,x_n)) + B(y)$$

where

$$H(x_1,...,x_n) = \min_{i=1,...,n} [A_i(x_i)]$$

Thus, for this example:

$$P_{m1,m2}(y_1,y_2,z) = \min(1,1 - \min_{i=1,2} R_i(y_i) + C(z))$$

$$= \min(1,1 - \min[\max_{i=1,2,3} [Q(i)\wedge D_i(x_1,x_2,x_3)],R_2(y_2)])$$

where C is the membership function for "patient is admissible".

The preceding procedure is sufficient to handle inferences when degree of presence of symptoms is considered. It remains to consider handling of the generalized SC with unequal contribution of antecedents.

In order to incorporate weighting of antecedents, we proceed from the statement:

QV's are A

where Q is a kind 1 quantifier to:

Q_1 (Q_2V's) are A

following the work of Yager.[22] This second type of statement can be interpreted as, for example:

(At least n) (important) objectives are satisfied by x

where Q_1 is "at least n" and Q_2 is "important." It is then necessary to determine the truth value of the proposition P. This can be done by considering P to be true if there exists some subset C of V so that:[42]

1.   The number of elements in C satisfies Q, or
2.   Each element in C satisfies the property A.

We then want to find the degree to which P is satisfied by C, which will be denoted $V_p(C)$. The overall validity is then:

$$V(P) = \max_{C \in 2^A} V_p(C)$$

For our application, there are two cases to consider, depending on whether Q_1 is a kind 1 quantifier or a kind 2 quantifier. First, consider Q_1 is kind 1, as before. The required result is then:[42]

$$V(P) = \max\left\{ [Q_1\left(\sum_{i=1}^n c_i \wedge q_{2i}\right) \wedge \min_{i=1,\ldots,n} [a_i^{c_i \wedge q_{2i}}] \right\}$$

where Q_1 is a kind 1 quantifier, $c_i \in \{0,1\}$ indicates membership status, q_{2i} is the weighting factor for the ith antecedent, and a_i is the degree of presence of the ith finding. Using an example from the EMERGE system for these values, we have:

	q_{2i} (weighting factor)	a_i (degree of presence)
Sweating	.4	.9
Dizziness	.3	.7
Nausea	.3	.5

The weighting factor is part of the rule and is determined through expert consultation. The degree of presence is determined for all consultations and is entered by the user. The substantiation value must then be determined by generating all subsets for $C \in 2^A$ to get the overall validity factor.

VI. CONCLUSION

The EMERGE system was extensively tested retrospectively and prospectively using traditional logic and crisp presence and absence of symptoms, and it performed extremely well.[42] The system, however, lacked the ability to record nuances and did not intuitively appear to follow the same processes as the human reasoning process. Preliminary analyses indicate that the incorporation of techniques of approximate reasoning allow more accurate evaluation of borderline cases, where these nuances are most important. EMERGE is currently undergoing extensive evaluation to compare the approximate reasoning version with the original binary logic version.

REFERENCES

1. **Shortliffe, E. H.**, *Computer-Based Medical Consultion, MYCIN*, Elsevier/North-Holland, New York, 1976.
2. **Winston, P.**, *Artificial Intelligence*, Addison-Wesley, Reading, MA, 1977.
3. **Miller, R. A., Pople, H. E., and Myers, J. D.**, INTERNIST-1, An experimental computer-based diagnostic consultant for general internal medicine, *N. Engl. J. Med.*, 307, 468, 1982.
4. **Pople, H. E., Myers, J. D., and Miller, R. A.**, DIALOG: A model of diagnostic logic for internal medicine, *Proc., IJCAI*, 4, 848, 1975.
5. **Reggia, J., Tabb, D. R., Price, T. R., Banko, B., and Hebel, R.**, Computer aided assessment of transient ischemic attacks, A clinical evaluation, *Arch. Neurol. (Chicago)*, 41, 1248, 1984.

6. **Reggia, J. A.,** A production rule system for neurological localization, *Proc. Symp. Comput. Appl. Med. Care,* 2, 254, 1978.

7. **Swartout, W.,** A digitals therapy advisor with explanations, *Proc., IJCAI,* 5, 819, 1977.

8. **Weiss, S. and Kulikowski, C.,** EXPERT: A system for developing consultation models, *Proc., IJCAI,* 6, 942, 1979.

9. **Yu, V. L., Wraith, M. S., Clancey, W. J., et al.,** Antimicrobial selection for meningitis by a computerized consultant: a blinded evaluation by infectious disease experts, *J. Am. Med. Assoc.,* 241(12), 1279, 1979.

10. **Fink, P. K., Lusth, J. C., and Duran, J. W.,** A general expert system design for diagnostic problem solving, *IEEE Trans. Pat. Anal. Mach. Intell.,* PAMI 7(5), 553, 1985.

11. **Reggia, J. A., Perricone, B. T., Nau, D. S., and Peng, Y.,** Answer justification in diagnostic expert system. I. Supporting plausible justifications, *IEEE Trans. Biomed. Eng.,* BME 32(4), 268, 1985.

12. **Kulikowski, C. A.,** Artificial intelligence methods and systems for medical consultation, *IEEE Trans. Pat. Anal. Mach. Intell.,* PAMI 2(5), 464, 1980.

13. **Shortliffe, E. H., Buchanan, B. G., and Feigenbaum, E. A.,** Knowledge engineering for medical decision making: a review of computer-based clinical decision aids, *Proc., IEEE 67,* 9, 1207, 1970.

14. **Szolovits, P.,** *Artificial Intelligence in Medicine,* Westview Press, Boulder, CO, 1982.

15. **U.S. Department of Health, Education and Welfare,** *Report on Artificial Intelligence in Medicine,* Colorado, 1982.

16. **Shortliffe, E. H.,** A model of inexact reasoning in medicine, *Math. Biosci.,* 23, 251, 1975.

17. **Shortliffe, E. H., Scott, A. C., and Bischoff, M. B.,** ONCOCIN, An expert system for oncology protocol management, *Proc., IJCAI,* 81, 876, 1981.

18. **Duda, R. O. and Gashnig, J. G.,** Knowledge-based expert systems for the age to come, *Byte Mag.,* 238, September, 1981.

19. **Zadeh, L. A.,** Fuzzy sets, *Inf. Control,* 8, 338, 1965.

20. **Zadeh, L. A.,** Fuzzy sets as a basis for a theory of possibility, *Fuzzy Sets Syst.,* 1(1), 3, 1978.

21. **Yager, R. R.,** Measurement of Preparing on Fuzzy Sets and Possibility Distribution, *Proc. 3rd Int. Sem. Fuzzy Set Theory,* 211, Linz, Austria, 1981.

22. **Yager, R. R.,** General multiple-objective decision functions and linguistically quantified statements, *Int. J. Man-Mach. Stud.,* 21, 389, 1984.

23. **Gupta, M. M. and Sanchez, E., Eds.,** *Approximate Reasoning in Decision Analysis,* Elsevier/North-Holland, Amsterdam, 1982.

24. **Tanino, T.,** Fuzzy preference orderings in group decision making, *Fuzzy Sets Syst.,* 12, 117, 1984.

25. **Zimmermann, H. J. and Zysno, P.,** Decision and evaluations by hierarchical aggregation of information, *Fuzzy Sets Syst.,* 10, 243, 1983.

26. **Czogala E. and Zimmermann, H. J.,** The aggregation operations for decision making in probabilistic fuzzy environment, *Fuzzy Sets Syst.,* 13, 223, 1984.

27. **Zadeh, L. A.,** Making computers think like people, *IEEE Spectrum,* 26, August 1984.

28. **Czogala, E.,** On distribution function description of probabilistic sets and its application in decision making, *Fuzzy Sets Syst.,* 10, 21, 1983.

29. **Anderson, J., Bandler, W., Kohout, L. J., and Trayner, C.,** The design of a fuzzy medical expert system, in *Approximate Reasoning in Expert Systems,* Gupta, M. M., Kandel, A., Bandler, W., and Kiszka, J. B., Eds., Elsevier/North-Holland, 1985, 689.

30. **Whaler, T. and Schott, B.,** Alternative logics for approximate reasoning in expert systems, *Int. J. Man-Mach. Stud.,* 19, 57, 1983.

31. **Adlassnig, K. P.,** A survey on medical diagnosis and fuzzy subsets, in *Approximate Reasoning in Decision Analysis,* Gupta, M. M. and Sanchez, E., Eds., 203, Elsevier/North-Holland, Amsterdam, 1982.

32. **Yager, R. R.,** Approximate reasoning as a basis for rule-based expert systems, *IEEE Trans. Syst. Man, Cybern.,* SMC 14(4), 636, 1984.

33. **Kerne, E. E.,** The use of fuzzy set theory in electrocardiological diagnostic, in *Approximate Reasoning in Decision Making,* Gupta, M. M. and Sanchez, E., Eds., 277, Elsevier/North-Holland, Amsterdam, 1982.

34. **Zadeh, L. A.,** The role of fuzzy logic in the management of uncertainty in expert systems, in *Fuzzy Sets Syst.,* 11, 199, Elsevier/North-Holland, Amsterdam, 1983.

35. **Adlassnig, K. P.,** A fuzzy logical model of computer-assisted medical diagnosis, *Math. Inform. Med.,* 19(3), 141, 1980.

36. **Esogbue, A. O. and Elder, R. C.,** Management and valuation of a fuzzy mathematical model for medical diagnosis, in *Fuzzy Sets Syst.,* 10, 223, Elsevier/North-Holland, Amsterdam, 1983.

37. **Vila, M. A. and Delgado, M.,** On medical diagnosis using possibility measures, in *Fuzzy Sets Syst.,* 10, 211, Elsevier/North-Holland, Amsterdam, 1983.

38. **Hudson, D. L. and Estrin, T.,** EMERGE, A data-driven medical decision making aid, *IEEE Trans. Pat. Anal. Mach. Int.,* PAMI 6, 87, 1984.

39. **Hudson, D. L. and Estrin, T.,** Microcomputer-based expert systems for clinical decision making, *Proc. Symp. Comp. Appl. Med. Care,* 5, 976, 1981.

40. **Hudson, D. L. and Estrin, T.,** Derivation of rule-based knowledge from established medical outlines, *Comput. Biol. Med.,* 14, 3, 1984.
41. **Lewis, C. E., et al.,** *The Atlas, UCLA EMCRO Criteria Mapping System,* University of California, Los Angeles, 1976.
42. **Hudson, D. L., Cohen, M. E., and Deedwania, P. C.,** EMERGE — An expert system for chest pain analysis, in *Approximate Reasoning in Expert Systems,* Gupta, M. M., Kandel, A., Bandler, W., and Kiszka, J. B., (Eds.), 705, Elsevier/North-Holland, Amsterdam, 1985.
43. **Hudson, D. L. and Cohen, M. E.,** Development of decision making rules for transportable, microcomputer-based expert systems in medicine, *J. Clin. Eng.,* 9(4), 301, October-December 1984.
44. **Hudson, D. L. and Cohen, M. E.,** Approximate reasoning in a diagnostic program for chest pain analysis, *Proc., NAFIPS,* 249, 1986.
45. **Zadeh, L. A.,** A computational approach to fuzzy quantifiers in natural languages, *Comp. Mach. Appl.,* 9, 149, 1983.
46. **Hudson, D. L. and Cohen, M. E.,** Approaches to management of uncertainty in an expert system, *Int. J. Intelligent Sys.,* 3(1), 45, 1988.
47. **Hudson, D. L. and Cohen, M. E.,** Fuzzy logic in a medical expert system in *Fuzzy Computing: Theory, Hardware Realization and Applications,* Gupta, M. M. and Yamakawa, T., Eds., 273, Elsevier/North-Holland, Amsterdam, 1988.

Chapter 12

FESS: A REUSABLE FUZZY EXPERT SYSTEM

Lawrence O. Hall and Abraham Kandel

TABLE OF CONTENTS

I. INTRODUCTION

Expert systems are computer programs that emulate the reasoning process of a human expert or perform in an expert manner in a domain for which no human expert exists. They typically reason with uncertain and imprecise information. There are many sources of imprecision and uncertainty. The knowledge that they embody is often not exact, in the same way that a human's knowledge is imperfect. The facts or user-supplied information is also uncertain.

An expert system is typically made up of at least three parts: an *inference engine*, a *knowledge base*, and a *global* or *working memory*. The knowledge base contains the expert domain knowledge for use in problem solving. The working memory is used as a scratch pad and to store information gained from the user of the system. The inference engine uses the domain knowledge together with acquired information about a problem to provide an expert solution.

Expert systems have modeled uncertainty and imprecision in various ways. MYCIN[30] uses certainty factors, while CASNET[33] uses the most significant results of tests. We apply the well-established theory of fuzzy sets and fuzzy logic to the problem of modeling imprecision and uncertainty in expert systems. This provides a theoretical foundation for this crucial portion of an expert system. We also show that it is the case that fuzzy techniques can be used to provide a valid model of human reasoning.

The principles of software engineering do not appear to have been followed by many expert system developers. We have followed those principles. Expert systems have typically been developed, and then the development process has given rise to practical advice on building one. We developed a theory of building a fuzzy expert system, a methodology, and then applied it.

An expert system building methodology has been developed along with the necessary tools and theory to use fuzzy sets and logic in the reasoning process. A coherent description of the expert system is provided by the use of relations in all aspects of it. We have used our methodology to develop a fuzzy expert system, called **Fess**, which incorporates these described attributes.

II. RELATIONAL KNOWLEDGE REPRESENTATION

There are many different forms in which knowledge may be represented in an expert system. Several different forms have been used very successfully for certain types of applications. The most pervasive form of knowledge representation is the rule-based form. That is because it is a natural form and easy to understand, as well as quite powerful. It enforces very little structure. Data-driven production rules are the most widely used form of rules. We desired a knowledge representation form that will provide great flexibility along with some structure to aid in its understanding.

Since it was desired a build a reusable expert system, its knowledge representation scheme must provide at least the power of a rule-based system. On the other hand, for some applications, it would be useful to allow a scaling down or structuring of the knowledge representation scheme. All knowledge representation schemes can be described as relations. For example, in a semantic net two nodes are either related by an arc (usually weighted) or they are not. So a natural fuzzy relation suggests itself. We chose fuzzy relations as a powerful and flexible knowledge representation scheme.

The combination of rules and frames in an expert system provides some definite structure to an expert system and has been found to be quite powerful. Along those lines it was speculated that the combination of relations and frames may also be very powerful. This would appear true, since rules can be easily described in a relational format. Frames also

Invariant Frame

properties	transitive
How to find domain values	ask-user
method of calculating relational values	Lukasiewicz
instance creation method	get implication
parent	none

Instance Frame

instance type	implication
name	10
premise	sky is blue
conclusion	rain is not falling
clause information	FRAME B
a priori strength	0.95
premise value	0.9

FIGURE 1.   Invariant and instance implication frames.

allow inheritance properties to come into play which can be important in some expert system applications.

A. FRAME STRUCTURE AND RELATIONS

The frames used are broken up into two general classes. The classes are instance frames and invariant frames. The invariant frames contain the information that does not change about the relations. It will contain the number of elements in a relational-tuple. The properties of the relation such as transitive, reflexive, symmetric, etc. will also be in the invariant frame. A slot will contain information on how to calculate the fuzzy value of the relation; there will be a slot which contains information on how to find the attribute values, and one for the domains of each attribute of the relational-tuple. The name of the relation type is also included here. Invariant frames may also have a parent invariant frame associated with them. This provides a standard sort of inheritance mechanism. They may inherit properties that their parent frames possess.

In the instance frame, the unique name of a relation is contained. A link to its invariant parent frame is also included. It also contains the values for each attribute of the relation, the value for the relation, and the *a priori* relation certainty, if one exists. An example invariant frame and an instance of it are shown in Figure 1.

Frames are also used to store information about all the conclusions in the expert system. This includes intermediate and final conclusions. These frames indicate what relations give evidence about the conclusion and what relations use the conclusion.

sun is shining and there are no clouds ⇒ it is not raining, 0.95

This is read as:

IF sun is shining and there are no clouds THEN it is not raining.

The 0.95 indicates that this relation usually holds.

FIGURE 2.   Example implication relation.

The relations that are used in the system are simply n-tuples with some value assigned to them when each field in the relation is filled. The value of the relation will lie in the rational interval [0,1]. The relations are therefore not binary valued, but fuzzy valued. A relation may have an *a priori* determined certainty of less than one which is given by the expert. This means that the maximum value for the relation when completely satisfied will be the strength provided by the expert. All knowledge is represented in some type of relation. This includes meta-knowledge.

Implication relations are the primary relations used for the inference procedure. While other types of relations can be used within the system to perform inference, fuzzy implication relations were chosen for initial system validation. They are also very flexible relations which can be used for any application. A simple implication relation is shown in Figure 2.

In addition to the implication relations, another important type of relation in Fess is called a factual relation. Factual relations may be gathered before the expert system is called upon to solve a problem, or they may be obtained by querying the system user. The factual relations are fuzzy relations, which are usually two-place. A two-place relation is one with two fields which are filled in to determine the value of the relation. Factual relations correspond to the clauses in the implication relations. They provide information about the truth of clauses and also the conclusions of the implication relations. Factual relations are denoted as $R_f(X,Y) \in [0,1]$ for a fact:

$$X \text{ is \{quantifier\} } Y$$

The quantifier is optional and the fact may be assigned a numerical truth value rather than quantified by linguistic descriptors. Actual facts may read:

The sky is somewhat gray.
OR
The age is about 30.

In the case of the second fact phrase the system will have a function associated with the term *about 30*. Given an age, the function will provide a fuzzy belief value about how well the age matches *about 30*. In the same way other imprecise statements may be handled by Fess.

One important aspect of Fess is how it matches imprecise clauses with others which involve the same concept. Fess is able to determine the value of a clause if it has a fact which matches the clause in the phrase before the *is* and after it except for the optional quantifier(s). That is, if we are trying to determine the value of the clause, X *is quantifier$_1$* Y, and we know the value of the clause X *is quantifier$_2$* Y we can do a fuzzy match based on the quantifier to find the truth value for the unknown clause. When the quantifiers are the same, the solution is obvious. An example shows what happens when the quantifiers are different.

Example 1 — Let clause$_1$ $\equiv$ *the temperature is cold* be the clause whose value we wish to determine. Let clause$_2$ $\equiv$ *the temperature is almost cold* be known to be true with a value of 1.0. We now use the appropriate translation function, in this case $\mu_{truth}(C) \leftarrow \mu_{almost}(C) *0.9$, to determine the value for clause$_1$. In this case we apply μ_{almost} to clause$_2$ to find $\mu_{truth}($clause$_1) = 0.9$. Note that our double left arrow operator is not symmetrical.

The mechanism previously described provides a powerful and convenient method for quickly determining clause values without badgering the system user with what would be perceived as redundant questions. In the case that information has been previously put into the fact base, it allows flexibility of expression, making the system more user amiable.

In a manner consistent with the Fess knowledge representation formalism, our factual relations are represented in frames. The factual frame consists of the type of the relation, name, actual fact, and a numerical truth value for the fact. These are called instance frames. These frames are created as facts come into the system. For each domain in which facts may occur, there is in variant frame created. These are created when the knowledge base is loaded since we can discern the domains of all clauses in implication relations at this time. The invariant frame consists of the type, domain, instance creation procedure, and possibly a link to a frame with which it shares properties.

There are three important types of relations that involve the conclusions of implication relations. Each conclusion is involved in a relation with the names of the implication relations, if any, in which it is in the premise. A conclusion may be in no premise if it simply causes an action to be taken. These relations have a variable number of elements in them. Each conclusion is also in a relation with the implication relations which act upon it. Those relations serve to confirm or deny the conclusion. We actually have two relations in use here. One for the implication relations, which act to confirm the conclusion, and one for the implication relations, which act to confirm the opposite of the conclusion, thereby denying it. These relations may also be viewed as lists and are stored and manipulated as such. They look like the following:

$$R_p(C_i, name_1, \ldots , name_j)$$

The p denotes the relation with implication relations which act on the positive form of the conclusion C_i. It should be noted that a NIL name may exist, denoting a relation with value zero, or we may have only one item in the relation with the conclusion. The relation for the negative form of the conclusion R_n is viewed in the same manner. It will always be the case that one of R_p and R_n will be nonzero. The use of the conclusion relations will be fully discussed in the next section.

The third important type of relation which involves conclusions is the contextual relation. This is a unary relation on each conclusion whose value is the context of the conclusion. By context, we mean whether it is used as a relation to start with in a forward chaining system, provides an intermediate result (the largest group), is a startstop relation for beginning operation in backward chaining mode, or represents a final conclusion, a problem solution. The context of the relations becomes evident in the process of knowledge acquisition. The expert needs only to be queried about those that are used to begin and end processing. All others are intermediate. The use of contexts for the implication relations allows the simple use of the system for diverse applications by keeping all information about the knowledge external to the inference and control sections of the system. It should be noted that these context relations provide a form of metaknowledge.

It was noted earlier that implication relations may have a premise which consists of any combination of clauses conjuncted and disjuncted together. We require that they be fully parenthesized for clarity. A complex relation may look like the following:

IF (((X1 is Y1) and (X2 is almost Y2)) or (X3 is not Y3)))

THEN X4 is almost Y4

This relation is easy to read and requires no difficult translations. The relations may be added to the knowledge base simply by using a text editor, although a more sophisticated knowledge acquisition system is desirable.

Variables are represented by a percent sign (%) before the variable word. Clauses with variables in them are matched to all possible matches in the working memory. With consistency enforced across the premise clauses, this enables the system to infer several related items with the application of a single implication relation. Variables may also be in the conclusion of the implication relation and are given the value resulting from the premise matches. Variable use is implicit in the following discussion.

III. THE FESS FUZZY INFERENCE MECHANISM

Fuzzy relations are a very flexible representation formalism which can be used to model any knowledge formalism. The frames that are used to contain the relations do not detract from this ability. One formalism was chosen for implementation and testing. One of the beauties of Fess is that the representation and inference formalisms could be changed without significant modifications, if any, to other parts of the system. Implication relations are used for inference chaining in the expert system. The implication relation consists of two main parts: a premise and a conclusion. The premise is made up of a number of clauses. Each clause in a premise looks like:

$$X \text{ is } Y$$

where X and Y represent variable linguistic statements. Y may contain a fuzzy quantifier such as *almost, very*, etc. The premise consists of one to n clauses, with $n < \infty$. A typical clause is the following: *the plant is almost green*. The clauses are connected by the conjunction and disjunction operators $\wedge, \vee$. Any t-norm, t-conorm pair may be used.[11]

The conclusion is a set of one or more clauses connected by the linguistic connective *and*, which is used as a uniform conjunctive operator. That means that each clause in the conclusion holds with the same strength as the overall conclusion. Each relation is given, by the expert, some *a priori* strength which we call a certainty. This certainty may be represented linguistically with terms, such as *usually, sometimes,* or *occasionally*. It may also be represented with discrete numerical values in the rational-valued interval [0,1], into which the linguistic terms are currently translated.

We use a technique somewhat like generalized modus ponens[11] to determine that a conclusion holds with some true value in the interval [0,1]. A conclusion that has a truth value of 0 is considered false, 0.5 as the midpoint is considered to indicate neither truth nor falsity, and 1 is considered to represent the crisp true. Intermediate values have varying degrees of truth and falsity. In the interval [.5,1], as the conclusion becomes more true its certainty approaches 1. Likewise, as the value of the conclusion certainty decreases in the interval [0,.5], the conclusion is seen as increasingly false.

We are interested in determining a method for operating akin to modus ponens, but in the context of fuzzy logic. Our problem may be stated in the following, letting *cf* denote the certainty of the relation, *a* denote the truth value of the premise, and *b* denote the truth value of the conclusion. We have an equation which looks like the following:

$$cf = a \rightarrow_i b$$

We have values for *a* and *cf*; we must find a value for *b* that is consistent with the ith fuzzy implication operator.[19] There are many different implication operators which may be used in fuzzy implication relations. We choose an implication operator, the ith in this case, and calculate *b* our confidence in the conclusion. There are cases in which no value for *b* can be calculated. These have been referred to as *out of bounds* conditions by Bandler and Kohout.[4] They are a practical problem that an expert system must be able to solve.

a

$a \rightarrow b$	0	.1	.2	.3	.4	.5	.6	.7	.8	.9	1
0											0
.1			O	U	T					0	.10
.2			O	F					0	.11	.20
.3		B	O	U	N	D	S	0	.13	.22	.30
.4							0	.14	.25	.33	.40
.5						0	.17	.29	.38	.44	.50
.6					0	.20	.33	.43	.50	.56	.60
.7				0	.25	.40	.50	.57	.63	.67	.70
.8			0	.33	.50	.60	.67	.71	.75	.78	.80
.9		0	.50	.67	.75	.80	.83	.86	.88	.89	.90
1	≥ 0	1	1	1	1	1	1	1	1	1	1

$$b = \frac{(a \rightarrow b) + a - 1}{a}$$

FIGURE 3. Modus ponens for the Kleene-Dienes-Lukasiewicz operator.

It was speculated that an *out of bounds* condition denoted some inconsistency or error within the expert system knowledge or operation. Toward this end records were kept of such occurrences in the initial prototype of Fess. An examination of some occurrences showed that this was not always the case.

In Figure 3, modus ponens for the Kleene-Dienes-Lukasiewicz implication operator is shown. For this operator, given an *a priori* certainty for a relation of 0.7 and a certainty in the premise which lies anywhere in the range [0,0.2], false to almost false, we find that we are indeed *out of bounds*. This possibility does not, intuitively, seem to be unreasonable. Clearly, if such cases can happen with some legitimacy, a methodology for dealing with them must be developed. We deal with the problem by ignoring the relation which causes the problem. This means when an *out of bounds* value is obtained for a conclusion in an implication relation R_i, the relation is left out of our inference chain. We will either establish the relative truth or falsity of the conclusion of R_i by other relations or will be obliged to find a reasoning path which does not make use of the conclusion contained within the offending relation.

There are quite a few fuzzy implication operators. They each have different properties. The question arises about the best properties for a fuzzy implication operator. First, it must certainly match the classical implication operator at the crisp corners. All of them do this. There are two other properties that are important to an expert system inference. They involve the case when the certainty factor for an implication relation, R_i, is equal to one and when the values for the conclusion are obtained along the diagonal with the out of bounds region. In the first case it is desired that the value for the conclusion not be one except in the crisp case. In other cases it should be less than one. This property is provided by the Lukasiewicz implication operator. Along the diagonal, which for our purposes are the certainty, premise pairs (1,0), (0.9,0.1), (0.8,0.2), . . . ,(0,1), we would like the conclusion values to be nonzero in general. In fact, the conclusion certainties should be decreasing as the relation certainty increases. For example, when the premise value is 0.7 and the relation certainty is 0.3, we desire that our conclusion not be totally false. This property is provided by the Kleene-Dienes implication operator. This property is not, it should be noted, as important as the first one mentioned. The Gaines 43 and Modified Gaines 43[4] do have both these properties and, hence, are most useful as implication operators in fuzzy expert systems.

The reasons that these properties are desirable is given in the following. In the first case, it is misleading to label a conclusion completely certain when the premise which is involved in its verification has a low degree of truth. Note that it is the case in classical logic that when the premise is false, and the overall certainty of the implication relation is true, that the conclusion may be considered completely true or false. Given the lack of supporting evidence we choose to represent it as completely false. The second property indicates that it is desired that some certainty exist in a conclusion, given a premise which has some positive degree of certainty and an implication relation with a positive certainty. That it should decrease with the decreasing certainty in the relation is required, since there is no justification for a conclusion with a higher certainty than the relation it is contained within.

There is one other property which was found important in generating modus ponens. The lower bound must be taken for modus ponens generation. An expert system often must take a conservative approach in its reasoning, since this is the typical approach of experts. Therefore, we must know our lower bound from the function which we use to generate modus ponens.

IV. FESS EVIDENCE COMBINATION

For an expert system which is intended to be used in different applications, the combination of evidence method must be easily changed. This is accomplished in the Fess system by having the combination method(s) confined to a specific code unit. New methods may be plugged in without affecting the rest of the system. This feature is one of the advantages provided by the language in which Fess was developed.

Each conclusion in Fess is seen as consisting of two parts: the positive part, *X is Y*, and the negative part, *X is not Y*. Our certainty for a conclusion depends upon how much we believe in its component parts. A strong belief in the negative part with a small belief in the positive part of the conclusion will make us certain that the negative part of the conclusion may be considered true to some degree. For each conclusion the system maintains a measure of belief and a measure of disbelief in its conclusion frame. The measure of disbelief may be described as the measure of belief in the negative part of the conclusion or the negation of the conclusion.

The accrual method of evidence combination is the most generally applicable of the various methods available.[16] It was chosen for the initial implementation of Fess. The method of calculating the belief and disbelief measures will be detailed in the following.

For each conclusion, denoted by C_i, there will be some relation(s), denoted by R_j, which provide the evidence necessary to decide whether to believe the positive or negative parts of the conclusion, or that no decision can be made with the evidence. With each relation R_j there will be an associated certainty factor, denoted by $CF_j(C_i)$, which indicates the strength of the evidence about conclusion C_i provided by the relation. The certainty factors of the relations used on a conclusion C_i accrue to provide us with our measure of belief in the following manner.

A. DEFINITION 1
The **measure of belief** in a conclusion C_i is

$$MB(C_i,t) = MB(C_i,t - 1) + (1 - MB(C_i,t - 1)) \star CF_j(C_i)$$

where t is the current number of relations that have been used to determine C_i and $MB(C_i,t - 1)$ denotes the belief value before the current relation was used, $MB(C_i, -1) = 0$.

For convenience allow $MB(C_i)$ to denote the value for the measure of belief function at the current t, the number of relations used to determine the belief in the conclusion. The measure of disbelief is calculated in a corresponding manner shown as follows.

B. DEFINITION 2

The **measure of disbelief** in a conclusion C_i is

$$MD(C_i,t) = MD(C_i,t - 1) + (1 - MD(C_i,t - 1)) \star CF_j(\neg C_i)$$

where t is the current number of relations that have been used to determine C_i and $MD(C_i,t - 1)$ denotes the belief value before the current relation was used, $MD(C_i, - 1) = 0$.

Again, for convenience, we allow $MD(C_i)$ to denote the value for the measure of disbelief function at the current t, the number of relations used to determine the disbelief in the conclusion. It is important to note that the measure of belief and measure of disbelief are calculated separately. Therefore, it is possible that $MB(C_i) \neq 1 - MD(C_i)$. To obtain relations which act on both the positive and negative part of a conclusion may require a bit more effort by the knowledge engineer and engender a very small number of additional queries of the system user. However, the clarity of reasoning provided by such a system is well worth the trade-off.

Given that measures of belief and disbelief in a conclusion have been calculated, we must come to an overall certainty for the conclusion. The overall certainty lies in the interval [0,1]. The value 0.5 indicates neither belief nor disbelief in the interval [0.5,1.0] our belief increases as the certainty approaches one, and in the interval [0.5,0] our disbelief increases as the certainty approaches zero. The overall certainty of a conclusion C_i is calculated as shown in the following definition.

C. DEFINITION 3

The overall certainty of a conclusion is defined as:

$$OC(C_i) = 0.5 - 0.5 \star (MD(C_i) - MB(C_i))$$

The overall certainty, just defined, has the required characteristics which were previously discussed. We now have the mechanisms for determining whether to believe or disbelieve a conclusion based on one or more pieces of evidence about it. Some strategy about the number of relations to use to determine a conclusion, and when to stop trying to determine a conclusion, are necessary for an expert system implementation using this combination of evidence scheme.

All applicable relations may be applied to determine a conclusion. This is a straight-forward scheme, but it is not very efficient. If there are many relations, and we use some that give us a strong measure of belief and no measure of disbelief, should we continue processing? Unless we have an unusual application for the expert system, we do not. In fact, to continue will require a great deal of extra processing time, and an extremely slow expert system is not likely to be accepted. This is clearly an area in which different approaches may be suitable for distinct problem areas. Again, in Fess, the algorithm which governs the determination of a conclusion is confined to a specific section and can be easily changed. The current algorithm used in Fess to determine a conclusion will be described in the following text.

If more than one implication relation that acts on a conclusion exists, we will use a minimum of two relations to determine the conclusion. We alternate in using relations which act to increase our measure of belief and our measure of disbelief regarding the current conclusion. This process continues until we run out of relations which act to increase MB or MD, or we have tried two of each type of relation and the overall certainty has not changed more than some value Δ. If we find the overall certainty has changed more than the Δ, we will continue the process until we run out of relations or, after attempting to increase our MB and MD, the overall certainty has not changed by Δ. If there are no more relations providing evidence upon the MB or MD, but we have not used at least two relations, or our overall certainty changed by more than Δ, we will continue using relations to determine the overall certainty. This continues until two relations have been used, no more exist which act on this conclusion, or the overall certainty is not changing by more than Δ.

The algorithm just described, while somewhat complex, provides for efficient search and conclusion determination. It attempts to ensure that both the measure of belief and disbelief are fully determined so that an erroneous reasoning path is not followed.

V. CHAINING THROUGH THE FUZZY RELATIONS

We have previously discussed finding the value of the conclusion of an implication relation by the use of fuzzy modus ponens. What is done with the conclusion when it has been determined with some truth value? It is used in the continuance of the reasoning process. The reasoning process halts when the system has come to a conclusion which provides the information that the user of the expert system seeks or there are no more relations available to determine an answer for the user.

Fess may operate in a limited forward chaining or a backward chaining mode. The mode is chosen by the system with the use of context relations which will be discussed later. When the system is operating in the forward chaining, it begins by choosing an initial implication relation for processing. An evaluation function chooses the best implication relation. If the conclusion is determined to be certain above a threshold, it is used to guide the choice of the next relation to be processed. The conclusion, a portion of it or a result from it, will be in the premise of the relation which is evaluated as the best one for processing. This process of determining a conclusion certainty, and proceeding to choose a new relation with some trace of the former conclusion in its premise, will continue until a conclusion is determined to have a certainty below the preset threshold, or a final conclusion, expert answer, is determined with a truth value above the appropriate threshold.

If a conclusion along the path is found to have a certainty below the positive threshold, and is not conclusively false, then other relations, which act upon the conclusion, are used. This process continues until the conclusion certainty is above the true threshold, or below the false threshold, or there are no more relations available to determine it. In this case, another path must be found, if possible.

In this system, the chaining is more directed than in the OPS style,[14] match, recognize and act cycle. The system would not be applicable in the case of a planning or design problem.

If Fess is operating in a backward chaining mode, the first implication relation chosen will be one which contains a final conclusion. That is, a relation whose conclusion, when found to be above a threshold, causes the system operation to halt and the user to be supplied with a problem solution. The same evaluation function mentioned above chooses the *best* relation for processing. In order to determine the fuzzy truth value of the conclusion, the premise must be evaluated. Typically, some of the clauses in the premise will have their values supplied by the user, while other clauses will be the conclusions of some implication relations. We will discuss the use of the implication relations to find clause values since this is where the name backward chaining comes from.

The best relation to determine the clause will be chosen by the *implication relation evaluation function*. In order to determine it there will usually be a clause which requires another relation or set of relations to be used. This backward chaining process will continue until a relation, which has all clauses supplied by the user or factual relations, is evaluated.

When this occurs, certainty values are found for the various intermediate conclusions, and these are propagated along the chain to give us a truth value for our final conclusion which is our main interest. If the path does not hold up, and the system cannot determine a truth value for our final conclusion above some preset certainty threshold, then it will try to determine the certainty of the next best final conclusion. This process continues until a final conclusion, with a certainty above the truth threshold, is found or there are none left to try.

This was a brief discussion of forward and backward chaining in Fess. The basic building blocks of a fuzzy relational reasoning scheme for a general expert system has been presented before. One of the real benefits of the methodology we have presented is that it is relatively simple, yet powerful and flexible. Its implementation is not overly slow and creaking, while the results it provides allow a human expert reasoning with uncertainty to be emulated, as will be shown later.

VI. SINGLE OR MULTIPLE CONCLUSION SEARCH

During the operation of Fess the user will be asked whether a single conclusion configuration or a multiple conclusion configuration is desired. If a single conclusion configuration is chosen, the system will attempt to find a reasoning path which leads to one conclusion with a certainty above a threshold. In the multiple conclusion configuration, the system will attempt to find all possible conclusions above the specified certainty threshold. The single conclusion configuration provides faster inferencing and should be used whenever the problem is amenable.

In forward chaining mode and a single conclusion configuration, Fess will search for a path from the start relations to a final conclusion which has a certainty (overall certainty) above a threshold. The first one that it finds is presented to the user with its certainty as the system answer to the problem. In backward chaining mode, Fess presents the user with the first conclusion that it is able to establish with the a certainty above a threshold. The certainty of the conclusion is also provided. In both inference modes it simply searches for the first path which provides a conclusion above the specified certainty threshold. After this is accomplished, processing is halted thus, possibly, leaving other valid conclusions undiscovered.

In a multiple conclusion configuration and forward chaining mode, Fess will search for **all** possible paths that provide a conclusion which is above a threshold. It will examine all the start relations to determine if they lead to conclusion that is certain to its satisfaction. The search for all of the paths is quite time-consuming. In backward chaining mode, Fess will try each final conclusion to determine if it can establish the conclusion with a certainty above the threshold. In both inference modes Fess is searching for all possible reasonable answers to the problem. This requires that many more relations be evaluated than in the other configuration. It also requires some efficient method for deciding when to give up the search, otherwise all relations will be tried. This leads us to our next topic on halting path search in this configuration.

A minimal threshold of overall certainty is set in which a conclusion certainty must exceed in order for processing along the path to continue. If the evidence about a conclusion along a path cannot support it above a minimal threshold corresponding to *slightly true*, then we abandon the path on which the conclusion lies. This has been shown to be an effective, yet simple, method of pruning paths in our search space.

Now a summary of how Fess halts the inference process will be presented. In forward chaining mode and a single conclusion configuration, processing is halted when a conclusion is found and presented to the user or all paths have been tried unsuccessfully. In forward mode and a multiple conclusion configuration, processing halts when all conclusions with certainties above a threshold are found and presented to the user, or all paths have been tried unsuccessfully. In backward chaining mode and a single conclusion configuration, processing is halted when a conclusion with certainty above a threshold has been presented to the user or none could be found. In multiple conclusion configuration and backward chaining mode, processing halts when all conclusions with certainty above a threshold are presented to the user or no conclusions could be found which were above the threshold. In this case the user is informed that the system is *stuck*. The certainty threshold is fuzzy and corresponds, linguistically, to *fairly certain*.

VII. SUMMARY

The concept of a reusable expert system which makes use of fuzzy reasoning techniques and a design methodology has been incorporated into Fess. A reusable expert system is one in which **no** domain knowledge is incorporated in the inference engine. Any necessary domain knowledge is contained solely in the knowledge base(s). Only the incorporation of a new knowledge base is necessary for the system to be applied to a new application area. Limited explanation capabilities are incorporated in the system. It will basically provide the user with a trace of the rules fired.

Modula-2 worked out well as a development language. It provides good structures, which enabled a modular and structured system to be developed. Fess has proved easy to extend and update. The one necessity with the Modula-2 language is a good symbolic debugger, which was available on our VAX* development system. The language also provides a nice mechanism, in process modules, to distribute the expert system.

The use of fuzzy sets and logic in the system provide a viable uncertainty handling method. This provides the system with the ability to be successfully applied to problems which have uncertainty associated with them. It has been applied to several different domains in which diagnosis of some problem is the objective. The performance has been good. The relatively straightforward fuzzy reasoning mechanism has proven up to the task of representing uncertainty in these problems. Overall, Fess is robust and capable of providing expertise on a diverse set of problems.

REFERENCES

1. **Aiello, N., Bock, C., Nii, H. P., and White, W. C.,** Joy of AGEing: an introduction to the AGE-1 system, Report HPP-81-23, Computer Science Department, Stanford University, CA, 1981.
2. **Aikens, J. S.,** Prototypical knowledge for expert systems, *Artif. Intell.,* 20, 163, 1983.
3. **Bandler, W.,** Representation and Manipulation of Knowledge in Fuzzy Expert Systems, presented at Workshop on Fuzzy Sets and Knowledge-Based Systems, Queen Mary College, University of London, England, 1983.
4. **Bandler, W. and Kohout, L. J.,** The four modes of inference in fuzzy expert systems, *Cybernetics and Systems Research 2,* Trappl, R., Ed., North-Holland, Amsterdam, 1984.
5. **Bennett, J. S. and Engelmore, R. S.,** Experience using EMYCIN, in *Rule-Based Expert Systems,* Buchanan, B., and Shortliffe, E., Eds., Addison-Wesley, Reading, MA, 1984, 314.
6. **Bonissone, P. P. and Brown, A. L., Jr.,** Expanding the Horizons of Expert Systems, *Proc. Conf. Expert Syst. Knowledge Eng.,* Gottlieb Duttwailer Institut, Zurich, April 1985.
7. **Buckley, J. J., Siler, W., and Tucker, D.,** A fuzzy expert system, *Fuzzy Sets Syst.,* 20(1), 1, 1986.
8. **Buckley, J. J., Siler, W., and Tucker, D.,** FLOPS, A fuzzy expert system: applications and perspectives, in *Fuzzy Logics in Knowledge Engineering,* Negoita, C. V. and Prade, H., Eds., Verlag TUV Rheinland, Germany, 1986.
9. **Cheng, P. and Hall, L. O.,** A Comparison of Point-Valued and Interval-Valued Reasoning under Uncertainty, *N. Am. Fuzzy Inf. Processing Soc. Conf.,* San Francisco, June, 1988.

---

*   VAX is a trademark of the Digital Equipment Corporation.

10. **Davis, R., Buchanan, B. G., and Shortliffe, E. H.,** Production rules as a representation for a knowledge-based consultation program, *Artif. Intell.,* 8, 15, 1977.
11. **Dubois, D. and Prade, H.,** *Fuzzy Sets and Systems: Theory and Applications,* Academic Press, New York, 1980.
12. **Erman, L. D., Fennel, R. D., and Reddy, D. R.,** System organizations for speech understanding: implications for network and multiprocessor computer architectures for A.I., *IEEE Trans. Comput.,* C 25(4), 414, 1976.
13. **Erman, L. D.,** The HEARSAY-II Speech-understanding system: integrating knowledge to resolve uncertainties, *Comput. Surveys,* 12(2), 213, 1980.
14. **Forgy, C. L.,** *OPS5 User's Manual,* Tech Report CMU-CS-81-135, Department of Computer Science, Carnegie-Mellon University, Pittsburgh, 1981.
15. **Gaines, B. R.,** Foundations of fuzzy reasoning, in *Fuzzy Automata and Decision Processes,* Gupta, M., et al., Eds., North-Holland, New York, 1977.
16. **Hall, L. O.,** A Methodological Approach to a Re-usable Fuzzy Expert System, Ph.D. dissertation, Florida State University, Tallahassee, 1986.
17. **Hall, L. O. and Kandel, A.,** *Designing Fuzzy Expert Systems,* Verlag TÜV Rheinland, Germany, 1986.
18. **Anon.,** A Fuzzy Expert System Based on Relations, *1986 Int. Symp. Multiple-Valued Logic,* Blacksburg, VA, May 1986, 252.
19. **Hall, L. O.,** The choice of ply operator in fuzzy intelligent systems, *Fuzzy Sets Syst.,* 34, 135, 1990.
20. **Hayes-Roth, F., Waterman, D. A., and Lenat, D. B., Eds.,** *Building Expert Systems,* Addison-Wesley, Reading, MA, 1983.
21. **Jones, P. L. K.,** REVEAL: an expert systems support environment, in *Expert Systems Principles and Case Studies,* Forsyth, R., Ed., Chapman and Hall, New York, 1984, 131.
22. **Kandel, A., Ed.,** Special issue on expert systems, *Inf. Sci.,* 37(1—3), December 1985.
23. **Kandel, A.,** *Fuzzy Mathematical Techniques with Applications,* Addison-Wesley, Reading, MA, 1986.
24. **Kacprzyk, J. and Yager, R.,** Emergency-oriented expert systems: a fuzzy approach, Tech. Report MII-213/247, Machine Intelligence Institute, Iona College, New Rochelle, NY, 1982.
25. **Kuipers, B.,** A frame for frames: representing knowledge for recognition, in *Representation and Understanding: Studies in Cognitive Science,* Bobrow, D. G. and Collins, A., Eds., Academic Press, New York, 1975, 151.
26. **Ogawa, H., Fu, K. S., and Yao, J. T. P.,** An Expert System for Damage Assessment of Existing Structures, *Proc. 1st Conf. Artif. Intell. Appl.,* IEEE Computer Society, Miami, FL, 1984.
27. **Ogawa, H., Fu, K. S., and Yao, J. P. T.,** Knowledge Representation and Inference Control of SPERILL-II, *ACM Conf. Proc.,* November 1984.
28. **Pople, H. E., Jr.,** Knowledge-based expert systems: the buy or build decision, in *Artificial Intelligence Applications for Business,* Reitman, W., Ed., Ablex, Norwood, NJ, 1984.
29. **Rich, E.,** *Artificial Intelligence,* McGraw-Hill, New York, 1983.
30. **Shortliffe, E. H.,** *Computer-Based Medical Consultation: MYCIN,* Elsevier/North-Holland, New York, 1976.
31. **Valverde, L. and Trillas, E.,** On Modus Ponens in Fuzzy Logic, *15th Int. Symp. Multiple-Valued Logic,* Kingston, Ontario, 1985.
32. **Waterman, D. A.,** *A Guide to Expert Systems,* Addison-Wesley, Reading, MA, 1986.
33. **Weiss, S. M., Kulikowski, C. A., Amarel, S., and Safir, A.,** A model-based method of computer-aided medical decision-making, *Artif. Intell.,* 11, 145, 1978.
34. **Wenstop, F.,** Applications of Linguistic Variables in the Analysis of Organizations, Ph.D. thesis, University of California, Berkeley, 1975.
35. **Whalen, T. and Schott, B.,** Goal-directed approximate reasoning in a fuzzy production system, in *Approximate Reasoning in Expert Systems,* Gupta, M. et al., Eds., North-Holland, New York, 1985.
36. **Yager, R. R.,** Robot planning with fuzzy sets, *Robotica 1,* 41, 1983.
37. **Zadeh, L. A.,** The role of fuzzy logic in the management of uncertainty in expert systems, *Fuzzy Sets Syst.,* 11, 199, 1983.
38. **Zadeh, L. A.,** Common sense knowledge representation based on fuzzy logic, *Computer,* 16(10), 61, 1983.

Chapter 13

DESIGN FOR DESIGNING: FUZZY RELATIONAL ENVIRONMENTAL DESIGN ASSISTANT (FREDA)*

Vasco Mancini and Wyllis Bandler

TABLE OF CONTENTS

* This chapter is dedicated to Edith Ludowyk Gyömröi.

I. THE CREATIVE DESIGN PROCESS

When Freud investigated the dynamic unconscious, which predominates in so much of our mental life, he distinguished between the *primary* and the *secondary* mental processes. In dreams, the primary process calls up the exciting, bizarrely structured material from the depths of the unconscious, which the secondary process partly resists and partly transforms into what the dreamer perceives.[6] In vision, the primary process seizes the meaningful *gestalt* of the object it cares about. The secondary process subdivides this and starts to organize the parts into relationships. The meaning of all of this, beyond the merely rational, must again be supplied by the primary process and so on, with many interchanges, as can be traced in Arnheim.[1]

In artistic creation the primary process makes use of loose, undifferentiated structures lacking clear boundaries, failing to distinguish between opposites, and lacking a clear articulation of space and time. It is syncretistic in its grasp of the whole object. The secondary process, on the contrary, aims at well-defined, differentiated structures, abstracts smaller elements from the concrete object, and matches them one by one. It uses an analytic mode. Both are necessary and interactive in the artistic creation but, as has been noted with children, the analytic mode of the secondary process can re-repress the other process, with disastrous consequences.[5]

All creative designing consists of an alternation between these two processes. The primary process turns up imagery from the unconscious, which the secondary process manipulates to make more explicit and analytic. While this is going on, or after it has been completed (often long after), the primary process turns up fresh material. This material is then reworked by the secondary process and, once more, inspiration comes from the primary process, and so on, an indefinite number of times. The alternation between the two creative processes occurs in various rhythms. There is a slow, tidal movement of emphasis from the one to the other and back, and there are rapid, almost undetectable, oscillations.

One of the present authors recollects that, in the midst of dealing with the challenge of designing a museum, he went for a drive through the Tuscan countryside. He experienced a sequence of thoughts, which are excerpted here. It will be apparent, as Freud noted concerning the recital of dreams, that the primary thoughts, when recalled, have already been clothed in words and partially rationalized by the secondary process.

Inward Movement 7 — The first time I went to the Uffizi (was I 13?), I had no idea what to expect. I was concerned that the owners of the place, who lived there, had all moved out of the rooms temporarily. The pictures seemed to live in a world of their own and, yet, to be quite alert and waiting, as if to resume some intimate conversation with friends. And, yet, they looked like someone one knew going about their lives in workshops or restaurants, except for an air of fancy and unashamed revelation, the more so because some of them lived in my classroom also, but so lifeless there and forlorn, like photographs of dead people. I decided that it must be the building and its immediate surroundings that provided that exalted quality, that heroic surrender.

Outward Movement 7 — The window reaches the floor and is open. The great thickness of the wall is revealed. The floor is made up of *cotto*, as at home, except that here it is very smooth and even, and redder. The white heavy cotton curtain moves slightly; the air is still, the light is secret and uneven; the room is large, and the ceiling high. There are three portraits in the room; two of them, on opposite walls, seem to mirror one another as if they were having a dialogue. The ceiling is flat and supported by wooden beams placed at a distance from one another such that a *cotto* tile, larger than the floor one, spans the interval. The beams are decorated with intricate geometric designs in color. The ceiling enrages me . . . it is all wrong.

Inward Movement 8 — The white, freshly laundered sheets would be brought in and piled on the bed. After dinner I would pretend to be very sleepy, and would slip in under the white mountain, like crushed sails, smelling of sun and fresh air. Eventually the women would come in and start folding the sheets. They would stand by the side of the large bed, one at each end and, having got hold of the four corners of one, they would throw it up into the air. The sheet would inflate, unruffling, and stand suspended for a moment like a white cupola. I would try to fight off sleep. When properly folded, it would be placed on the other bed, my parents', in a neat growing pile. Meanwhile, the women would converse softly about secret, terrible, and beautiful happenings in the village, and their muted voices echoed down the cascading corridors of cupolas. When I woke up next morning, I felt a great sense of loss . . .

Outward Movement 8 — The ceiling was flat and held up by parallel beams at a distance so that a *cotto* tile would bridge it, and both beams and tiles had been whitewashed and seemed to be lighted from within, as the shutters were closed. I would hold my right arm high and with my finger write in between the beam lines, as on a page of my notebook. I would spend a lot of time at it, since my discovery of writing was fresh. But this had to be a secret pleasure, because they would laugh and make fun of me if observed.

Unfortunately, the reverie came to an untimely end with the secondary process claiming control and starting the organization of the material in its own fashion. It began by abstracting and focusing upon certain associations, to which it then tried to give rational content:

Light — revelation

Light — painting visible

Painting — ultimate surrender

Intimate — secret

Buried under — depth perception

After some conflict, the two processes compromised upon a conviction that a deep enough level of meaning had been reached and a generative kernel established. The secondary process described it like this: a spatial configuration of vaulted spaces within vaulted spaces, a fluid circulation network that would allow short-linked relations among particular objects to be experienced, and a complex use of natural and artificial lighting to allow the individual displayed objects to reveal themselves. The primary process accepted that its qualitative concerns could be so met.

Eventually an iconic representation was produced that expressed and confirmed visually and globally these certainties. The secondary process would hammer at the representation, again and again, until the generative force of the kernel had found mutually acceptable concretization.

II. HOW CAN THE COMPUTER HELP? WHAT SORT OF SYSTEM DO WE WANT?

Up to this time, computers have been brought into the design process very late in the game, when the most important creative and imaginative work has already been done *or omitted*. The current use of technology and the current mental set reinforced by the ''problem-solving'' attitude, make it easy all too often for genuine creativity to be bypassed altogether. The machine lends itself, of course, to utterly explicit and completely algorithmic operations, that is, to be the more superficial levels of the secondary process alone. Emphasis on these tends to obscure the true creative challenge.

We propose to bring the computer *within* the creative process, as close to the primary process phases as possible. This aim and the consequent systemic structure and content are in sharp contrast with those of existing software.

We are not claiming that a computer system can perform the primary process function. Quite the contrary, we believe these to be beyond mechanistic abilities forever, not only at the present day. What we do seek to provide is a receptor for the fruits of the primary process, which will encourage the designer to confide his fresh ideas and enable him to return to them at will. Positive aid, where the system offers its own suggestions, will be confined to the secondary processes at as deep a level as we can manage.

Ambitious though it is, our object is to facilitate the cooperation of the two processes by providing the following features:

Feature 1 — In primary receptive mode, the system makes the expression of the ideas evolved by the primary process as undistracting as possible. Confident that his conceptions are accepted as meaningful and valid, the designer is to be spared the superfluous inhibition brought about by a feeling of estrangement from the recipient. The initial global ideas can be expressed to the machine in any combination of three notations, *graphic, verbal,* and *relational,* which, by their familiarity, encourage spontaneity. The special handling of each of the three notations, requiring interpretation followed by manipulation, is discussed in the next section.

Feature 2 — In promoting the secondary process, the system can be used either as an active or a passive agent:

- It will "volunteer" suggestions; that is, having itself performed a version of some of the processing, it can offer the results to the user for acceptance, rejection, or modification.
- It will, of course, also allow convenient secondary processing entirely on the part of the user.

For example, in connection with the enormously important *articulation* of an initial global vision into, for example, elementary solids, there can be available the system's own attempts, the designer's attempts, and any desired synthesis of these.

Feature 3 — By a rather sophisticated network filing subsystem (part of the *relational zone* touched upon earlier and to be discussed in the next section), the system facilitates free movement among the growing number of notes and sketches. There also exists alternation between primary and secondary activity on the part of the user who can, without hindrance, drop one of the activities, then resketch, and later tie the new work into the existing network. Fidelity to the first vision (which is the essence of the whole effort) is thus encouraged by the ease with which the later representations can be compared with the earlier ones. If the designer can decide upon *labels* to give to certain aspects of his conception, which he feels to be essential, the system can even issue reminders of them on subsequent occasions. These arrangements also encourage recourse to primary processing at all the later stages of the designer's work. These steps are those concerned with such matters as furnishing style, materials of decoration, adjustment of proportions, and, in general, whenever aesthetic considerations enter.

Feature 4 — A learning component is planned for the system to be installed after an effective working version has first been constructed. Its main function will be to acquaint itself with typical actions of an individual user, and to promote these to prominence in developing its own suggestions. This will diminish the amount of explanation the user must give to the system, as discussed in the next section, and, therefore increase the tranquillity of his or her self-expression. At the same time it will increase the frequency with which the system's suggestions are agreeable to the user. The effect desired is the emulation of a human assistant who gets to know the idiosyncratic ways of the principal.

III. THE SYSTEM AT WORK

We have tried to establish as paramount importance that the designer grasps a generative kernel of the design task, which is the grounds for his conviction and security, and, in effect, his knowledge — perceptions, reactions, feelings, and ideas. If and when this finds a syncretistic representation, visual or plastic, it carries great clarity and force; think, for example of Utzon's famous initial sketch of the Sydney Opera House.

This is the *sine qua non* of the design process. From then on we consider the two fundamental stages which have to be reached and reconciled in order to arrive at a successful conclusion. One is the geometric, volumetric, and spatial organization of the proposed artifact: the definition and arrangement of its parts. The other is the explicit elaboration of the structure of the activities which will go on inside the building: this is an aspatial structure, a sort of mirror image of the spatial one. We will refer to the first as *geometrization*, to the second as *expression of the activity structure*.

Both are developments of suggestions implicit in the generative kernel of the first expression. Both must, at the end, accord with one another and with the initial vision; only then will they feel right to the designer. "The formal levels ought to represent a system of functional levels," as Norberg-Schulz[15] puts it.

Let us now examine how our system proposes to assist the designer in achieving these two goals. We assume, without further comment, that it contains sophisticated drafting manipulative software, of the general sort already in existence,[4] and we concentrate instead on its innovative features.

A. RECORDING OF THE INITIAL GLOBAL VISION

For completeness, we suppose the designer's initial idea to be embodied in a sketch, with perhaps a couple of subsketches, a few arrows showing special relationships on the sketches, and all bearing some annotations. This can all be drawn and entered into the machine with light pens on the screen, or on a special tablet, or (experimentally) with a special directly three-dimensional drawing apparatus. To make the entry as unfussy as possible, no extra requirements are put upon the designer, except only that the three kinds of information be kept distinct, for example, by the use of three differently colored pens for the graphic, the verbal, and the relational elements. Transgressions can be cleared up later.

B. DEBRIEFING PHASE

The great freedom of the initial entry must be tempered at a later moment by the explanation to the system of some of the entered information. If the sketch was entered to a two-dimensional receptor, then the three-dimensional lie of its strokes will need to be clarified. The relational arrows may need additional labels. The verbal notes will need to be divided into those to be kept verbatim and those subject to semantic interpretation; the latter will need to be re-entered by typing. The nuisance of this phase is mitigated by its postponement from the time of primary entry and also, perhaps, by the fact that, in clarifying things to the system, the user may also be clarifying them to himself.

C. GEOMETRIZATION

The computer system now has an internal representation, in three dimensions, of the designer's original sketch. It allows him to divide this into parts, substitute elementary solids for each of these parts, alter their mutual positions, experiment with their proportions and their hierarchical relationships, seek out their symmetries, view them from various angles, and, in general, perform upon them all the usual geometrical transformations and investigations until he reaches a configuration which he accepts.

If requested, the system will try its own hand at some of this; for example, it can fit one, two, . . . , *n* elementary solids to the initial or any subsequent sketch. Each such suggestion is subject to the user's acceptance, rejection, or retainment for modification. Continuing the example, the solids may be taken from a variety of repertories: classical elementary, rectangular parallelipipeds, polyhedra of different sorts, all from simply or multiply parametrized families of the kinds discussed in Mitchell.[14]

Of course, as the design advances, the parts themselves require articulation, elaboration, and detailing, to conform with their meaning. The same system facilities can be reapplied but, in addition, there are available the relational ones discussed in the next item.

As always, the assistance is interactive, so that its active and passive uses can be alternated at will.

D. RELATIONAL MANIPULATION

Whether between parts of the drawings or of the eventual artifact, or between activities of the building's users, or among properties and qualities of aesthetic, social, functional or psychological nature associated with the project, *relations* abound throughout the design process. We take the most general case, in which these may be either *fuzzy* or *crisp*, and apply very general methods,[2,3,17,18] displaying the results as *digraphs*, and further manipulating the latter by elaboration of existing methods. This zone of the system has a number of important uses, of which we discuss only one, in the next item.

E. ANALYSIS AND EXPRESSION OF THE ACTIVITY STRUCTURES

Relations among activities, of whatever sort the designer may be concerned with at the time — contiguity, temporal succession, complementarity, mutual observability or visual exclusion, etc. — lend themselves neatly to representation and manipulation as diagraphs, with prolific annotations. This partial visualization of the activity structures conduces, among other things, to the allocation and mutual arrangement of spaces in such a way as to meet the relational demands. The relational zone of our system supports thinking out the relations in these terms, and adds graph-untangling procedures, dual-graph floor planning, and so forth. In addition, it makes available a library of examples and furnishes continual reminders of the designer's own notes on qualities and visual features.

F. MATCHING OF THE VISUAL-FORMAL AND THE ACTIVITY-FUNCTIONAL STRUCTURES, AND OF BOTH WITH THE ORIGINAL CONCEPTION

As the repeated differentiation and refinement of the geometric and the activity structures proceeds (as in Sections C and E), they continually require mutual compliance. The system facilitates constant comparison in adjacent windows, and the annotations attached to the relational representations not only suggest some of the parameters for the readjustments of the geometry, but also demand figuration reflecting important qualities. The system, which never forgets the notation of a requirement, is ready at any time to offer reminders for an evaluation of the results. It can also always run through the successive steps the designer has taken, so that errors of judgment can be corrected.

IV. FEASIBILITY AND METHODOLOGY

Clearly, the system discussed here presupposes the existence of present-day, state-of-the-art, three-dimensional CAD systems, notebook systems, and the like, some of which it can utilize. Its chief advance upon them is central: the divorce between the "hot" (primary) and the "cool" (secondary) communication from the user to the computer, and the facilitation of both. All of this is twofold. First, there is the receptivity of the machine to the nonanalytical presentations of the primary process, preserving them, as they first flow, until the user is in the mood to process them secondarily. Second, there is the readiness of the machine to respond to the user's wishes — to "understand" his or her elaboration and clarification and sometimes to anticipate them. These abilities are those which will give the user confidence and the enlivened feeling of working in a sympathetic medium.

They can be obtained, in our view, with certain techniques on which much work has been done, but which are only beginning to find their way into general use. These are the *fuzzy* methods stemming from Zadeh,[17,18] in their relational version.[2,3] Very widely applicable, these methods are particularly appropriate here because of two almost contrasting reasons. Ideas, as well as performance specifications and other parameters of the initial building program and of the initial inspiration can, on the one hand, be left in imprecise form during the early stages of work, awaiting later clarification. On the other hand when, as often happens, they have been prematurely overspecified, they can be varied within wide limits as the design process goes along. Adaptations of one subsystem to another can also be handled by the relational component using fuzzy relational products[3] and continuing the fuzzy relational data base technology pioneered by Zemankova and Kandel.[19]

In what sense is this system more "expert" than what already exists? It is mainly in the interworking between itself and the designer, to whom it seems to be making "efforts" to "understand" what he means before he has made it explicit and, indeed, to assist him in doing so. For the system to deserve this impression requires more than a display of tact on its part. It demands the adroit exploitation of the extensive knowledge base, made possible by the intensive development of the relational component. What we consider to be the core of the design task is the interplay between the primary and the secondary processes. In attempting to promote this interplay, we have had to push the expert system capability beyond what is operationally expected of present day CAD programs, into a realm where the object is no longer constructed by the assembly of parts, but is articulated by the differentiation of a global image.

ACKNOWLEDGMENT

Research is partially supported by National Science Foundation Grant IST-8604575 and by Florida High Technology and Industrial Council Grant UPN-85100316.

REFERENCES

1. **Arnheim, R.,** *Visual Thinking,* University of California Press, Berkeley, 1969.
2. **Bandler, W.,** Representation and manipulation of knowledge in fuzzy expert systems, *Int. J. Syst. Res. Inf. Sci.,* 1, 113, 1985.
3. **Bandler, W. and Kohout, L. J.,** Fuzzy relational products as a tool for analysis and synthesis of the behaviour of complex natural and artificial systems, in *Fuzzy Sets: Theory and Applications to Policy Analysis and Information Systems,* Wang, P. P. and Chang, S. K., Eds., Plenum Press, New York, 1980, 341.

4. **Crosley, M. L.,** *The Architect's Guide to Computer-Aided Design,* John Wiley & Sons, New York, 1988.
5. **Ehrenzweig, A.,** *The Hidden Order of Art: A Study in the Psychology of Artistic Imagination,* University of California Press, Berkeley, 1967.
6. **Freud, S.,** *The Interpretation of Dreams, Standard Edition of the Complete Psychological Works of Sigmund Freud,* Vol. 4, Basic Books, New York, 1965 ff.
7. **Glancey, J.,** Eternal values, *RIBA J.,* January 28, 1986.
8. **Kaplan, B.,** Comment: on the rational reconstruction of intuition in the design process, in *Emerging Methods in Environmental Design and Planning,* Moore, G. T., Ed., MIT Press, Cambridge, MA, 1970.
9. **Klir, G. J.,** *Architecture of Systems Problem Solving,* Plenum Press, New York, 1985.
10. **Kohout, L. J.,** *Perspectives on Intelligent Systems: A Framework for Analysis and Design,* Abacus Press, Cambridge, MA, 1987.
11. **Mancini, V.,** An Architect's View of Systems Architecture, presented at *3rd Int. Conf. Syst. Res. Inf. and Cybern.,* Baden-Baden, West Germany, August 19—24, 1986.
12. **Michalski, R. S. and Chilausky, R. L.,** Learning by being told and learning from examples, *Int. J. Policy Anal. Inform. Syst.,* 4, 1980.
13. **Mitchell, W. J.,** *Computer-Aided Architectural Design,* Petrocelli, New York, 1977.
14. **Mitchell, W. J.,** Formal representations: a foundation for computer-aided architectural design, *Environment and Planning B: Planning and Design,* 13, 133, 1986.
15. **Norberg-Schulz, C.,** *Intentions in Architecture,* MIT Press, Cambridge, MA, 1966.
16. **Padovan, R.,** A necessary instrument?, 54, April; Measuring and counting, 54, May; Theory and practice, 54, June; *The Architect,* 1986.
17. **Zadeh, L. A.,** Fuzzy sets, *Inf. Control,* 8, 338, 1965.
18. **Zadeh, L. A.,** A theory of commonsense knowledge, in *Aspects of Vagueness,* Skala, H. J., Termini, S., and Trillas, E., Eds., D. Reidel, Dordrecht, Netherlands, 1984, 257.
19. **Zemankova-Leech, M. and Kandel, A.,** *Fuzzy Relational Data Bases,* Verlag TÜV Rheinland, Köln, Germany, 1984.

Chapter 14

ON THE DESIGN OF A FUZZY INTELLIGENT DIFFERENTIAL EQUATION SOLVER

Menahem Friedman and Abraham Kandel

TABLE OF CONTENTS

I. INTRODUCTION

In this chapter we present a fuzzy (''soft'') expert system operating as a fuzzy intelligent differential equation solver (FIDES). The main purpose of this project is to furnish a finite element program users with an expert system that will release them from the necessity of providing the finite element triangular mesh by themselves. Creating such a mesh could be quite complicated and tiresome even for an expert, pending on the complexity of the problem domain of solution.

Users with almost no knowledge of finite element theory and techniques are often prevented from using an intelligent finite element programming package, due to their unwillingness and occasionally inability to prepare and cope with a large and complex input data. Even an expert is not exactly thrilled having to manually form a finite element mesh for domains other than rectangular, circle, etc. In addition to spending their valuable time on putting the mesh together, users are quite often capable of inserting some subtle errors. Such errors in a large and complicated grid could backfire a little or a lot on the final solution. If that solution is luckily completely out of range, the error could (not easily) be spotted and removed. Otherwise, it could stay there forever, inflicting small but damaging deviations from the numerical solution.

FIDES was developed in order to eliminate the necessity of creating the grid and to encourage engineers and scientists to consider a finite element approach more often, particularly when it is advantageous to their specific problems. It is also aimed at significantly saving on the user's time, even at the expense of somewhat adding to the computing costs. In the era of supercomputers such as CRAY YMP, it is the expert's time that should have a clear priority.

FIDES users have to specify only the most essential basic data that uniquely define their problem. This involves interactively inserting the domain boundary — given as an ordered list of points, the boundary conditions, and the differential equation coefficients. This relatively small set of α numerical characters is necessary and sufficient for defining the problem. FIDES then provides a suitable finite element triangular mesh and sets a finite element package — MANFEP[1] — to work. The final result is a numerical solution to the user's problem. The system is based on the conceptual structure of fuzzy expert systems[9] which is closely related to the theory of fuzzy sets as developed by Zadeh.[7] For a detailed exposition of fuzzy set theory and its applications, the reader is referred to.[5,6]

MANFEP, which is a two-dimensional finite element programming package designed for solving second order self-adjoint elliptic partial differential equations, can treat linear and various types of nonlinear problems with any combination of Dirichlet, Neumann, and mixed boundary conditions. Its efficiency and applicability have been broadly demonstrated,[2-4] and the package is naturally a major component of FIDES.

A detailed formulation of the problem is given in Section II. FIDES structure is discussed in Section III. and its applicability is demonstrated in Section IV.

II. FORMULATION OF THE PROBLEM

The general problem treated by FIDES is given by the 2-D elliptic partial differential equation:

$$\frac{\partial}{\partial x}\left(a\frac{\partial}{\partial x} + b\frac{\partial}{\partial y}\right)\phi + \frac{\partial}{\partial y}\left(b\frac{\partial}{\partial x} + c\frac{\partial}{\partial y}\right)\phi + d\phi + p = 0 \tag{1}$$

defined over a general bounded domain D with a boundary B. The equation coefficients are in $C^1(D)$ and satisfy $ac - b^2 > 0$ over D.

The boundary is composed of three sections B_1, B_2, B_3 on which the most commonly used boundary conditions are given:

$$Dirichlet: \phi = f \text{ on } B_1 \tag{2}$$

$$Homogeneous\ Neumann: \frac{D\phi}{Dn} = 0 \text{ on } B_2 \tag{3}$$

$$Mixed: \frac{D\phi}{Dn} + \sigma\phi = h \text{ on } B_3 \tag{4}$$

where

$$\frac{D\phi}{Dn} \triangleq \left(\frac{\partial\phi}{\partial x}\ \frac{\partial\phi}{\partial y}\right)\begin{pmatrix} a & b \\ b & c \end{pmatrix}\begin{pmatrix} n_x \\ n_y \end{pmatrix} \tag{5}$$

f, σ, h are continuous functions specified on B_1, B_3, respectively, and n_x, n_y — the x,y components of the outward normal to the boundary.

FIDES (via MANFEP) can also solve the general eigenvalue problem defined by:

$$\frac{\partial}{\partial x}\left(a\frac{\partial}{\partial x} + b\frac{\partial}{\partial y}\right)\phi + \frac{\partial}{\partial y}\left(b\frac{\partial}{\partial x} + c\frac{\partial}{\partial y}\right)\phi + d\phi = E\phi \tag{6}$$

with homogeneous Dirichlet or Neumann boundary conditions, and boundary value problems given by Equations 1—4 where $d\phi$ is replaced by a nonlinear term of Thomas-Fermi type.[8]

The user supplies the input data, first the coefficients a, b, c, d, p, f, σ, h which could be either constants or functions. The boundary is then defined as an ordered list of pairs of numbers $(x_i,y_i)_{i=1}^N$ that creates a closed polygon. Each polygon side that is not a straight segment must be followed by a pair of functions:

$$x = f_i(t),\ t_{i_1} \leq t \leq t_{i_2}$$
$$y = g_i(t),\ t_{i_1} \leq t \leq t_{i_2} \tag{7}$$

that specify the boundary between the vertices i, $i + 1$. This portion of the boundary is eventually replaced by an approximating polygon, so that the whole boundary B is finally represented by one closed polygon.

For each vertex i, the user is expected to specify the boundary condition that must hold between vertices i and $i + 1$. However, if no boundary condition is inserted, a "natural boundary condition" (homogeneous Neumann) will automatically hold.[1]

Once the input is inserted, FIDES automatically replaces the domain by a triangular finite element mesh and sets MANFEP to solve Equations 1—4.

III. THE ARCHITECTURE OF FIDES

FIDES code is artificially intelligent since it understands the objects it reasons with. The relationship between the numerical and the symbolic codes is based on the need to interpret massive amounts of data. This is done by making the numerical code "Smarter" just by adding the symbolic interface, and thus we have a system where efficient symbolic code is replacing parts of the numerical code in order to achieve a higher level of performance.

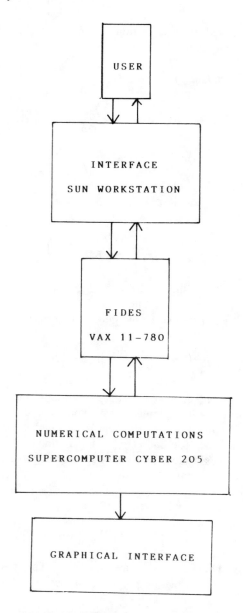

FIGURE 1.   Present architecture of the system.

Figure 1 shows the architecture of FIDES. The user interface runs at the moment on a SUN workstation. The expert system is written in LISP and runs on a VAX 11/780. The super-computer CYBER 205 is used for the numerical computations. Hence, FIDES is a distributed system that runs on different machines with different operating systems and is certainly not a conventional expert system, at least not in the clerical sense.

A small set of FIDES rules, which is given as follows, demonstrates in Figure 2 a first stage of subdivision of three different subpolygons that could participate in composing (along with other subpolygons) the original closed polygon. The four-sided polygon (Figure 2A) is divided into triangles which are transferred to another knowledge base for further investigation by FIDES. The long fuzzy rectangle (Figure 2B) is presented by an eight-sided polygon which will now be properly subdivided. The seven-sided polygon (Figure 2C) is

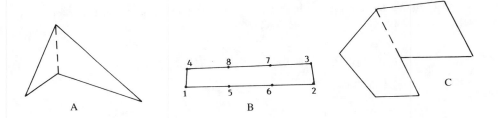

FIGURE 2.   Subdivision. (A) A concave four-sided polygon; (B) a long fuzzy rectangle; (C) a seven-sided polygon.

replaced by four-sided and five-sided polygons. On each of these polygons further subdivision will be performed.

Rule 1.   If the object has three sides, then the object is triangle.
Rule 2.   If the object has four sides, then the object is four-sided.
Rule 3.   If the object is four-sided and the object has all angles equal to 1.57, then the object is rectangle.
Rule 4.   If the object is rectangle and the object has largeratio, then the object is narrowrec.
Rule 5.   If the object is rectangle and the object has smallratio, then the object is standrec.
Rule 6.   If the object has more than four sides, then the object is polygon.
Rule 7.   If the object is four-sided and the angle is greater than 1.7, then the object is square.

A. THE FUZZINESS IN FIDES

Any conclusion related to a given object (subpolygon) is naturally associated with a certainty factor α, $0 \leq \alpha \leq 1$. For example, consider the question whether a given four-sided polygon is a "rectangle". We must first decide which subpolygons are perfect rectangles. One alternative is to define a four-sided subpolygon as a perfect rectangle if and only if each of its four angles is 90°. We then propose a *rectangle fuzzy function*, i.e.:

$$\chi_R(x) = \frac{1}{1 + A\sum_{i=1}^{4}\left(\phi_i - \frac{\pi}{2}\right)^2}, \; A > 0$$

defined for *any* four-sided polygon x with angles ϕ_i, $1 \leq i \leq 4$. If $\phi_i = \frac{\pi}{2}$, $1 \leq i \leq 4$, then x is a perfect rectangle. Otherwise, it is a rectangle with certainty $\alpha = \chi_R(x)$. We thus constructed a rectangle fuzzy set:

$$R = \{(x,\chi_R(x)) \mid x \text{ is a four-sided polygon}\}$$

Similarly we may define a "narrow" fuzzy set:

$$N = \{(x,\chi_N(x)) \mid x \text{ is a four-sided polygon}\}$$

whose general member is again any four-sided polygon. This time the grade of membership is given by $\chi_N(x)$ — a fuzzy function that determines just how narrow x is by calculating ratios of opposite sides of x. A four-sided polygon is considered *perfectly narrow* if the minimum of all these ratios exceeds a given threshold σ and $\chi_N(x)$ is defined accordingly, i.e., $\chi_N(x) = 1$ for such objects.

Now, let us try to determine whether a given four-sided object is a narrow rectangle. We define a third fuzzy set NR — the intersection of N and R:

$$NR \overset{\triangle}{=} N \cap R = \{(x, \chi_{NR}(x)) \mid x - four\text{-}sided\ polygon\}$$

where

$$\chi_{NR}(x) = \min\{\chi_N(x), \chi_R(x)\}$$

A perfect narrow rectangle must, therefore, be a four-sided polygon with four right angles whose large ratio of opposite sides exceeds the threshold σ.

Thus, in general, each "intermediate" conclusion CI that *identifies* a given object is associated with a certainty factor α. This factor is computed by a fuzzy function, whose parameters characterize the geometry of the object. The pair (CI, α) is then transferred to the blackboard which is the logical part of the inference engine, provided that α exceeds some threshold α_0. When the inference of the knowledge base is completed, the blackboard contains a sequence of pairs $\{(CI_i, \alpha_i)\}_{i=1}^n$.

Let CF be a "final" conclusion which is fired by the sequence $\{CI_{i_j}\}_{j=1}^m$ of intermediate conclusions with certainty factors $\{\alpha_{i_j}\}_{j=1}^m$ (all of which are placed on the blackboard). Then, CF is placed on the blackboard with the certainty factor β_i, where:

$$\beta_i = \min\{\alpha_{i_1}, \ldots, \alpha_{i_m}\}$$

If the blackboard finally consists of the final pairs $\{(CF_i, \beta_i)\}_{i=1}^k$, then the object is recognized with the certainty factor CF_l for which:

$$\beta_l = \max\{\beta_1, \ldots, \beta_k\}$$

For example, let a four-sided object be a *rectangle* with certainty factor 0.73, a *square* with certainty factor 0.62, and possess a *largeratio* with certainty factor 0.80. Then, the final conclusions are as follows:

1. *Narrowrec* (a narrow rectangle) with certainty $0.73 = \min\{0.73, 0.80\}$
2. *Square* (a standard four-sided polygon) with certainty factor 0.62

The object is thus recognized as a narrowrec and is further subdivided as in Figure 3B. If we would simply exchange the certainty factors 0.73 and 0.62, the object would be recognized as a square and would be subdivided as in Figure 3A.

IV. EXAMPLES

A. A BOUNDARY VALUE PROBLEM OVER A CONCAVE POLYGON

Laplace's equation with Dirichlet and homogeneous Neumann boundary conditions is solved over a domain D — a concave polygon — represented by an ordered list of its vertices (0,0), (1,0), (2,.6), (2,2), (.8,2), (.8,.8), (0,2.) and shown in Figure 3.

The fineness of the finite element grid is a parameter supplied by the user. Three grids — coarse, semifine, and fine — are presented by Figures 3A—C and consist of 8, 17, and 27 elements, respectively. Obviously, a finer mesh would provide a better approximation to the exact solution. The boundary conditions are

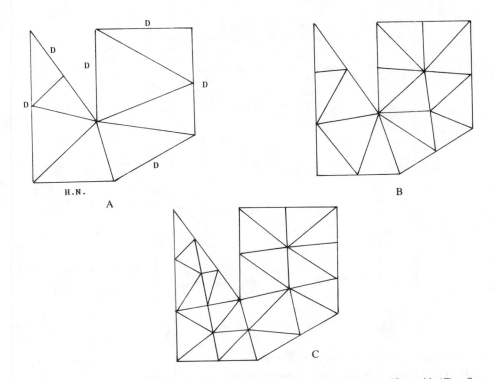

FIGURE 3. Concave polygon — various grids. (A) A coarse grid; (B) a semifine grid; (C) a fine grid.

$$\frac{\partial \phi}{\partial n} = 0 \quad \textit{between nodes} \ (0,0) \ \textit{and} \ (1,0) \tag{8}$$

$$\phi = c^x \ \textit{cosy elsewhere} \tag{9}$$

Dirichlet, homogeneous Neumann, and mixed are denoted by D, H.N., M, respectively. The exact solution to Laplace's equation given in Equations 8 and 9 is e cosy.

Once the grid is determined by FIDES, its partial differential equation solver — MANFEP — approximates the exact solution using different polynomials over the various triangles. The degree of these polynomials is another parameter supplied by the user and should not exceed four.

A comparison between the three approximating solutions related to the various grids, using cubics and the exact solution, is given in Table 1.

B. A BOUNDARY VALUE PROBLEM OVER A DOMAIN WITH A BOTTLENECK

The second-order partial differential equation:

$$\frac{\partial^2 \phi}{\partial x^2} + \frac{\partial^2 \phi}{\partial y^2} + (x^2 + y^2)\phi = 0 \tag{10}$$

is solved over the domain D (Figure 4) whose boundary is represented by the ordered list of points (2,0), (0,1), (0,.2), (−2.1,.2), (−2.1,0), and (0,0). This domain is composed of the right upper quarter of the ellipsoid:

TABLE 1
Laplace's Equation — Various Grids

(x,y)	Coarse	Semifine	Fine	Exact
(.5,0)	1.64763	1.64880	1.64877	1.64872
(.5,1)	0.89059	0.89039	0.89079	0.89081
(1,.5)	2.38712	2.38616	2.38524	2.38532

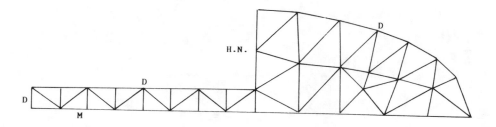

FIGURE 4.   A domain with a bottleneck.

$$\frac{x^2}{4} + y^2 = 1 \tag{11}$$

and a long narrow rectangle attached to it at the bottleneck $x = 0, 0 \le y \le .2$. Equation 10 is solved with the following boundary conditions:

$$M: \quad \frac{\partial \phi}{\partial n} + \phi = 1, y = 0 \tag{12}$$

$$H.N.: \quad \frac{\partial \phi}{\partial n} = 0, x = 0 \tag{13}$$

$$D: \quad \phi = cos(xy) \; elsewhere \tag{14}$$

FIDES recognizes the bottleneck near the origin (0,0) and separates D into two different sections, namely, a quarter of an ellipsoid and a long and narrow rectangle. It then replaces each section *independently* by a *proper* set of triangles, annexes them smoothly, and transfers the complete grid (Figure 4) to MANFEP.

In this example, second order polynomials are used to approximate the exact solution, cos(xy) and a comparison is given in Table 2.

C. A COMPLEX DOMAIN

Figure 5 shows the triangularization of a complex domain whose boundary is composed of straight and curved sections. The curved part consists of two parabolas, a circle, and an ellipsoid. In spite of the lack of any *real* bottleneck, FIDES is *aware* of the existence of almost closed subdomains as the circle of radius 1 centered at (5.0) and the ellipsoid with axes a = 2.32, b = 1.96 centered at (3,7). It replaces them separately and independently by their finite element grids, as it does to *any other section*. The complete net is then transferred to MANFEP.

TABLE 2
Curved Boundary and Bottleneck

(x,y)	Numerical	Exact
(.5,.5)	0.972	0.969
(1,.5)	0.884	0.878
(0,.1)	1.000	1.000
(−1.1,.1)	0.994	0.994

FIGURE 5.   A complex domain triangularized by FIDES.

ACKNOWLEDGMENT

Research is partially supported by National Science Foundation Grant IST 8405953 and by Florida High Technology and Industrial Council Grant UPN 85100316.

REFERENCES

1. **Friedman, M., Richards, D., and Wexler, A.,** MANFEP — University of Manitoba Finite Element Program, Polytechnic of Central London, 1974.
2. **Friedman, M., Rosenfeld, Y., Rabinovitch, A., and Thieberger, R.,** Finite element methods for solving two dimensional Schroedinger equation, *J. Comput. Phys.,* 26(2), 1978.
3. **Goshen, S., Friedman, M., Thieberger, R., and Weil, J. A.,** Models for the hydrogen atom confined within crystalline quartz, *J. Chem. Phys.,* 79, 1983.
4. **Friedman, M., Rabinovitch, A., and Thieberger, R.,** The influence of an electric field on a hydrogen atom confined to boxes of different shapes, *A. Phys.,* A 316, 1984.
5. **Kandel, A.,** *Fuzzy Techniques in Pattern Recognition,* Wiley Interscience, New York, 1982.
6. **Kandel, A.,** *Fuzzy Mathematical Techniques with Applications,* Addison Wesley, Reading, MA, 1986.
7. **Zadeh, L. A.,** Fuzzy sets, *Inf. Control,* 8, 338, 1985.
8. **Friedman, M., Rosenfeld, Y., Rabinovitch, A., and Thieberger, R.,** Finite element methods for solving 3-D Thomas-Fermi equation, accepted by *J. Comput. Phys.,* September 1986.
9. **Hall, L. O. and Kandel, A.,** *Designing Fuzzy Expert Systems,* Verlag TÜV Rheinland, Köln, Germany, 1986.

Chapter 15

MILORD: A FUZZY EXPERT SYSTEMS SHELL

R. López de Mántaras, J. Agusti, E. Plaza, and C. Sierra

TABLE OF CONTENTS

I. INTRODUCTION

This paper describes MILORD, an expert systems building tool containing a knowledge elicitation module and two inference engines (forward and backward) with uncertain reasoning capabilities based on fuzzy logic. MILORD allows the user to express the degree of certainty by means of expert-defined linguistic statements and provides the possibility to choose among three different calculi of uncertainty corresponding to three different models of the *and*, *or* and *implication* connectives.

The switching between the two engines is transparent to the user. MILORD has two types of control strategies: one consists of a lookahead technique that allows the user to detect, in advance, whether or not the linguistic certainty value of a conclusion will reach a minimal threshold acceptance value. The other concerns the selection of rules according to several criteria. MILORD also contains a limited, but useful, explanation module as well as a rule editor, not described in this chapter.

II. THE KNOWLEDGE REPRESENTATION

The knowledge base consists of facts and rules. The facts are LISP atoms associated with a linguistic certainty value. A nonevaluated fact will have the value *nil* and, therefore, is very fast to check if a given fact is known, i.e., if a certainty value has been assigned to it.

Every rule has a set of conditions which, when evaluated with a certain degree of linguistic certainty, leads to a conclusion whose degree of linguistic certainty depends on the degrees of the conditions. The rules are externally represented as follows:

(Rule rule — number (If conditions) [vc] (Then conclusions))

where [vc] is the linguistic certainty value of the rule.

In order to enable a fast access to the rules, MILORD translates the preceding list into the following internal representation that uses the LISP property lists:

$$\text{Rule-N} \rightarrow \text{VAL}[vc] \text{ IF } (p_1,\ldots,p_N) \text{ THEN } (c_1,\ldots,c_M)$$

where *VAL*, *IF*, and *THEN* are properties of the atom *rule*. The access to the conditions and conclusions of a rule is then an access to the properties of an atom.

The internal representation of the rules builds, for each conclusion, a property list which is the list of rules that deduce this conclusion, together with the linguistic certainty value of each rule, i.e.,

$$\text{Conclusion} \rightarrow \text{Rules } ((\text{rule}_1 \text{ vc}_1)\ldots(\text{rule}_k \text{ vc}_k))$$

where the rules in this list are listed in decreasing order of their linguistic certainty values. This ordering will be used by the lookahead control strategy that will be described later.

III. FORWARD, BACKWARD, AND THEIR COMBINATION

The forward reasoning starts with a set of given facts and its goal is to deduce a hypothesis whose linguistic certainty value reaches a given acceptance threshold. If the forward reasoning gets to a hypothesis whose certainty value is below the threshold, the backward reasoning is called in order to try to increase this certainty value by considering, through a lookahead process, other rule-paths that would conclude the same hypothesis with a higher certainty.

A. THE LOOKAHEAD PROSPECTION TECHNIQUE

MILORD applies a prospection process from the hypothesis toward the external (non-deducible) facts in such a way that at any time it checks if the certainty value of the hypothesis can reach the acceptance threshold value. If not, it will consider a new hypothesis. Let us now briefly describe such a process with the following default operators, and for the *and*, *or* and → connectives, to perform the calculus of uncertainty (although the lookahead process is independent of the operators used):

$$v(A \text{ and } B) = \min(v(A),v(B))$$

$$v(C_{R1} \text{ or } C_{R2}) = \max(v(C_{R1}),v(C_{R2}))$$

$$v(C) = \min(v(R),v(P))$$

where A and B are conditions of a same premise, C_{R1} and C_{R2} represent the same conclusion deduced by the two rules R1 and R2, and C is the conclusion of rule R whose premise is P.

The preceding operators are used, respectively, in the evaluation of the satisfaction of the premise, in the combination of several rules with the same conclusion, and in the propagation of the uncertainty from the premise to the conclusion of a rule.

The lookahead process in the backward reasoning starts assuming that all the nonevaluated conditions of the rules leading to the same conclusion, have the highest linguistic certainty value among the ordered set of linguistic values defined by the expert. This allows to compute the highest possible certainty value that this conclusion could reach. If this value is higher than the acceptance threshold, the backward reasoning proceeds asking the user to assign a linguistic certainty value to the nonevaluated, nondeducibile conditions one by one. Each time a condition gets its value, it is propagated to the conclusion using the preceding formula, and if its certainty value is still higher than the threshold, the process proceeds asking for the value of the next nondeducible condition and so on until either the certainty value of the conclusion falls below the threshold (in which case MILORD calls back the forward reasoning mode to deduce another hypothesis), or all the nondeducible conditions have been assigned a certainty value. As far as the deducible conditions are concerned, the lookahead process is applied recursively to each one of them, as described, and its certainty value is also propagated toward the conclusion in order to keep checking if its certainty value is higher than the threshold, in which case the process resumes. If not, the forward reasoning mode will try to deduce a new hypothesis.

If the user initially gives a set of hypotheses, instead of a set of facts, MILORD calls the backward reasoning mode with one of the hypotheses and tries to validate it with a linguistic certainty value higher than the threshold, using exactly the same process described previously. If it fails, it tries another hypothesis, and so on until either one of them succeeds or all of them fail.

B. THE RULE SELECTION CRITERIA

The set of criteria to select rules has to be easily modifiable because the efficiency of any criterion depends on each particular application. In MILORD it is very easy for the user to modify or introduce criteria. The selection among a given set of criteria can, in some cases, be done automatically. For example, if a knowledge base only contains rules which have a single conclusion, any criterion based on the number of conclusions would not be considered. The criteria that, in addition to metarules, are available in MILORD are

1. The order of the rules
2. The linguistic certainty values

3.   The number of conditions
4.   The number of conclusions
5.   The rule most recently used
6.   The rule containing the most recently deduced fact in its premise

Furthermore, the user can combine several criteria according to a given priority. For example:

R1: Condition$_1$, condition$_2$ $\Rightarrow$ [absolutely-true] conclusion$_1$
R2: Condition$_2$, condition$_3$ $\Rightarrow$ [almost-true] conclusion$_2$
R3: Condition$_4$ $\Rightarrow$ [quite-true] conclusion$_1$

The extreme values corresponding to the following ordered criteria are

1.   Maximum certainty value: absolutely-true
2.   Maximum number of conclusions: 1
3.   Minimum number of conditions: 1

In this case the system will try to select a rule, among the applicable ones, having a certainty value equal to "absolutely-true", and having one condition and one conclusion. If there is no rule satisfying these criteria, it will drop the last one (number of conditions) and so on until one or more rules are obtained. If several rules have been obtained, the user can use the rest of the criteria to end up with only one rule. In our example, after dropping the last criterion, the selected rule is R1.

IV. THE MANAGEMENT OF UNCERTAIN REASONING

The numerical approaches to the representation of uncertainty imply hypotheses of independence, mutual exclusiveness, etc. about the information they deal with. On the other hand, they oblige the expert and the user to be unrealistically precise and consistent in the assignment of such numerical values to rules and facts. Furthermore, these approaches are computationally expensive.

Our approach is based on a linguistic characterization of the uncertainty and follows the work of Bonissone.[3] The linguistic certainty values are terms defined by the expert. The internal representation of each term is a fuzzy number on the interval [0,1] characterized by a parametric representation for computational reasons.

MILORD has been parametrized in order to perform three different calculi of uncertainty operating on the expert defined term set of linguistic certainty values.

A. THE CALCULUS OF UNCERTAINTY

It can be shown[5] that triangular norms (t-norms) and triangular conorms (t-conorms) are the most general families of two-place functions from [0,1] $\times$ [0,1] to [0,1], that satisfy the requirements of conjunction and disjunction operators, respectively.

A t-norm $T(p,q)$ performs a conjunction operator, on the degrees of certainty of two or more conditions in the same premise, satisfying the following properties:

$$T(0,0) = 0$$
$$T(p,1) = T(1,p) = p$$
$$T(p,q) = T(q,p)$$
$$T(p,q) \leq T(r,s) \text{ if } p \leq r \text{ and } q \leq s$$
$$T(p,T(q,r)) = T(T(p,q),r)$$

A t-conorm $S(p,q)$ computes the degree of certainty of a conclusion derived from two or more rules. It is a disjunction operator satisfying the following properties:

$$S(1,1) = 1$$
$$S(0,p) = S(p,0) = p$$
$$S(p,q) = S(q,p)$$
$$S(p,q) \leq S(r,s) \text{ if } p \leq r \text{ and } q \leq s$$
$$S(p,S(q,r)) = S(S(p,q),r)$$

The propagation function $P(p,r)$, giving the certainty value of the conclusion of a rule as a function of the certainty value of the premise and the certainty value of the rule itself, satisfies the properties of a t-norm.

For suitable negation operators $N(x)$[13], t-norms and t-conorms are dual in the sense of DeMorgan's law.

Some usual pairs of dual t-norms and t-conorms are

$T_0(x,y) = \begin{cases} \min(x,y) \text{ if...etc.} \\ 0 \end{cases}$ $S_0(x,y) = \begin{cases} \max(x,y) \text{ if...etc.} \\ 1 \end{cases}$

$T_1(x,y) = \max(0, x + y - 1)$ $S_1(x,y) = \min(1, x + y)$ (Luckasiewicz)

$T_{1.5}(x,y) = x \cdot y/[2 - (x + y - xy)]$ $S_{1.5}(x,y) = (x + y)/(1 + xy)$

$T_2(x,y) = x \cdot y$ $S_2(x,y) = x + y - xy$ (Probabilistic)

$T_{2.5}(x,y) = x \cdot y/(x + y - xy)$ $S_{2.5}(x,y) = (x + y - 2xy)/(1 - xy)$

$T_3(x,y) = \min(x,y)$ $S_3(x,y) = \max(x,y)$ (Zadeh)

It can be shown that they are ordered as follows:

$$T_0 \leq T_{1.5} \leq T_2 \leq T_{2.5} \leq T_3$$

$$S_3 \leq S_{2.5} \leq S_2 \leq S_{1.5} \leq S_1 \leq S_0$$

In MILORD we have implemented the pairs (T_1, S_1), (T_2, S_2), and (T_3, S_3), following the experimental results obtained by Bonissone,[3] which consisted of applying nine t-norms to three different term sets. Bonissone analyzed the sensitivity of each operator with respect to the granularity (number of elements) in the term sets and concluded that only the t-norms T_1, T_2, and T_3 generated sufficiently different results for term sets that do not have more than nine elements. On the other hand, according to the results of Miller[9] concerning the span of absolute judgment, it is unlikely that any expert or user would consistently qualify uncertainty using more than nine different terms.

V. THE LINGUISTIC CERTAINTY VALUES

MILORD allows the expert to define the term set of linguistic certainty values which constitutes the verbal scale that he and the users will use to express their degree of confidence in the rules and facts, respectively. Recent psychological studies[1] have shown the feasibility of such verbal scales. " . . . A verbal scale of probability expressions is a compromise between people's resistance to the use of numbers and the necessity to have a common numerical scale," according to Beyth-Marom.[1] " . . . People asked to give numerical estimations on a common-day situation err most of the time and in a nonconsistent way. Furthermore, they are unable to appreciate their judgment imprecision (errors are by far bigger than the maximum error accepted as possible by the subjects themselves). Nevertheless, judgments embodied in linguistic descriptors appear consistent in this same situation."[6]

Each linguistic value is represented internally by a fuzzy interval (fuzzy number), i.e., the membership function of a fuzzy set on the real line, or, more precisely, on the truth

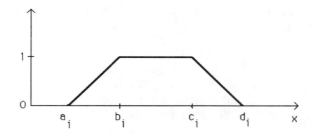

FIGURE 1.   The trapezoidal function.

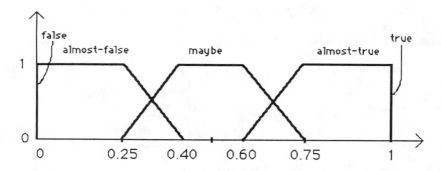

FIGURE 2.   Five elements representation.

space represented by the interval [0,1]. These membership functions can be interpreted as the meanings of the terms in the term set. The conjunction and disjunction operators applied to these functions will produce another membership function, as a result that will have to be matched to a term in the term set, in order to keep the term set closed. This can be done by a linguistic approximation process that will be described later (see Bonissone[2] for an extensive study of the linguistic approximation process).

A. A DEFAULT TERM SET AND ITS REPRESENTATION
Although the expert can define its own term set together with its internal representation, MILORD provides the following default term set:

{False, almost-false, maybe, almost-true, true}

Each term T_i is represented by a membership function $\mu_i(x)$, for x in the interval [0,1]. For computational reasons, each membership function is represented by four parameters $T_1 = (a_i, b_i, c_i, d_i)$, corresponding to the following trapezoidal function:
The five element default term set has the following representation:

$$False = (0,0,0,0)$$
$$Almost\text{-}false = (0,0,.25,.40)$$
$$Maybe = (.25,.40,.60,.75)$$
$$Almost\text{-}true = (.60,.75,1,1)$$
$$True = (1,1,1,1)$$

corresponding to the following functions in Figure 2.

In order to be able to evaluate the t-norms T_1, T_2, T_3 and the t-conorms S_1, S_2, S_3 on the elements of the term set, we have applied the following formulas according to the arithmetic rules on fuzzy numbers.

Given two fuzzy intervals $l = (a,b,c,d)$ and $l' = (a',b',c',d')$, we have the following:

$$l+l' = (a+a', b+b', c+c', d+d')$$
$$l-l' = (a-d', b-c', c-b', d-a')$$
$$l*l' = (aa', bb', cc', dd')$$
$$\text{Min}(l,l') = (\min(a,a'), \min(b,b'), \min(c,c'), \min(d,d'))$$
$$\text{Max}(l,l') = (\max(a,a'), \max(b,b'), \max(c,c'), \max(d,d'))$$

B. THE LINGUISTIC APPROXIMATION

A linguistic approximation process is performed in order to find a term (linguistic value) in the term set whose "meaning" (membership function) is the closest (according to a given metric) to the "meaning" (membership function) of the result of the conjunction or disjunction operation performed on any two linguistic values of the term set. This allows us to maintain, closed, the operations for any t-norm and t-conorm. The problem is, therefore, that of computing a distance between two trapezoidal membership functions. In order to do so, we have adopted a simple solution consisting of the computation of a weighted Euclidean distance of two features of the functions: the first moment and the area under the function. The next figure shows the results obtained with the selected t-norms T_1, T_2, and T_3 on the default term set of Figure 2.

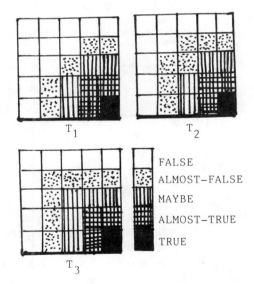

VI. THE KNOWLEDGE ELICITATION MODULE OF EXPERTISE TRANSFER

This section describes elicit-analyze-refine (EAR), a system and ancillary methodology for aiding knowledge engineers in the early phases of knowledge base design. That is to say, we focus on the top half of Figure 3 because we are convinced that much of the difficulty in knowledge acquisition lies in the fact that the expert cannot easily describe how he views a problem, because he may not distinguish between the facts or beliefs and the factors which influence his decision making. Much of the expertise lies in the way an experienced person views the problem, and this is a psychological issue that can be dealt in terms of the personal

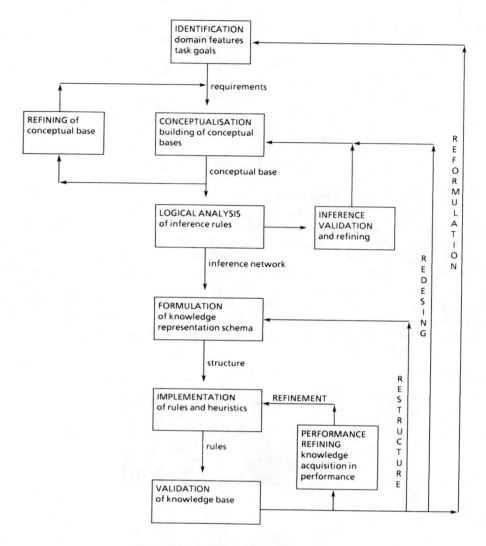

FIGURE 3. Knowledge acquisition stages.

construct psychology. The psychology of Kelly[7] views a human as a scientist classifying and theorizing about his world and basing his theory on the hypothesis that everybody has his own model of the world made up of personal constructs.

Based on this claim, the system conducts a dialogue with a domain expert eliciting relevant constructs and interactively detecting constructs poorly or ambiguously defined. Such constructs are fed back to the dialoguing expert for further refining. In this way, the expert is forced to investigate how he thinks about the problem at hand. This process builds up a repertory grid relating domain constructs with domain elements to which they apply. These relations are expressed by a contrastive set of linguistic labels, and are represented by possibility distributions, e.g., are of the form: E_i is $Q^k j \ C_i$, where E_i is an element, C_j is a construct pole, and $Q^k j$ is the linguistic label relating them.

In the second stage, a logical analysis of the repertory grid shows the implication strength between the constructs and this allows us to generate an initial set of fuzzy rules, forming an inference network. This network is then presented to the expert who points out his disagreements and enters a refining stage that uses several techniques of the personal construct theory (PCT). The elicitation dialogue is also based on the PCT and on recent implementations of psychological analysis systems.[4,10,12]

EAR produces a validated rule set for knowledge base building through the following three-stage cycle:

1. Interactive elicitation and analysis (EINA) is a program to assist in the conceptualization by building a repertory grid through a guided dialogue.
2. Subjective inference logical analysis (ALIS) is a program that generates a tentative rule-set represented as an inference network.
3. Inference validation and refining environment (EVR) is a program containing several techniques for disagreement resolution (laddering, concrete explanations, etc.). At this point if disagreement remains, it is possible to go back to point 1. If no disagreement remains, the cyclic process ends.

A. ELICITATION

To enter the elicitation stage the expert must characterize the context with a minimal set of elements (i.e., cases, examples, diagnostics, etc.) pertaining to the domain of expertise. Next, the elicitation process builds up groups of elements according to their similarities and dissimilarities with respect to the constructs already present.

The refining mode carries out two indistinguishability analyses: one over the domain elements and another over the already elicited constructs. Its main feature is that the interactive analysis is fed back to the expert in such a way that the incremental building of the repertory grid, and the validation/refinement of the repertory grid are the same process. Construct analysis shows the expert the most similar constructs, and he may point out his disagreements. If two constructs are similar, it means that they structure similarly the domain elements, and if they had to be more different than the domain context, they should be enlarged with new elements that are still missing. Therefore, the expert is asked to supply a concrete explanation of his disagreement, i.e., a counterexample embodied in a new element that will increase the construct distinction, as well as the representativeness of the domain context with regard to the real expertise domain.

Element analysis shows the most similar elements and, if disagreement arises, the expert is requested to supply a new concrete explanation, i.e., a new construct that distinguishes these too-similar elements. Undue indistinguishability between elements reveals a poor discrimination power of the set of elicited constructs.

By defining a new construct, the expert introduces a new distinction over the context elements in a process to build an opposite characterization of the set of domain elements for the task at hand. Both interaction modes form the incremental constructing process of the repertory grid. The set of fuzzy relations between domain elements and constructs constitutes the fuzzy repertory grid and the set of fuzzy predicates applying to a construct; for example, a type 2 fuzzy set constitutes the representation of the construct.[11]

B. INFERENCE ANALYSIS

In the second stage, ALIS elucidates the implicational relationship holding between constructs. As constructs are represented by type 2 fuzzy sets, we apply a type 2 semantic entailment (an extension of ordinary fuzzy set entailment) to model subjective inferences. The analysis outcome is an inference network, a digraph where nodes stand for construct poles and weighted arcs stand for implicational strength. The inference network is fed back to the expert for validation in the EVRI stage.

C. INFERENCE VALIDATION AND REFINING ENVIRONMENT

This stage implements a set of techniques and aids for validation and refining founded in PCT. The EVRI environment is the turnover of the developmental EAR cycle, for it focuses on disagreement resolution. Expert disagreements about the inference network may arise for different reasons (ambiguous or polysemic constructs, insufficient elements characterization, domain context incompleteness), and they are handled in different ways:

Counterexample proposal — Disagreement with a rule is justified by the expert stating a counterexample that is incorporated in the repertory grid as a new concrete explanation. As before, concrete explanation has a global repercussion and may modify other rules in addition to the intended one. Counterexamples may be new elements, in the case of lack of representativeness of the current context, or new constructs, in the case of insufficient element characterization.

Revision of assignment values — The expert may have used different criteria in estimating the linguistic evaluations, applying constructs to elements, in a nonconsistent way. This is solved by editing the repertory grid to revise the assignment linguistic values.

Revision of contrastive sets — Disagreements may also arise for inappropriate or poorly discriminating contrastive sets. Contrastive sets can then be drastically changed or augmented with new linguistic labels in order to achieve a finer discrimination over the domain. A revision of the linguistic values of the associated constructs is finally conducted.

Laddering techniques — Concept ambiguity is resolved splitting a construct into two or more constructs by asking *how* and *why* questions. These new constructs are added to the repertory grid. *Why* questions lead to superordinate, more abstract constructs whereas *how* questions lead to subordinate, more concrete constructs.

VII. CONCLUDING REMARKS

We have described some aspects of the MILORD system and, in particular, its management of uncertainty. The most relevant features of our approach are the representation of uncertainty by means of expert-defined linguistic statements and the use of the certainty values to guide the search tree by means of a lookahead prospection technique.

The main advantage of this approach is that once the linguistic values have been defined by the expert, the system computes and stores the matrices corresponding to the different conjunction and disjunction operations on all the pairs of terms in the term set. Later, when MILORD is run on a particular application, the propagation and combination of uncertainty is performed by simply accessing these precomputed matrices.

The gain in speed, with respect to the most common numerical approaches, is remarkable; for example, a rule with N conditions in its premise will need N-1 accesses to a matrix to obtain the linguistic certainty value of the premise, and one additional access to combine this value with that of the rule itself in order to obtain the linguistic certainty value of the conclusion.

The easiness for the expert and the user in expressing linguistically their confidence in the rules and facts is also a remarkable feature.

The EAR cycle facilitates the knowledge engineering process, decoupling knowledge acquisition from implementation, and sticks to systematic refinement, requiring concrete explanations for the resolution of disagreements. The decoupling is obtained creating a refinement cycle prior to knowledge base implementation and prototype testing. This decoupling also allows a structured way in which a group of experts may develop their individual perspectives and, furthermore, using several content-free conversational procedures, engage in a process of discussion and negotiation for reaching a meaningful consensus. The system is being used for designing a knowledge base medical diagnosis system. In this experience, we have cited the insight provided by the different perspectives of the elicited data, the easiness in eliciting the individual conceptual repertoires, and generating tentative inference networks.

In the near future, the resulting knowledge base will be submitted to other experts for a final validation, and research has to be done in order to be able to implement a consensus process on a second stage over the inference network.

REFERENCES

1. **Beyth-Marom, R.,** How probable is probable? A numerical taxonomy translation of verbal probability expressions, *J. Forecasting,* 1, 257, 1982.
2. **Bonissone, P. P.,** The Problem of Linguistic Approximation in System Analysis, Ph.D. dissertation, EECS Dept., University of California, Berkeley, in University Microfilms International Publications 80-14, Ann Arbor, MI, 1979, 618.
3. **Bonissone, P. P. and Decker, K. S.,** Selecting uncertainty and granularity: an experiment in trading-off precision and complexity, *KBS Working Paper,* General Electric Corp. Res. Develop. Center, Schenectady, NY, 1985.
4. **Boose, J. H.,** *Personal construct theory and the transfer of human expertise,* in Advances in Artificial Intelligence, O'Sheea, F., Ed., Elsevier/North-Holland, Amsterdam, 1984.
5. **Dubois, D. and Prade, H.,** Criteria aggregation and ranking of alternatives in the framework of fuzzy set theory, in *TIMS/Studies in the Management of Science,* Vol. 20, Elsevier, New York, 1984, 209.
6. **Freksa, C. and Lopez de Mantaras, R.,** A learning system for linguistic categorization of ''soft'' observations, *Actes Colloq. Assoc. Rec. Cognit.,* Université de Paris-Sud, Orsay, 1984, 331.
7. **Kelly, G. A.,** *The Psychology of Personal Constructs,* W. W. Norton, London, 1955.
8. **Lopez de Mantaras, R., Agusti, J., Cortes, U., and Plaza, E.,** *Fuzzy Knowledge Engineering Techniques in Scientific Document Classification,* Int. Symp. Methodol. Intell. Syst., Knoxville, TN, October 1986.
9. **Miller, G. A.,** The magical number seven plus or minus two: some limits on our capacity for processing information, in *The Psychology of Communication,* Penguin Books, New York, 1967.
10. **Plaza, E.,** *Sistema interactivy d'explicitaciö de constructes personals usant semäntica difusa,* Master's thesis, Facultat d'Informática de Barcelona, 1984.
11. **Plaza, E. and López de Mántaras, R.,** *Knowledge Acquisition and Refinement Using a Fuzzy Conceptual Base,* Proc. NAFIPS Workshop Fuzzy Expert Syst. Decision Support, Georgia State University, Atlanta, October 1985.
12. **Shaw, M. and Gaines, B. R.,** New directions in the analysis and interactive elicitation of personal construct systems, *Int. J. Man-Mach. Stud.,* 13, 86, 1980.
13. **Trillas, E.,** Sobre funciones de negacion en la teoria de conjuntos difusos, *Stochastica* (Journal of the Universitat Politecnica de Barcelona, Spain), 3(1), 47, 1979.

Chapter 16

MEDICAL DECISION MAKING USING PATTERN CLASSIFICATION TECHNIQUES FOR ESTABLISHMENT OF KNOWLEDGE BASES

M. E. Cohen and D. L. Hudson

TABLE OF CONTENTS

I. INTRODUCTION

Automated decision-making aids have found wide application in medicine.[1,2] These systems have been based on a variety of techniques, including traditional artificial intelligence knowledge-based systems[3] utilizing frames[4] or rules[5] for the knowledge base, data base query systems,[6] Bayesian-based decision-making systems,[7,8] as well as other types of pattern classification.[9] All of these techniques have a place in decision making, and in fact some techniques are better suited to a given application. In this chapter, a decision-making system based on pattern classification techniques is described.

The idea of fuzzy expert systems was present in a rudimentary form in early medical expert systems,[10] although this terminology was not used. More recently, medical expert systems have been developed which explicitly use techniques from fuzzy set theory and fuzzy logic,[11-14] based on general advances in the use of fuzzy logic in expert systems.[15,16]

The field of pattern recognition and classification had its roots in the early days of artificial intelligence research.[17,18] The term pattern recognition was used in conjunction with a wide variety of techniques, including two general fields: classification of data by discerning inherent patterns and recognition of pattern in visual images.[19-22] The first of these is the field of interest here and is designated by both the terms pattern recognition and pattern classification.

Pattern classification itself encompasses a variety of techniques,[23] including parametric statistical techniques;[24] discriminant analysis;[25] Bayesian classification;[26,27] maximum likelihood;[28] and nonparametric techniques such as supervised learning,[29-31] unsupervised learning,[32] nearest neighbor,[33] and clustering.[34] Clustering algorithms have been expanded to include clusterings of fuzzy data sets.[35,36] The method described here is a supervised learning approach to pattern classification based on a new class of multidimensional orthogonal polynomials developed by Cohen.[37]

Pattern classification involves a number of steps, including problem definition, feature extraction, classification, and verification. The objective of the method is to divide data into two or more categories; thus, the problem must be clearly defined in these terms. As a supervised learning approach is employed, it is necessary to initially have a set of data of known classification. The data are then randomly divided into two sets: a training set and a test set. The training set is used to determine the separating criteria, while the test set is used to ascertain the accuracy of the results. During the problem definition phase, parameters are identified which are pertinent to the problem at hand. Data must then be collected to ascertain values for these parameters. As in all data analysis approaches, an increase in the number of parameters to be analyzed necessitates collection of data for more cases.

In the second phase, feature extraction, some or all of the parameters identified in phase 1 are selected as features to be used in classification. The features are collected into a structure known as a feature vector. The number of features included determines the dimensionality of the resulting hypersurface.

Once the features have been selected, classification is begun. Values are assigned to the feature vectors using data in the training set, with weighting factors attached to each of the features. The weighting factors are then adjusted iteratively with the aid of the multidimensional Cohen polynomials until separation of the data into correct categories is achieved. If separation does not occur, the nonseparable feature vector is discarded, and a new vector is randomly selected to replace it. At the end of the classification phase, a dividing hypersurface (for the two category case) or hypersurfaces (for the multicategory case) are obtained.

The hypersurface obtained from the classification phase is then used directly to classify data in the test set. Results are then compared with independently obtained classification to ascertain the effectiveness of the dividing hypersurface. The ultimate goal is to obtain a reliable hypersurface which can be used to classify future data for which correct categorization is unknown.

This technique can be used for two purposes. The first is for classification of data sets into categories, as explained before. The second is to use data of known classification to determine which parameters are important in the decision-making process and also to determine the relative importance of these parameters by assigning weights to them. This information can be used to reduce the amount of data collected and, subsequently, to design an expert system which collects only necessary information. Derivation of information directly from data circumvents the long and tedious process of establishing a knowledge base through expert consultation, a process which must be repeated for each new application.

The classification method can operate on fuzzy data as well as crisp data. The nature of the data is determined in the problem definition phase. The algorithm itself operates directly on either categorical or continuous variables. However, the categorical variables must have an ordering. The decision in the two-category case is obtained by substituting the values for the feature vector into the equation for the hypersurface. If the result is positive, category 1 is assumed, negative implies category 2, and a value of zero is inconclusive. However, rather than interpreting the result as a crisp decision, the relative absolute magnitude can be considered as a degree of membership in the category. The classification method can be extended to the multicategory case in two ways. The most direct approach is to obtain hypersurfaces for all combinations of classes, although this method results in a combinational explosion if a large number of classes are involved. Alternately, a hypersurface value can be computed for each category, with classification assigned to the class yielding the highest numerical value.

This method has been applied to a number of medical applications,[38-47] although it is generally applicable to any type of data. It will be illustrated here in a medical application: a multicategory problem involving classification of exercise testing data to determine the extent of coronary artery disease.

In the next section, details of the classification method are given. In the following sections, the method is illustrated in conjunction with the application just described.

II. CLASSIFICATION METHODOLOGY

The classification method is based on the potential function approach to decision function generation. The multivariant potential function is defined by:

$$P(\overline{x},\overline{y}) = \sum_{i=1}^{\infty} \lambda_i^2 \, f_i(\overline{x})f_i(\overline{y}) \qquad (1)$$

where $f_i(\overline{x})$ are orthonormal functions, $\overline{x}$ and $\overline{y}$ are n-dimensional feature vectors, and λ_i, $i = 1,2, \ldots$ are real numbers. The equation which represents the decision hypersurface is

$$D(\overline{x}) = \sum_{i=1}^{\infty} c_i(f_i(\overline{x})) \qquad (2)$$

where c_i, $i = 1,2, \ldots$ are unknown weighting factors determined in an interative procedure using the training set. For a two-class problem, the decision is made according to:

$$D(\overline{x}) > 0 \text{ class 1}$$

$$D(\overline{x}) < 0 \text{ class 2}$$

$$D(\overline{x}) = 0 \text{ indeterminant} \qquad (3)$$

The iterative procedure works in the following manner. The basic iterative equation is

$$D_{k+1}(\bar{x}) = D_k(\bar{x}) + r_{k+1}P(\bar{x},\bar{x}_{k+1}) \qquad (4)$$

where $D_0(\bar{x})$ is assumed to be zero, and:

$$D_1(x) = \begin{cases} P(\bar{x},\bar{x}_1) & \text{if } \bar{x}_1 \in w_1 \\ -P(\bar{x},\bar{x}_1) & \text{if } \bar{x}_1 \in w_2 \end{cases} \qquad (5)$$

where w_1 represents class 1 and w_2 represents class 2. For the two-class problem:

$$r_{k+1} = \begin{cases} 1 & \text{for } \bar{x}_{k+1} \in w_1 \text{ and } D_k(\bar{x}_{k+1}) \leq 0 \\ -1 & \text{for } \bar{x}_{k+1} \in w_2 \text{ and } D_k(\bar{x}_{k+1}) \geq 0 \end{cases} \qquad (6)$$

It then remains to choose the orthogonal function $f_i(\bar{x})$. In previous work of the authors,[38] a new class of one-dimensional Cohen orthogonal polynomials were used. The general form of this polynomial is

$$F_n(\lambda,a_i;x) = \sum_{k=0}^{n} \frac{\prod_{i=0}^{n-1} (\lambda + a_i + a_k)x^{a_k}}{\prod_{j=0}^{k-1} (a_j - a_k) \prod_{s=1}^{n-k} (a_{k+s} - a_k)} \qquad (7)$$

where

$$\prod_{j=0}^{k-1} (a_j - a_k) = (a_0 - a_k)(a_1 - a_k)...(a_{k-1} - a_k) \quad k \geq 1$$

$$= 1 \text{ for } k = 0$$

$$\prod_{s=1}^{n-k} (a_{k+s} - a_k) = (a_{k+1} - a_k)(a_{k+2} - a_k)...(a_n - a_k) \quad n \geq k$$

$$= 1 \text{ for } n = k$$

The recurrence relation for this polynomial is

$$(a_n - a_0)F_n(\lambda,a_i;x) + (\lambda + a_n + a_0)F_{n-1}(\lambda,a_{t+1};x) -$$

$$(\lambda + a_0 + a_{n-1})F_{n-1}(\lambda,a_i;x) - (a_{n-1} - a_0)F_{n-2}(\lambda,a_{t+1};x) = 0 \quad n \geq 2$$

$$F_1(\lambda,a_i;x) = [(\lambda + 2a_0)xa^0 - (\lambda + a_0 + a_1)xa^1)]/(a_1 - a_0)$$

$$F_0(\lambda,a_i;x) = x^{a0} \qquad (9)$$

The orthogonal relationship is

$$\int_0^1 x^{\lambda-1}F_n(\lambda,a_i;x) F_m(\lambda,a_i;x) dx = 0 \qquad n \neq m$$

$$= 1/(\lambda + 2a_n) \quad n = m \qquad (10)$$

Multidimensional orthogonal polynomials were then generated using the expression:

$$\Theta_1(x_1,x_2,\ldots,x_n) = \emptyset_{j_1}(x_1)\emptyset_{j_2}(x_2),\ldots,\emptyset_{j_n}(x_n) \tag{11}$$

where $\emptyset_{j_i}(x)$ is the one-dimensional orthogonal polynomial defined in Equation 7. $P(\bar{x},\bar{x}_k)$ is then defined by:

$$P(\bar{x},\bar{x}_k) = \sum_{l=1}^{m} \Theta_l(\bar{x})\Theta_l(\bar{x}_k) \tag{12}$$

where m is the degree of the resulting hypersurface and determined experimentally.

It should be noted that the polynomials used are orthogonal, rather than orthonormal. This is permitted since all comparisons are made to zero, and the normalizing constant serves only as a scaling device and does not affect the class decision.

In the method described, computational complexity is reduced by the use of the new multidimensional polynomial developed by Cohen. It takes the form:

$$C_n(x_1,x_2,\ldots,x_m) = \frac{m!}{n!(m-n)!} + \sum_{k=1}^{n} \frac{(-1)^k(m-k)!}{(n-k)!(m-n)!}$$
$$\sum_{i_k=k}^{m} \sum_{i_k-1=k-1}^{i_k-1} \cdots \sum_{i_2=2}^{i_3-1} \sum_{i_1=1}^{i_2-1} \sum_{p=1}^{k} \frac{x_i^{a(n,i_p)}[a(n,i_p)+v_{i_p}]}{v_{i_p}} \tag{13}$$

$$\int_0^1 \int_0^1 \cdots \int_0^1 C_nC_r x_1^{v_1-1} x_2^{v_2-1} \cdots x_1^{v_m-1} dx_1,\ldots,dx_m = \begin{cases} 0 & n \neq r \\ A & n = r \end{cases} \tag{14}$$

where m is the dimensionality of the data, $a(n,i_p),i_p = 1,\ldots k$ are parameters which may be arbitrarily selected, A is the normalization constant, and $v_i, i = 1,\ldots,m$ are assigned values corresponding to the components of the first feature vector.

In this chapter, we make the assumption $a(n,i_p) = 1$. $P(\bar{x},\bar{x}_k)$ is then defined by:

$$P(\bar{x},\bar{x}_k) = \sum_{i=1}^{\infty} \Theta_i(\bar{x})\,\Theta_i(\bar{x}_k) \tag{15}$$

where $\Theta_i(\bar{x})$ are the components of the Cohen polynomials, which themselves form an orthogonal set:

$$\int_{-1}^1 \cdots \int_{-1}^1 \Theta_n(x_1,\ldots,x_l)\,\Theta_r(x_1,\ldots,x_l)dx_1,\ldots,dx_m = \begin{cases} 0 & n \neq r \\ B & n = r \end{cases} \tag{16}$$

where B is the normalization constant.

III. IMPLEMENTATION

The pattern recognition method described in Section II. is implemented on the VAX 11/750 computer. It is written in FORTRAN 77 and contains a number of features to facilitate its use. Figure 1 shows a systems diagram.

A. AUTOMATIC FORMATTER

This program allows the user to input the format of the new data file, and then reformats it in accordance with the format expected by the pattern recognition programs. At this stage, the user indicates the number of possible parameters, the number of classes into which the data should be divided, and the variable which determines the classification. A disk file is then written to be used by the training set selector.

B. TRAINING SET SELECTOR

This program creates the training set by starting with two randomly selected feature vectors and adding additional vectors one at a time. At each stage, the training set is tested to make sure that separation occurs. The user can indicate the size of the training set desired. Its size usually falls between one third and one half of the available pattern vectors.

C. PATTERN RECOGNITION ALGORITHM

This algorithm implements the details of the classification method described in the previous section. It receives as input the training set and proceeds iteratively to produce a separating hypersurface. The result of this program is an algebraic equation which represents this hypersurface.

In the algorithm outlined in Section II. algebraic equations must be maintained through all stages of the computation. The result itself is an equation. In order to accomplish this, the program must be capable of performing algebraic operations, rather than straight-forward numerical computation.

If multidimensional polynomials are generated from one-dimensional polynomials, the following procedure is used. The required sets of polynomials of one variable are stored in 2 two-dimensional arrays: $con(i,j)$, $icon(i,j)$ — where $con(i,j)$ contains the coefficient of term j in polynomial i and $icon(i,j)$ contains one plus the power of x corresponding to that coefficient. If a coefficient is zero, it is not stored. This representation permits compact storage of the polynomials.

The polynomial coefficients are generated through the use of the appropriate recurrence relation. The coefficients for the first two terms are initialized with all remaining terms generated recursively.

The multiplication of term i in polynomial m by term j in polynomial n to yield term k in polynomial l is accomplished by the two operations:

$$con(l,k) = con(m,i)*con(n,j)$$

$$icon(l,k) = icon(m,i) + icon(n,j) - 1 \qquad (17)$$

In order to generate the multivariant polynomials, the following procedure is followed:

1.   An initial set of one-dimensional polynomials is computed using the appropriate recurrence relation, and the representation in Equation 17.
2.   Combinations are then made according to Equation 11.

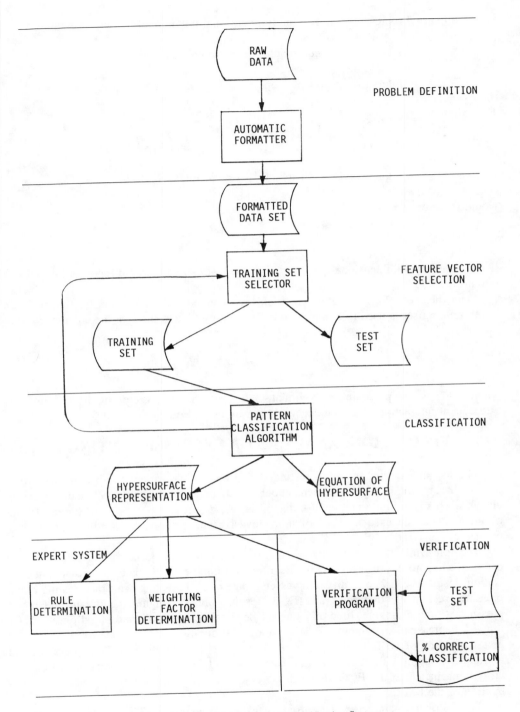

FIGURE 1.   Flow of Pattern Classification System.

The internal representation of a multidimensional polynomial is the following:

phi(i,j): coefficient for polynomial i, term j

iphi(i,j,k): power plus one of x_k for polynomial i, term j

For example:

$$phi(1,1) = 3, \; phi(1,4) = -6, \; phi(1,7) = 8$$

$$iphi(1,4,2) = 8, \; iphi(1,4,3) = 5, \; iphi(1,7,2) = 15$$

would represent:

$$f(x) = 3 - 6x_2^7 x_3^4 + 8x_2^{14}$$

If the Cohen multidimensional polynomial is used, it can be stored directly in the phi and iphi arrays.

Once the chosen orthogonal polynomial is represented in this structure, the first equation in the training set is used to compute initial weighting factors through the equation:

$$D_1(\overline{x}) = K(\overline{x}, \overline{x}_1) \sum_{i=1}^{m} \emptyset_i(\overline{x}) \emptyset_i(\overline{x}_1) \tag{18}$$

where $\emptyset_i(\overline{x})$ is the preceding polynomial, and x_1 is the first vector in the training set. Subsequent interactions are then accomplished by using Equations 4 and 6.

IV. APPLICATION: ANALYSIS OF EXERCISE TESTING DATA

The use of exercise treadmill testing (ETT) to classify patients with coronary artery disease has resulted in disagreement about which parameters should be included in the analysis. This is an important test for diagnosis, because it is noninvasive. If the results are accurate, it can be used in place of more invasive and potentially harmful tests such as cardiac catheterization.

The following data were collected for exercise testing subjects who also had cardiac catheterization to confirm diagnosis: identification number, age, date of cardiac catheterization, results (0, 1, 2, or 3 diseased vessels), date of ETT, resting heart rate, resting blood pressure, resting electrocardiogram (ECG) results, time of angina, heart rate and blood pressure at time of angina, time of 1 mm ST depression, heart rate and blood pressure at 1 mm ST depression, maximum ST depression, total duration of ETT, heart rate and blood pressure at end of ETT, total duration of chest pain, total duration of ST depression, and reason for stopping ETT. Seventy cases were analyzed.[37,41,42] An ST depression is a cardiac arrhythmia in which the ECG signal drops below the baseline following ventricular contraction of the heart.

The following parameters were used as features: maximum ST depression, percentage changes in heart rate and in blood pressure from beginning to end of test, and product of heart rate and blood pressure at end of test.

Four classes were used: class 0 — patients with no coronary artery disease; class i — patients with i diseased vessels, i = 1,2,3. Each class was compared separately to each of the other classes.

V. RESULTS

For the application described in Section IV. the results for the new Cohen multidimensional polynomial were compared to other classification methods. The first was the method previously developed by the authors for the single-dimensional Cohen orthogonal functions. A special case of the one-dimensional Cohen orthogonal polynomial, $g_n^{5,2}(x)$, is used to demonstrate the one-dimensional approach.[38] This is a special case of Equation 7. The new multidimensional method is represented by $C_n(x)$. The second method used for comparison was standard nonlinear statistical discriminant analysis.

The classification results are shown in Tables 1 and 2 for coronary artery disease. For two cases, equivalent results were obtained by the multidimensional polynomials in fewer iterations. In the remaining two cases, more accurate results were achieved with the multidimensional polynomial. In all cases, the pattern recognition approach produced superior results to nonlinear statistical discriminant analysis.

Sample equations for class 0 vs. 1 for the ETT data are shown below for the one-dimensional polynomial $g_n^{5,2}(x)$ and the multidimensional polynomial $C_n(x)$:

$$D_g = -2.2 + 2.9x_1 + 3.3x_2 + 4.6x_3 + 2.9x_4 - 21.5x_3x_4 + 11.4x_2x_4$$
$$+ 1.9x_1x_4 - 19.4x_1x_2 + 7.9x_1x_3 - 1.6x_2x_3$$

$$D_c = 5.3 - 7.4x_1 + 2.4x_2 - 23.3x_3 + 39.8x_4 - 60.0x_3x_4 + 31.1x_2x_4$$
$$+ 5.3x_1x_4 - 52.1x_1x_2 + 28.9x_1x_3 - 6.8x_2x_3$$

where

x_1: Maximum ST depression
x_2: Percentage change in heart rate from beginning to end of test
x_3: Percentage change in blood pressure from beginning to end of test
x_4: Percentage of heart rate and blood pressure

VI. EXPERT SYSTEM DESIGN

The procedure in Section V. reduced the relevant parameters from 16 to 5. This information can then be utilized to design a prospective decision-making model to collect the required information. An important advantage of this method is that the information which has been inferred directly from the data can be combined with heuristic rules representing expert decision-making strategies. For example, the physician may wish to include rules of the form:

IF   Reason for stopping ETT is severe chest pain
THEN   Artery blockage is probably severe
IF   Patient exercises less than 2 min
THEN   Artery blockage is severe

A sample consultation run of the expert system is shown in Figure 2. Note that questions pertaining to parameters derived from the classification procedure are asked first. Then questions pertaining to heuristic rules are asked. All yes/no rules can be answered instead with a degree of presence between 0 and 10, inclusive. The results from all questioning is evaluated using approximate reasoning techniques for rule evaluation.[48]

TABLE 1
Number of Classified Incorrectly, ETT

Number of diseased vessels	$g_n^{.5,.2}(x)$	$C_n(x)$	BMDP	Training set	Size of sample
0 vs. 1	2	2	4	7	14
0 vs. 2	4	3	4	14	23
0 vs. 3	5	4	6	31	49
1 vs. 3	8	8	11	29	47

Note: BMDP.

TABLE 2
Number of Iterations, ETT

Number of diseased vessels	$g_n^{.5,.2}(x)$	$C_n(x)$
0 vs. 1	175	103
0 vs. 2	84	76
0 vs. 3	118	188
1 vs. 3	123	82

```
In the following consultation, if a yes/no response is indicated, a number
between 0 and 10, inclusive, may be entered instead to indicate a degree of
presence of the symptom.  A value of 0 corresponds to no; a value of 10
corresponds to yes.


Name of patient:  DD
Age (years):  73

Maximum ST depression in mm (xx.x):  2.4
Blood pressure before test (xxx/xxx):  145/92
Heart rate before test (xxx):  74
Blood pressure at end of test (xxx/xxx):  142/90
Heart rate at end of test (xxx):  78
Patient experienced chest pain (y/n):  3
Length of exercise in minutes (xx):  4

The patient has 3 vessel blockage with a certainty of 0.70.
The patient has 2 vessel blockage with a certainty of 0.78.
The patient has 1 vessel blockage with a certainty of 0.90.
```

FIGURE 2.   Sample Consultation.

VII. CONCLUSION

The pattern recognition system described is a useful method for the classification of diverse types of medical data. It has been shown to consistently produce more accurate results than standard techniques of statistical discriminant analysis. The system easily accommodates data of diverse types, including continuous data, crisp categorical data, as well as fuzzy data. Although the method has been illustrated in conjunction with a medical application, it is generally applicable to data in any field for which classification results are useful in the decision-making process.

The methodology has been applied successfully to a number of medical applications, including diagnosis of bacterial infections,[39] determination of the extent of coronary artery disease,[41,42] detection of liver and spleen disorders,[45] analysis of chromatographic data,[43,44] and analysis of survival rates for melanoma. In all these applications, the pattern classification algorithm has produced more accurate results than the standard method of nonlinear statistical discriminant analysis. The new Cohen multidimensional orthogonal polynomials provide straightforward classification for multivariate data, as well as flexible parameters for fine-tuning of the method to specific applications. The method determines importance of parameters, which results both in a reduction of the number of variables that must be collected and in a weighting of the relative importance of each parameter.

REFERENCES

1. **Shortliffe, E., Buchanan, B., and Feigenbaum, E.,** Knowledge engineering for medical decision making: a review of computer-based clinical decision aids, *Proc., IEEE,* 69, 1207, 1979.
2. **Kulikowski, C. A.,** AI methods and systems for medical consultation, *IEEE Trans. Pattern Anal. Mach. Intelligence,* PAMI 2, 464, 1980.
3. **Winston, P.,** *Artificial Intelligence,* Addison-Wesley, Reading, MA, 1977.
4. **Weiss, S. M., Kulokowski, C., Amanel, S., and Safir, A.,** A model-based method for computer-aided medical decision making, *Artif. Intell.,* 11, 145, 1978.
5. **Szolovits, P. and Pauker, S.,** Research on a Medical Consultation, *Proc. 3rd Ill. Conf. Med. Inf. Syst.,* 1976, 299.
6. **Blois, M. S., Sherertz, D. D., Kim, H., Tuttle, M. S., Erlbaum, M., Harrison, P., and Yamashita, D.,** RECONSIDER: An Experimental Diagnostic Prompting Program, *ACP Comput. Workshop,* 7, 1983.
7. **Ben-Bassat, M., Carlson, R. W., et al.,** Pattern-based interactice diagnosis of multiple disorders: the MEDAS system, *IEEE Trans. Pattern Anal. Mach. Intell.,* PAMI 2, 149, 1980.
8. **Ben-Bassat, M., Campbell, D. B., et al.,** Evaluating multimembership classifiers: a methodology and application to the MEDAS diagnostic system, *IEEE Trans. Pattern Anal. Mach. Intell.,* PAMI 5, 225, 1983.
9. **Patrick, E. A.,** Pattern recognition in medicine, *Syst. Man Cybern. Rev.,* 6, 4, 1977.
10. **Shortliffe, E. H.,** *Computer-Based Medical Consultation,* Elsevier/North-Holland, New York, 1976.
11. **Esogbue, A. O. and Elder, R. C.,** Measurement and valuation of a fuzzy mathematical model for medical diagnosis, in *Fuzzy Sets and Systems 10,* Elsevier/North-Holland, Amsterdam, 1983, 223.
12. **Adlassnig, K. P.,** A fuzzy logical model of computer-assisted medical diagnosis, *Math. Inf. Med.,* 19(3), 141, 1980.
13. **Adlassnig, K. P.,** A survey on medical diagnosis and fuzzy subsets, in *Approximate reasoning in Decision Analysis,* Sanchez, E., Ed., North-Holland, Amsterdam, 1982, 203.
14. **Vila, M. A. and Delgado, M.,** On medical diagnosis using possibility measures, in *Fuzzy Sets and Systems 10,* Elsevier/North-Holland, Amsterdam, 1983, 211.
15. **Zadeh, L. A.,** The role of fuzzy logic in the management of uncertainty in expert systems, in *Fuzzy Sets and Systems 11,* Elsevier/North-Holland, Amsterdam, 1983, 199.
16. **Yager, R. R.,** Approximate reasoning as a basis for rule-based expert systems, *IEEE Trans. Syst., Man, Cybernet.,* SMC 14(4), 636, 1984.
17. **Patrick, E. A. and Fattu, J. M.,** Artificial intelligence and pattern recognition, in *Consult-I,* Prentice Hall, Englewood Cliffs, NJ, 1983.
18. **Kulikowski, C. A.,** Pattern recognition approach to medical diagnosis, *IEEE Trans. Syst. Sci. Cybern.,* SSC 6(3), 173, 1970.
19. **Duda, R. O. and Hart, P. E.,** *Pattern Classification and Science Analysis,* John Wiley & Sons, New York, 1973.
20. **Tou, J. T. and Gonzales, R. C.,** *Pattern Recognition Principles,* Addison-Wesley, Reading, MA, 1974.
21. **Young, T. Y. and Calvert, T. W.,** *Classification Estimation and Pattern Recognition,* Elsevier, New York, 1974.
22. **Fu, K. S. and Rosenfeld, A.,** Pattern recognition and computer vision, *Computer,* 274, October 1984.
23. **Fu, K. S.,** A step towards unification of syntactic and statistical pattern recognition, *IEEE Trans. Pattern Anal. Mach. Intell.,* PAMI 5(2), 201, 1984.
24. **Morgena, S. D.,** Information theoretic covariance complexity and its relation to pattern recognition, *IEEE Trans. Syst., Man, Cybern.,* SMC 15(5), 608, 1985.
25. **Pino, J. A., McMurray, J. E., Jurs, P. C., Lavine, K. P., and Harper, A. M.,** Application of pyrolsis/gas chromatography/pattern recognition to the detection of cystic fibrosis heterozygotes, *Anal. Chem.,* 57, 295, 1985.

26. **Evans, D. H. and Caprihan, A.,** The application of classification techniques to biomedical data with particular reference to ultrasonic doppler blood velocity waveforms, *IEEE Trans. Biomed. Eng.,* BME 32(5), 301, 1985.

27. **Belforte, G., Bona, B., and Tempo, R.,** Conditional allocation and stopping rules in Bayesian pattern recognition, *IEEE Trans. Pattern Anal. Mach. Intell.,* PAMI 8(4), 1502, 1986.

28. **Flick, T. E. and Jones, L. K.,** A combinational approach for classification of patterns with missing information and random orientation, *IEEE Trans. Pattern Anal. Mach. Intell.,* PAMI 8(4), 482, 1986.

29. **Jellum, E., Bjornson, I., Nesbakken, R., Johansson, E., and Wold, S.,** Classification of human cancer cells by means of capillary gas chromatography and pattern recognition analysis, *J. Chromatogr.,* 217, 231, 1981.

30. **Wold, S., Johansson, E., Jellum, E., Bjornson, I., and Nesbakken, R.,** Application of SIMCA multivariate data analysis to the classification of gas chromatographic profiles of human brain tissues, *Anal. Chem. Acta,* 133(1), 251, 1981.

31. **Coomans, D., Broeckaent, I., Derde, M. P., Tassin, A., Massart, D. L., and Wold, S.,** Use of microcomputer for the definition of multivariate confidence regions in medical diagnosis based on clinical laboratory profiles, *Comput. Biomed. Res.,* 17, 1, 1984.

32. **Patrick, E. A., Fattu, J. M., Blomberg, D., Patrick, E. A., Jr., and Detterman, P.,** CONSULT-I, Network of two consult subsystems: CONSULT electrolytes, and CONSULT acid base, *Proc. Symp. Comput. Appl. Med. Care,* 9, 268, 1985.

33. **Keller, J. M., Gray, M. R., and Givens, J. A.,** A fuzzy k-nearest neighbor algorithm, *IEEE Trans. Syst., Man, Cybern.,* SMC 15(4), 580, 1985.

34. **Cheng, Y. and Fu, K. S.,** Conceptional clustering in knowledge organization, *IEEE Trans. Pattern Anal. Mach. Intell.,* PAMI 7(5), 592, 1985.

35. **Gu, T. and Dubuisson, B.,** A loose-pattern process approach to clustering fuzzy data sets, *IEEE Trans. Pattern Anal. Mach. Intell.,* PAMI 7(3), 366, 1985.

36. **Selim, S. Z. and Ismail, M. A.,** On the ocaloptimality of the fuzzy isodata clustering algorithm, *IEEE Trans. Pattern Anal. Mach. Intell.,* 8(2), 284, 1986.

37. **Cohen, M. E., Hudson, D. L., Touya, J. J., and Deedwania, P. C.,** A new multidimensional polynomial approach to medical pattern recognition problems, *Medinfo,* 86, 614, 1986.

38. **Cohen, M. E. and Hudson, D. L.,** Medical decision making utilizing techniques from pattern recognition, in *Approximate Reasoning in Expert Systems,* Gupta, M. M., Kandel, A., Bandler, W., and Kiszka, J. B., Eds., Elsevier/North-Holland, Amsterdam, 1985, 435.

39. **Cohen, M. E., Hudson, D. L., and Gitlin, N.,** Pattern Classification Using a New Orthogonal Function for Recognition of SBP, *Proc. AAMSI Congr.,* 1984, 114.

40. **Hudson, D. L., Cohen, M. E., and Gitlin, N.,** Pattern classification of patients with spontaneous bacterial peritonitis using new orthogonal functions with expansions to higher dimensions, *Proc. Symp. Comput. Appl. Med. Care,* 8, 112, 1984.

41. **Cohen, M. E., Hudson, D. L., and Deedwania, P. C.,** Use of pattern recognition techniques to classify exercise testing data, *Comput. Cardiol.,* 361, 1984.

42. **Cohen, M. E., Hudson, D. L., and Deedwania, P. C.,** Pattern Recognition Analysis of Coronary Artery Disease, *Proc., AAMSI Congr.,* 1986, 262.

43. **Cohen, M. E., Hudson, D. L., Gitlin, N., Mann, L. T., and Van den Bogaerde, J.,** Pattern recognition analysis of ascitic fluid chromatograms for patients with liver disorders, *Proc., Symp. Comput. Appl. Med. Care,* 9, 263, 1985.

44. **Cohen, M. E., Hudson, D. L., Gitlin, N., Mann, L. T., Van den Bogaerde, J., and Leal, L.,** Knowledge representation and classification of chromatographic data for diagnostic medical decision making, *Artif. Intell. Appl.,* 481, 1985.

45. **Cohen, M. E., Hudson, D. L., Touya, J. J., Leal, L., Velasco, E., and Rahimian, J.,** Pattern Classification of Diseases with Liver and Spleen Involvement, *Proc., AAMSI Congr.,* 214, 1986.

46. **Cohen, M. E., Hudson, D. L., Mann, L. T., Van den Bogaerde, J., and Gitlin, N.,** Use of pattern recognition techniques to analyze chromatographic data, *J. Chromatogr.,* 384, 145, 1987.

47. **Cohen, M. E. and Hudson, D. L.,** The use of fuzzy variables in medical decision making in *Fuzzy Computing: Theory, Hardware Realization and Applications,* Gupta, M. M. and Yamakawa, T., Eds., North-Holland, Amsterdam, 1988, 273.

48. **Hudson, D. L. and Cohen, M. E.,** Approaches to management of uncertainty in an expert system, *Int. J. Intell. Syst.,* 3(1), 45, 1988.

Chapter 17

FUZZY EXPERT SYSTEMS FOR AN INTELLIGENT COMPUTER-BASED TUTOR

Lois Wright Hawkes, Sharon Derry, and Abraham Kandel

TABLE OF CONTENTS

I. INTRODUCTION

Computers have had a long association with the field of education. Some of the earliest uses were strictly in administration, but this was soon expanded to include games, simulations, and programs to provide individualized instruction. The latter are referred to as computer-aided instruction (CAI). Although many of these were little more than page-turners, there are a few notable exceptions.[18] In an attempt to make such programs *intelligent,* artificial intelligence (AI) techniques were incorporated into the designs. The resulting programs are called intelligent CAI (ICAI) or intelligent tutoring systems (ITS), with the former term referring to whole courses and the latter to the instruction or practice of a particular skill. Such systems typically have three main components:

1.   The problem-solving expertise, or the knowledge the system is to convey to the student
2.   The student model which houses the system knowledge concerning the student
3.   The tutoring strategies

Even though most of the main subdivisions of AI must be used to develop such systems, ITSs are still distinct from other AI applications. The reason for this is best described by Clancey[3] when he states that ITSs

    . . . must integrate these [AI techniques] to
— understand what another problem solver is doing by watching him perform a task (or looking at the results of his reasoning);
— in particular, recognize or simulate human problem solving;
— actively and systematically articulate problem-solving methods, not just passively respond to requests for information;
— in particular, explain reasoning such that it is learned and mimicked, not just accepted as a justification.
Of these criteria, the problem of interpreting a trace of human reasoning is what most directly distinguishes this research from other areas of AI. Student-modeling programs must deal with noisy data, shifting focus of attention, and goals or reasoning strategies that may be foreign to an ideal model of the world.

The inherent fuzziness of this task is obvious, however; to date, no other system but the one described here has incorporated fuzzy expert systems into the design. In the following we describe an ITS for training arithmetic problem-solving skills (TAPS), a system initially targeted to fourth to seventh grade children, but which is also applicable to adolescents and adults by simply changing the noun and verb lists of the problem templates. After giving an overview of TAPS and an outline of the other system components, the two fuzzy expert systems in the design will be described in detail, illustrating a powerful and unique application of this technology.

II. OVERVIEW OF GOALS OF TAPS

TAPS, a curriculum centered around an intelligent tutoring system, is designed to substantially speed development of *expert* performance on complex verbal arithmetic word problems. Although work currently is focused on grades four through seven, we note that the problem-solving training provided by the system is appropriate also for older students and that the level of problem difficulty can easily be varied. Such problems frequently are encountered on tests that people take for educational and career advancement, and also represent a broad range of realistic daily situations in which people encounter basic mathematics. However, the most recent Presidential report on education (*A Nation in Crisis*) reveals that many students leave school without verbal problem-solving skills. Also, studies in our laboratory indicate that even typical college students can experience great difficulty with arithmetic word problems. Lack of this basic problem-solving ability also is evident within our military services.[8] (See *New York Times,* July, 1983.) Clearly we are dealing with a significant educational issue.

Our work is attacking this educational problem on several fronts. First, through the application of artificial-intelligence technologies to microcomputer-based instruction, we aim to achieve a feasible means of offering quality one-on-one tutoring in problem solving to a large number of students. Individualized human tutoring is regarded by researchers as the most powerful teaching method available; its substantial impact on student performance is well documented.[2] Unfortunately, the solution of offering each child an intelligent, trained human tutor is too costly for society to bear. Therefore, to the extent that artificial-intelligence methods can be applied to simulate important features of human tutorial interaction, substantial impact on performance should be realized through an affordable microcomputer technology.

Moreover, to the extent that computer tutors can be designed to surpass their human counterparts on certain critical variables known to affect performance, an even greater impact may be possible. As with most ITS projects, a major part of our development effort is a cognitively-based program of research, which aims to achieve detailed scientific understanding of the problem-solving process itself and how it evolves through tutorial interaction. Much of this knowledge is being programmed into the TAPS system, along with the capability of using this knowledge in designing instructional interventions. Few human tutors would understand students and learning situations at the detailed level of analysis being programmed into this system.

Our approach is based on the identification of sets in problems and their interrelationships. These sets are grouped into relational schemas based on the relation that holds between them. For example, the following problem contains a 1-combine schema:

Rob has 3 marbles. Lisa has 2 marbles. How many do they have altogether?

The other types of schemas we are working with are *compare, change,* and *vary.* More complex problems are formed by chaining single schemas together (see Figures 3 and 4).

Basically the system works as follows. It plans a lesson based on its extensive knowledge of the student and the goals it has for him. A set of initial problem templates are chosen and instantiated. The student begins to solve the problems and, if assistance is deemed necessary, a tutoring strategy is determined based on knowledge of the student and the planner's chosen approach. After each session the inference engine portion of the student record updates the data on the student. The system currently runs on a Xerox 1186 LISP machine.

In the following we briefly discuss the planners, the interface, the problem generator, and the expert solutions module in order to provide the requisite background for the more detailed discussion of the student record and the intervention module (tutor) which are both represented as fuzzy expert systems (FES). Next, the necessary cooperation between these two FESs is described. The conclusions from this research are given in the last section.

III. PLAN-BASED ARCHITECTURE

A. PLAN-BASED VS. OPPORTUNISTIC SYSTEMS

The tutor is best characterized as a plan-based opportunistic system that provides practice and tutoring on the skills of expert performance in the arithmetic word-problem domain. In a recent extensive review of intelligent tutoring systems, Wenger[27] distinguished plan-based tutoring architectures from opportunistic teaching systems, noting that almost all extant intelligent tutors are based on opportunistic designs.[27] The basic philosophy of opportunistic tutoring is learn-by-doing.[19,23] Thus, all didactic intervention is triggered by circumstances that occur while a student is problem solving.

Despite the power and appeal of opportunistic tutoring, the ITS movement can be criticized for its lack of attention to such matters as instructional design theory and curriculum structure. Park, Perez, and Seidel[19] note that developers of intelligent tutors have largely ignored task analysis, which attempts to identify instructional objectives and the pedagogically relevant relations among them. Peachey and McCalla[20] argue that a serious weakness of most intelligent tutors is their lack of a global curriculum plan. Wenger[27] calls for development of systems that constrain the range of student experiences with thoughtful curriculum planning. Clearly there is need for tutoring architectures that effectively combine the strengths of both instructional strategies, and this chapter addresses the issue.

An early step in designing our tutor was to construct a model of the expert's knowledge, which delineates the knowledge components, or capabilities, that presumably are engaged during expert performance. Our model also describes the hierarchical interactions among model components, since some higher-order systems in the model are dependent upon prerequisite learning of other lower-order model components. An abbreviated version of the expert knowledge model resulting from our studies has been provided as Figure 1.

Given this expert model, individual student models are expressed in terms of the knowledge components of the expert model. Assessment procedures have been developed to help determine which of the expert components are either missing from a student's model, or should be represented as being weak, or *buggy*. Common misconceptions may be associated with the domain, and if these can be identified through research, misconceptions also may appear in some student models.

The aim of tutoring is conceived as the task of operating on the student knowledge model until it evolves into the expert knowledge model. The system tries to accomplish this by building an individualized plan with the instructional goals of adding missing knowledge components to the student model, strengthening existing ones, correcting buggy ones, or eliminating troublesome misconceptions. The structure of the expert knowledge model constrains the order in which these goals must be taught, since it specifies which lower-level components are required as prerequisites for higher-level learning and performance.

Observational research by Leinhardt and Greeno[14] has shown that experienced master teachers employ hierarchical planning in accomplishing their goals. At the most global level, the teacher is guided by a master operational plan called an *agenda*. This agenda assembles and organizes goals and operators, and can be tuned and revised during instruction. At a less global level, master teachers also plan lessons. These are characterized in general terms as *action schemata*, which use various lower-order *routines*. Routines are *small, socially scripted pieces of behavior that are known by both teachers and students,* and that are shared by many classroom activities.

We have designed our tutor to carry out three levels of instructional activity: planning an individual's route through a curriculum (the agenda), planning lessons (using action schemata and routines), and on-line tutorial intervention. At each of these instructional levels, data pertaining to student performance can be collected and made available for use by other levels.

B. CURRICULUM PLANNER

For each student, curriculum planning begins by establishing that student's current achievement goals. To determine these the student must be tested on skills in the knowledge model, beginning at the lowest prerequisite level. On each assessed skill, the student's level of expertise is classified as ranging from nonmaster (cannot perform skill), through novice master (performs accurately but slowly), to expert (performs rapidly and accurately). If expertise is attained on all skills at one horizontal level in the hierarchy, the testing continues

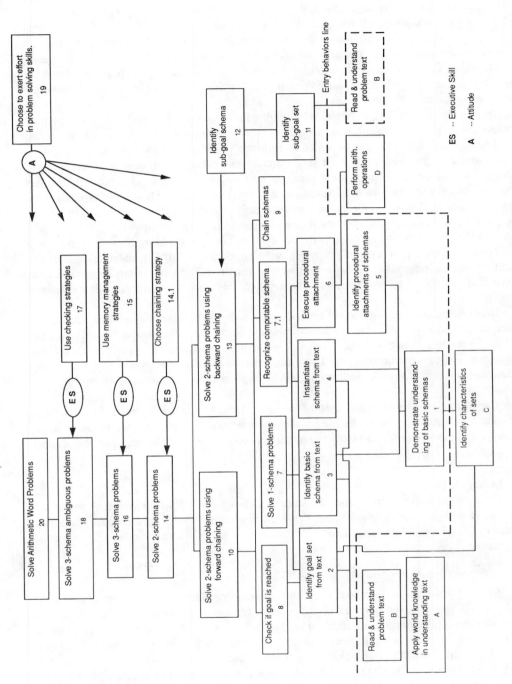

FIGURE 1. TAPS instructional hierarchy.

at the next horizontal level and continues until a level is reached where deficiencies are found.

Setting achievement goals is, at first, conceptually simple. At horizontal level 1 (Figure 1 is an abbreviated hierarchy, with skill box 1 actually representing a class of skills), the tutor sets a performance goal for each skill that is an increment above the student's current level. For example, if the student is performing at the level of novice mastery on the skill *understand change schema*, the tutor will set a goal for the student to reach an expert level on that skill. This goal-setting heuristic can be stated as follows: Curriculum Planning Rule 1 — for any assessed skill, the achievement goal for that skill is set at one level above the currently assessed achievement level.

In addition to setting goals for all eligible skills within a horizontal level, a planner should also set all possible instructional goals for moving vertically from one horizontal level to another. This idea can be stated as follows: Curriculum Planning Rules 2 and 3 — a performance goal can be set for any nonassessed higher-level skill that represents a combination of assessed lower-level prerequisite competencies, provided that all involved prerequisite competencies have been mastered at the novice level. The performance goal for the combined skill is set at the maximum possible achievement level that does not exceed the lowest performance goal attained for any single prerequisite competency. For example, if a student has achieved novice competency on skill 4.1 (instantiating change-schema problems) and an expert level of competency on prerequisite skill 6.1 (calculational attachments to change schemas), the achievement goal for the next-higher competency that combines these skills (e.g., instantiating change problems and attaching procedures) cannot exceed novice mastery.

Goal setting becomes more complicated as the student progresses through the curriculum, since the student will soon be exhibiting multiple levels of performance in different parts of the hierarchy. However, the original goal-setting heuristics can continue to manage movement through the curriculum. These heuristics constrain one another in a convenient manner. The vertical rules cannot set a goal too high because they are constrained by Rule 1.

The current achievement goals set by the curriculum planner are recorded in the student's permanent record and remain eligible for that student until they are changed by the planner. Goals are updated after each instructional session or if a lengthy time lag expires between sessions.

C. LESSON PLANNER

The lesson planner selects from the student's currently eligible achievement goals, and plans a tutorial (approximately 40 min) designed to accomplish that lesson goal. The lesson planner can call upon various tutorial routines, which have the capability of teaching these routines to the student.

D. SHOW ME ROUTINE

We have designed our tutor to permit implementation and testing of different instructional routines. To illustrate, the routine called Show Me has been developed to have various skill level variations. The purpose of Show Me is to reach the schema-chaining skill. The routine also promotes awareness of problem structure — a problem's underlying schemas and how they are joined through common sets. The routine uses a worked examples approach that has proved effective, whereby the student studies and models an expert solution.[6,25] A system expert illustrates for the student how to solve a word problem by constructing a diagrammatic tree representation. This follows what Anderson et al.[1] found concerning geometry proofs, i.e., "The graphical structure . . . seems to be the key to enabling students to understand the structure of a proof . . . ." The complete expert solution has the same structure as the partial one shown in Figures 3 and 4.

Following the expert's demonstration, the student is then presented with a different problem and asked to model the expert's performance by using the system's problem-solving tools provided by the interface. A partial student solution for *the apartment problem* is shown in Figure 3. In Show Me, the tools used to build a solution include a set of problem schema boxes (see Figure 2) that can be dragged by the student to the workspace and a set of labels (see Figure 3) and operators generated by the system's expert which the student may use if she chooses. If the tutor deems it necessary, tutorial assistance, individualized to the particular student, may be given (see Figure 4, for example). Depending upon the exact instructional goal of the exercise and the version of the drill being offered, the student may then be asked to compare her solution to an alternative expert solution.

Other decisions made by the lesson planner also determine the particular version of Show Me that will be implemented. Different versions are created, depending upon whether the system is intending to promote novice mastery or expertise. In general, when novice mastery is the goal, students receive easier problems that are isomorphic or very similar to those demonstrated by the expert. Also, the lesson provides a great deal of *scaffolding*[4,17] in the sense that more graphic reification is supplied during problem solving. In addition, the lesson planner informs the intervening monitor that the student is a novice, as the monitor also varies its coaching strategy accordingly. However, when the goal is development of expertise, there is less reliance on graphic scaffolding. Also, the coach is more withdrawn, to encourage more discovery learning and self-checking. Moreover, students are given more difficult problems that are structurally different from those demonstrated by the expert, for the purpose of promoting transfer. Also, the tutor may request that the student study and develop alternative solutions and strategies for the same problem.

Once the lesson is planned, the tutor presents a problem representing a middle level of difficulty, based on the range in the problem set that has been selected for that student. (Thus, the lesson can be modified simply by increasing or decreasing the difficulty of the next problem.) Then, control is passed to the intervening monitor, designed to observe the student's performance and make specific tutorial interventions.

The many instances where the planners' knowledge of the student is uncertain and fuzzy is obvious. Details of this are discussed in Sections VIII. and IX. which deal with the student record, which stores and manipulates such knowledge, and the tutor, which seeks to implement and assist with the plans.

IV. THE INTERFACE

A typical screen layout for the Show Me routine is shown in Figures 2, 3, and 4. Although the system is designed to be used with any age audience, the primary user of this system will be in the fourth to seventh grade age group. Thus, a text-based interface would be cumbersome, so we chose to use menus, windows, icons, graphics, and a mouse, supported by the observation that humans acquire graphic information at a substantially higher rate than text-based information.[21] The student can click on sets, schema icons, and operators to instantiate schemas. Anderson et al.[1] found that this approach decreases significantly the number of typographical errors.

Another observation concerning the interface[4,27] is that reification can be a very effective tool for tutoring. Graphics, especially the sophisticated version available on the Xerox LISP machine we are using, make the implementation of reification for our domain very effective. In the Show Me routine, various interface tools are used by the expert and the student to build problem solutions as trees.

Menu selections are used in the Help window to avoid time-consuming and complex natural language understanding. However, we allow the student to type in a question if

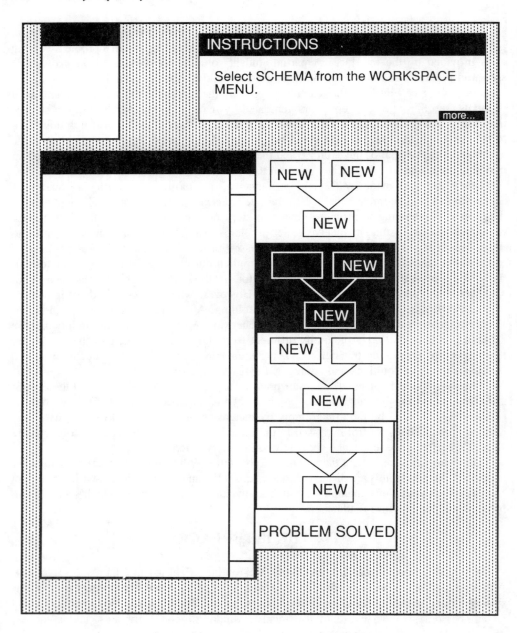

FIGURE 2.   Show me window — schema building boxes.

nothing on the menu captures his or her uncertainty. In the manner of Lesgold,[13] this will be saved and analyzed later and, if necessary, an entry added to the menu. Also, the format of how the messages will be presented will vary with the age of the student. For example, a fourth grade student may be more motivated by a familiar cartoon character giving a *bubble-style* message than a rectangular window-type message.

245

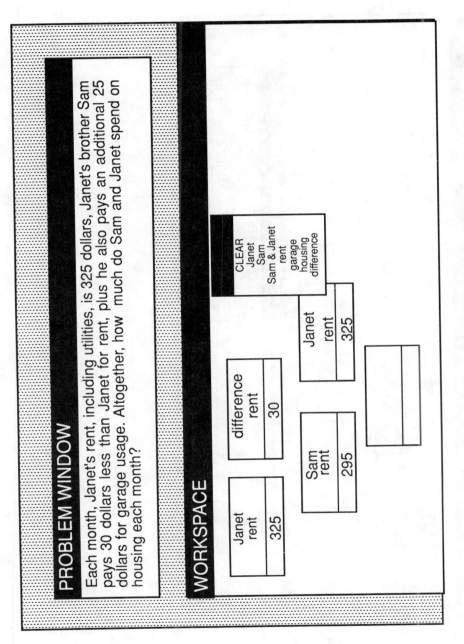

FIGURE 3. Show me window — label selection.

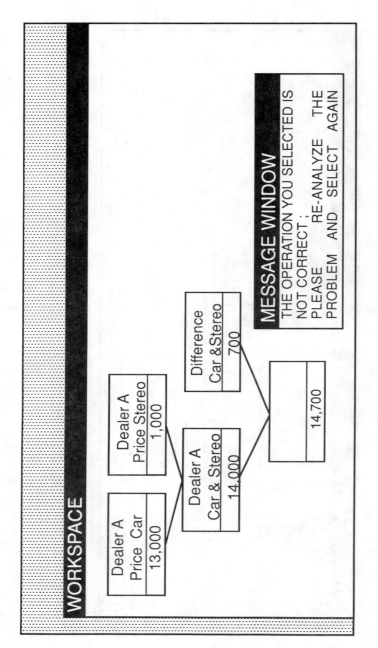

FIGURE 4.   Show me window — tutoring message.

V. PROBLEM GENERATOR

This implementation consists of templates to represent the large number of word problems that are needed for tutoring. Templates include semantic and syntactic parses, schemas used by the system, and other information about the problem set that can be represented in this format. Using instantiated templates and a system dictionary, the system is able to generate large numbers of problems. This approach will significantly increase the speed of the final system as well as decrease the amount of storage required. Problems can individualize the system by reflecting the student's interests and preferences. This should prove helpful in generating problems interesting to particular student groups, such as females, that are not highly motivated to do mathematics.

VI. EXPERT SOLUTIONS MODULE

The solutions module contains the major problem-solving expertise of the system. It constructs a tree of the solution or solutions (many problems have several solutions) from the semantic information and schemas passed to it from the problem generator. The tutor refers to this tree both as a guide in identifying where the student is in his progress and where he should be. The expertise involved in this module consists of the ability to identify how the various schemas or subtrees are interrelated. Interrelationship is established through sets which are used in more than one schema. Each schema consists of a set of nodes which can be represented as a subtree. If the root node of the schema subtree is not the goal of the problem, then it must be a leaf node of some higher order (in the tree) schema. Figure 5 shows the tree of solutions for the *car problem*.

The expert module is based on an object-oriented approach and is implemented using a hierarchical layering technique with the following levels: general problem-solving level, tree management and pattern matching level, application domain level, and data structure level. Key word paths are used to traverse the hierarchy. The intervening monitor works closely with the expert and the detailed tree it produces. It is the detailed information stored here which allows the tutoring of the student.

VII. STUDENT RECORD

A. FUZZY TEMPORAL RELATIONAL DATA BASE

The student record (SR) in this ITS is the student knowledge model, i.e., it represents the knowledge the system has about the student. In extant systems,[27] this information is generic in that only information on general student types is stored. In the TAPS system a holistic view of each individual student is stored, allowing the tutor to be highly individualized. This is accomplished through the use of a fuzzy temporal relational data base (FTRDB). This approach to knowledge modeling differs from its predecessors in several significant ways.

First, a relational representation is employed. The advantages of the relational model over other common data base representation techniques have been described by Tanimoto,[26] Rundensteiner,[22] and others. Unlike semantic nets, for example, when the size of a relational data base (RDB) increases, retrieval time does not increase, a significant issue in maintaining response times. Also, the RDB model is associated with a set of well-defined operators that permit extremely flexible data manipulation and facilitate both high-level retrieval and development of friendly query languages.

Second, we have incorporated into our RDB an extension that allows for each student's model to contain fuzzy data entries. This permits the representation of incomplete, imprecise, and vague information about the student, as well as precise data. An example of an imprecise

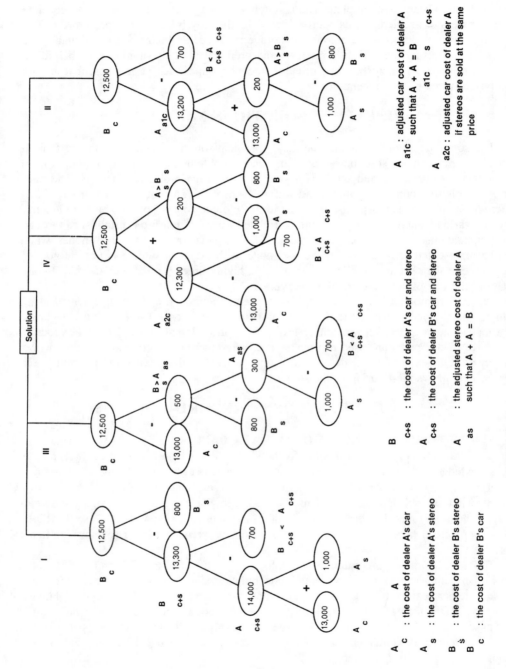

FIGURE 5. Expert solutions module solution tree for car problem.

NAME	GR LEVEL	INIT AGE	APTITUDE	GOALS
Elke	(3,4,5)*	8.2	(0.4/novice, 0.7/nov exp)**	(4,7,8)
Frank	(6,7)	10.3	(0.9/expert)	(9,10.1)

FIGURE 6. Possible domain values in a FRDB relation student. *Represents uncertainty; **represents a range of uncertainty.

entry would be a notation indicating that the student *often*, but not *always*, performs as an expert on two-schema problems (somewhat of an expert). An example of a vague entry resulting from incomplete data would be a notation indicating that the student is *probably* an older student. Although such information, as obtained, is often inherently inexact and vague, it can provide valuable guidance for global planning decisions, such as the selection of lesson goals and problems.

Our final enhancement to the fuzzy relational data base (FRDB) is a temporal extension, which addresses the issue of how to store information over time. This is accomplished by maintaining multiple tuples within each student relation, each tuple representing a holistic student evaluation at a different point in time. Historical data permits us to evaluate the student's development over the sequence of his sessions on an ITS. This is necessary for effective tutoring at the global level, because the student does not always have a consistent performance and a trend of behavior has to be developed over several sessions. A poor performance may represent a bad day and, thus, should not unduly bias selection of future instructional goals and problems. Storing information over time improves the global performance of the system as it enhances the system's understanding of the student and his behavior. It also makes a useful research tool, since it allows for comparison and analysis of student's performance changes. Thus, the actual effectiveness of the system as a tutor can be verified.

B. TYPES OF DATA

A heterogeneous data representation with weakly typed data domains is used. The various types of values that can be assigned to an element are:

1. Domain attribute not applicable, i.e., null value
2. Linguistic value
3. Known interval
4. Member of a discrete set of values (all equally possible)
5. Value somewhat ambiguous, as only partial information known

The relation in Figure 6 demonstrates some of these choices. A review of existing ITSs reveals that, to date, the binary approach has been used to indicate whether a student does or does not have a particular characteristic. The advantages of a nonbinary representation are obvious. Consequently, the choice of a fuzzy approach is both a positive and necessary enhancement for the student record.

An analysis of the kinds of information necessary in a SR supports the preceding five categories; however, no other existing ITS has this ability. The following examples illustrate typical items that may be stored. A teacher may want to store the information that a particular student is *very good* in type X problems but only *average* in type Y problems. Or, a teacher using the tutoring system might request a list of all first period students who are performing *very well* on two-schema problems but *below average* on three-schema problems. Also, our SR houses fuzzy linguistic information on the student's typical motivation level. It has been determined[12] that motivation is a significant component in a student's success or lack of it in word problems. Knowledge of the student's motivational level (from teacher input and system observation and inference) is used to determine the extent and type of tutoring offered.

By incorporating temporal knowledge in our SR, the tutor can detect trends in the student as well as inconsistencies in performance. Knowledge of the student's level of expertise for solving particular problem types is used to individualize the tutoring. We are also incorporating information on executive level skills. For example, we wish to promote such executive level skills as self-monitoring and good memory management. These are a few specific examples of the types of information that can be stored in the system. In addition, the design of the SR allows the researcher to augment these with other student characteristics that are determined to be relevant by using linguistic codes, ranges, fuzzy possibilities, and incomplete data.

C. STUDENT RECORD FUZZY EXPERT SYSTEM

Even though extant ITSs are designed to incorporate intelligence into the tutoring portion of the system, they have neglected to do so in the SR component of the system. In contrast, a human tutor, as well as making informed tutorial decisions, also is continually analyzing and making inferences concerning his or her uncertain or fuzzy knowledge about the student. The purpose of our SR is to have a similar capability. By designing the FTRDB to work with an inference engine we endow the student model with *intelligence* in the sense that it can evaluate and update the SR.

In the TAPS system after each session, the SR monitor evaluates the progress of the student in much the same way a teacher would in order to update her records on this student. Moreover, just as any decisions the human teacher would make must be based on uncertain information and, hence, be fuzzy, so also are the decisions of the SR expert system. The decisions as to progress made by the student are based on the current SR, information made available by the tutor FES, and patterns observed from analyses made on students and their solution protocols. The latter provides the *fuzzy* component in the production rule representation of the inference engine. The typical rule has the following structure (i.e., is a fuzzy relation between two fuzzy propositions):

If char **is** [fuzzy____adj] **and** observed____pattern **is** [fuzzy____pattern] **then** char **is** [fuzzy____adj].

For example, consider the case where the expert system must analyze the student's capability after a session on two-schema problems with one combine-vary, when the SR indicated that initially the student was *very poor* on these two-schema problems. The tutor indicates that the student has completed three of the ten problems with no assistance, two with a suggestion to *check your work,* two with mislabeling hints, one with *operator-incorrect* message, one with *bad-label* message (i.e., made explicit vs. *hint* above), and one unable to complete without tutor assistance. These are generic descriptions of the messages; each message is actually individualized to the particular student. The observed pattern is referred to as 3-c; 2-g; 2-h; 1-o-e; 1-bl-e; 1-w where:

c: correct, no assistance	g: general assistance
h: specific hint	o: operator
e: explicit	bl: bad label
w: wrong	

and where a hint is directed at the error but is not made *explicit*. For example, a hint for one mislabeled node in a schema would be: "Rob, before proceeding, review the last schema you built." In contrast, an explicit bad-label message is much more intrusive and results from the student's performance indicating she is having *significant difficulty* at this point (the tutor makes these decisions). In this case an applicable rule would be:

If 2-schema **is** very___poor **and** observed___pattern **is**
[(3,5)c; (2,4) g/h; (1,3) e; (1,2) w]
then 2-schema is weak___good,

where c + g/h + e + w = 10, i.e., the analysis is based on sets of 10 problems.

One argument against making the SR *intelligent* has been that the response time for ITSs must be realistic, if not real-time. However, designed in this way, the information needed by the tutor is read in and accessed by the tutor initially, and the analysis by the SR monitor is done after the student completes the session. Consequently, no overhead is added to the response time, yet the tutor gains significantly in the sensitivity and accuracy of the knowledge it has on each student.

Of interest to teachers and researchers are certain cumulative scores and averages. The SR monitor can easily calculate these in an off-line manner also.

Updating and statistical analyses are performed based on data currently in the SR and information already in the system. A second type of updating involves inferences made in a forward-chaining manner by the SR FES. Although considerable information is made available to the system through pretests and teacher evaluations, there are many instances where, for various reasons, such information is unavailable. The FES must then make fuzzy entries for as many of these as possible using fuzzy inference rules, known data and results provided by the tutor. For example, if the student has not been asked to solve three-schema problems previously, there will be no entry for this in the SR. A typical rule follows for such knowledge:

If 3-schema **is** null **and** observed___pattern **is**
[(0 − 1)c; (0 − 2) g/h: (3 − 5) e; (2 − 4) w]
then 3 − schema is rather___poor,

where c + g/h + e + w = 10.

Other information is supplied by the tutor based on error-analysis inferences made by the tutor. Such information is passed on directly to the SR manager for updating purposes. This is discussed in more detail in Section IX.

The general format of the rules are as given previously. Specific rules may have the [fuzzy___adj] and [fuzzy___pattern] portions modified with experience.

The power of such an implementation to enhance an ITS is obvious.

VIII. INTERVENTION MODULE — THE TUTOR

This module performs several complex tasks all geared toward providing individualized and intelligent guidance to the student. Initially, the intervening monitor obtains input from the lesson planner concerning the routine and variations that have been chosen for the lesson plan. There are specific forms of intervention associated directly with specific routines. Also, the monitor obtains information regarding the student's expected level of achievement for the target skill in order to set an intervention-frequency level. For example, if the student is working toward *expert novice* mastery, a higher intervention frequency is set than if the student is working toward *expert*. The tutor also has access to any other information from the SR necessary for this session, such as motivation for this problem type, previous success record, language level (i.e., choice of vocabulary) of prompts, *most frequent* error-types for this student, etc. Of course, the tutor is able to modify these levels if it decides the current choice is inappropriate.

A. LOWER LEVEL ERRORS

The monitor observes the user's progress in performing the lesson's problem-solving tasks by comparing student responses (using fuzzy production rules) to a tree of all acceptable solution paths generated by the solutions module. By comparing the student's solution moves to those constituting a solutions tree, the monitor is able to obtain on-line observations of such behaviors as how long the student takes to accomplish various parts of the task (a time measure), how frequently the student alters her response (a certainty measure), as well as correctness in schema choices, operands, operators, path, and memory-management decisions. The specific items of interest vary, depending upon the current performance goal and routine in use. Some of the matching to the tree is direct, for example, determining if the student schema values are identical to those of the tree. However, the decision as to the *correctness* of the schema labels is a fuzzy decision and is implemented through fuzzy production rules. This approach is consistent with our observations of student protocols that mislabeling is a significant cause of difficulty in problem solving. A particular node in a schema may have several correct as well as incorrect labels, some being *more correct* than others (see Appendix A, Part 2). Next we discuss other errors.

In developing this module, we conducted studies of the problem-solving errors and developed error categories and their possible causes (see Appendix A). Using these error categories and knowledge of the problem setting in which they occurred, the tutoring of these lower level errors can be implemented in a rather straightforward manner using fuzzy production rules. For example, the tutor has knowledge of the student's past error patterns, ability level for this problem type, and interruption history, and uses these to select a tutoring response through matching a fuzzy production rule. A typical such rule is (in English):

If student *often* selects wrong__operator **and** is rather__weak on combine__problems **and** student has not been interrupted recently
then suggest [closer__look__at__schema].

The actual wording of the instantiated suggestion will vary with the student, her reading level, and other prompts given during this session.

Tutoring these low level errors is done quite simply with knowledge from the SR and the solutions module. For example, a *bad value assignment* (see Appendix A) is easily observed from the tree. However, the response will vary with the student and the tutoring strategy for this student. If on the one hand, the student is a *very good* one, then this is probably simply a slip rather than a conceptual problem. If the tutor does not interrupt, then this is an opportunity for self-monitoring to come into play. If the student does not detect the error, at the end of the solution process the tutor can highlight this subtree and suggest, "Is this step what you intended?" and allow the student to redo the solution. On the other hand, if the student is a *very poor* one, then the tutoring strategy could be set at a much more intrusive level, for example, "Can Car A stereo plus Car B stereo produce a useful set?" The decision as to what message to give and in what format is the result of executing the most appropriate fuzzy production rule. The inference engine is forward chaining and may adjust its own knowledge depending on the response of the student to these prompts. Counts of the various error types are maintained to be used in matching the fuzzy production rules.

B. EXECUTIVE SKILLS ERROR MODELS

The executive skills (e.g., poor memory manager, no self-monitoring) error models are more esoteric and require a more sophisticated approach. It relies on the SR and also on previous psychological analyses of signs of the absence of certain executive level skills.

Such signs include time between actions, uncertainty, and accuracy of solution. These, when represented in various weighted combinations, indicate the absence or confusion of different executive skills. For example, poor memory management would include longtime spaces when trying to recall an unsaved value (i.e., a large value in the fuzzy [0,1] range), a significant degree of uncertainty in trying to recall a value (for example, selecting a value and then erasing it repeatedly). These combinations give a value in the range [0,1]. Each error model will have a cut-off level. The level itself is variable and is a function of the nature of the problem being solved, how long the student has worked on the problem, how close the student has come to a solution, and the history of interventions for the current problem and session. If the calculated value exceeds this level, we can assume that this error model is representative of the student's difficulties, and the tutor then applies the corresponding tutoring strategy (previously determined to be the most efficacious for this error model). If no conclusion can be made, then more monitoring is done and the values recalculated.

Some error models have been determined through analysis;[7] we expect to fine-tune these as well as add more. Currently, we are implementing the corresponding production rules with entries which are fuzzy intervals. These are more representative of the uncertainty inherent in executive skills. New error models can be added or inappropriate ones removed simply, as necessary with the object-oriented approach used for the overall design.

The tutoring module provides a novel application for FESs. The significant degree of uncertainty involved in modeling the human mind makes the fuzzy approach not only appropriate but indispensable.

IX. COOPERATION BETWEEN SR AND TUTOR FESs

It is obvious from the description of the SR expert system and the tutor that each depends significantly on the other. In order for the system to achieve its twin goals of individualization and intelligence in tutoring, the SR must feed the tutor its impression of the student and the tutor must return its knowledge of the student's progress to the SR expert system for evaluation and update. For efficient and real-time execution, these two subsystems must not duplicate their activities. By using a node labeling procedure on the solutions module tree, a succinct graph of nodes visited by the student together with tutoring messages can be passed from the tutor to SR for processing after the session. The SR FES can then do its analysis and updating off-line.

X. CONCLUSIONS

As expert systems move out of the strictly technical and scientific areas into more humanistic fields, the inherent uncertainty in measuring the human mind demands the use of fuzzy techniques. We have demonstrated the efficacy of the fuzzy expert system in the setting of an intelligent tutoring system for arithmetic word problems. As pointed out by Clancey,[3] these systems must constantly handle "noisy data, shifting focus of attention, and goals or reasoning strategies that may be foreign to an ideal model of the world." To deal with these complexities, we feel that future systems have no choice but to incorporate fuzzy expert systems into their designs in order to achieve individualized and intelligent tutoring. We propose that the TAPS system is a first step in that direction.

Appendix

DEFINITIONS OF ERROR CATEGORIES AND STUDENT MOVES

1. BAD VALUE ASSIGNMENT (FOUR TYPES)

Value for set required in solution is incorrect in otherwise correctly labeled scheme.

Transfer error — Set value is given in problem and entered into solution, but subject makes incorrect transfer into problem solution. This indicates carelessness, lack of self-monitoring, disorganization — general self-monitoring deficit.

Inference error — Set value is not given but can be inferred or easily calculated based on problem information. World knowledge may be required to infer appropriate value. However, value is determined through mental calculation or inference is incorrect and is entered into solution.

Propagation error — Set value entered into solution has been obtained from previous bad schema-based operation (e.g., bad schema, bad calculation).

Others — All others not listed previously are included here.

2. BAD LABEL

Schema is correct and present in solution path, but one set is imperfectly or imprecisely labeled, and this leads to problems.

Imprecise label — For example, instead of labeling a set as "cost of apartment for one month," the set may be labeled "cost of apartment". Self-monitoring skills are possibly implicated. This may indicate lack of understanding.

Other — Any labeling problem not covered previously is included here.

3. BAD SCHEMA (FIVE TYPES)

The student constructs a schema that is not on current solution path. Any schema is bad if it has at least one inaccurately labeled set.

Bad subgoal — The student sets subgoal to compute a set that is not on a current solution path. A *correct* schema may be formed for the unneeded set. Plan or problem-solving strategy may be implicated, as the student has developed a bad goal. The student may not understand the problem.

Irrelevant data — A schema not in solution is constructed because the student has failed to eliminate irrelevant problem information. This may result in extra schema, a hybrid schema that combines irrelevant with valid solution sets, or in the substituting of a wrong for a correct schema. Inability to recognize and eliminate information irrelevant to goal-based plan (planning skill) is implicated. If the student is working forward, he may spontaneously eliminate unnecessary schema at a later point.

Good subgoal — The student clearly has set a good subgoal that is on current solution path and is attempting to calculate that subgoal, but is combining wrong sets for calculating it. This error code may be combined with the type in Part 2, "Other." Schema knowledge may be implicated — the student does not possess a schema for calculating desired value, or the student might be distracted by irrelevant problem information or just being careless.

Strategy confusion — The student is on one solution path, but then *borrows* schema that is not on a current solution path but is correct for another one. The student may be changing strategies, or this may be a confusion error. This code is used only when there is confusion between two strategies. The student is having difficulty in developing problem representation and is being distracted by valid alternative schemas that are not on current path. This might be due to a reading error that causes problem to be comprehended incorrectly, or it may indicate that the student is having difficulty with problem-solving strategy.

Other — All schema errors that do not fit clearly in categories described previously are included here.

4. ARITHMETIC ERRORS
Calculation error — The correct schema and operation are indicated but the student makes an arithmetic error while computing unknown set value. The error may be a decimal error. The error is typically propagated through entire solution. Self-checking and memory-management skills possibly are implicated. In younger children, there is a possibility of an arithmetic bug.

Wrong attachment — Correct schema has been set up but a wrong operator is applied.

5. OMIT ERRORS
Omit step (schema) — The student is pursuing the correct solution strategy but one schema is omitted. Self-monitoring skills are possibly implicated. Planning skills are implicated in the sense that the subgoal has been left out.

Omit branch — The student omits an entire branch in the problem-solving strategy. This indicates serious comprehension failure and lack of comprehension monitoring. The student has failed to analyze the problem in depth, possibly indicating a lack of effort. The student may sense the answer is wrong.

6. ISOLATED BITS
Non-schema set — Names (writes down) or calculates set but is unable to relate it to other sets. May indicate lack or inaccessibility of schema knowledge.

Non-strategy schema — Names or calculates schema, but is unable to connect it to a strategy. Indicates lack of plan and planning ability.

7. HESITANCY
Long idle period (10 secs) with no action. If student successful, indicates lack of automaticity. If student not successful, may also indicate lack of understanding due to failure of problem-solving strategy.

8. WILD GOOSE CHASE
Three or more bad schemas connected together as solution strategy.

9. UNUSUAL PATH
Student pursuing highly unusual solution path, one that few people pursue and thus that represents an unusual way of seeing problem. If poor student, this is probably signal of difficulty, lacks understanding, etc.

10. GIVES UP
Student decides to quit or make wild stab at answer without completing solution strategy. If abandoning good strategy, may indicate lack of motivation, self-confidence, effort. If abandoning bad strategy, may indicate reality that cannot obtain answer in reasonable time.

Promising effort — Gives up before finishing good strategy.

Unpromising effort — Abandons unpromising effort.

11. BAD GOAL (2 SUBTYPES)
Computes set value and declares it as the answer when it is not.

Premature goal — On correct path, but does not completely finish solution. May implicate lack of self-checking and comprehension-monitoring skills, or ambiguity in problem goal statement. Always implies steps missisng at end.

Bad value — Valid goal set from valid strategy, but value incorrect, perhaps propagated through solution tree.

Bad strategy — Goal derived from bad strategy.

12. NEW START

Student begins new schema that is not immediately connected to any tree. Many new starts with few joins (below) may indicate lack of planning, skill, trial-and-error problem solving.

Same tree — Moves to new part of same solution tree.

Good shift — Abandons unpromising effort, starts anew.

Unnecessary shift — Abandons promising effort.

13. JOINS (COUNT)

Number of times student connects new start to another. Interpret in conjunction with start count, above.

14. TIME EXPIRES

15. NO EVIDENCE OF CHECKING OR ON-LINE MONITORING

Refuses prompted check.

Conducts unsuccessful checks.

Other

REFERENCES

1. **Anderson, J. R., Boyle, C. F., Corbett, A., and Lewis, M.,** Cognitive modelling and intelligent tutoring, Technical Report No. ONR-86-1, Carnegie-Mellon University, Pittsburgh, PA, 1986.
2. **Bloom, B. S.,** The 2 sigma problem: the search for methods of group instruction as effect as one-to-one tutoring, *Educ. Res.*, 13, 4, 1984.
3. **Clancey, W. J.,** Qualitative student models, *Annu. Rev. Comput. Sci.*, 1, 1986.
4. **Collins, A. and Brown, J. S.,** The computer as a tool for learning through reflection, in *Learning Issues for Intelligent Tutoring Systems,* Mandl, H. and Lesgold, A., Eds., Springer, New York, in press.
5. **Collins, A., Brown, J. S., and Newman, S. E.,** Cognitive apprenticeship: teaching the craft of reading, writing, and mathematics, in *Cognition and Instruction: Issues and Agendas,* Resnick, L. B., Ed., Lawrence Erlbaum Associates, Hillsdale, NJ, in press.
6. **Cooper, G. and Sweller, J.,** Effects of schema acquisition and rule automation on mathematical problem-solving transfer, *J. Educ. Psychol.*, 79, 347, 1987.
7. **Derry, S. J., Hawkes, L. W., and Tsai, C.,** A theory for remediating problem-solving skills of older children and adults, *Educ. Psychol.*, 22, 55, 1987.
8. **Derry, S. J. and Kellis, A.,** A prescriptive analysis of low-ability problem-solving behavior, *Instruc. Sci.*, 15, 49, 1986.
9. **Francioni, J. M. and Kandel, A.,** A software engineering tool for expert system design, *IEEE Expert,* Spring, 33, 1988.
10. **Hall, L. O. and Kandel, A.,** *Designing Fuzzy Expert Systems,* Verlag TÜV Rheinland GmbH., Köln, Germany, 1986.
11. **Kearsley, G.,** *Artificial intelligence and instruction: applications and methods,* Addison-Wesley, Reading, MA, 1987.
12. **Keller, J. M.,** Motivational design, in *Encyclopedia of Educational Media Communications and Technology,* 2nd ed., Greenwood Press, Westport, CN, 1985.
13. **Lesgold, A.,** Intelligent tutoring systems: practice environments and exploratory models, Address presented at Florida State University Department of Psychology, Tallahassee, March 1988.
14. **Leinhardt, G. and Greeno, J. G.,** The cognitive skill of teaching, *J. Educ. Psychol.*, 78(2), 75, 1986.

15. **Macmillan, S. A. and Sleeman, D. H.,** An architecture for a self-improving instructional planner for intelligent tutoring systems, *Comput. Intell.,* 3(1), 1987.
16. **Negoita, C. N.,** *Expert Systems and Fuzzy Systems,* Benjamin/Cummings, Menlo Park, CA, 1985.
17. **Palincsar, A. S.,** The role of dialogue in providing scaffolded instruction, *Educa. Psychol.,* 21(1 & 2), 73, 1986.
18. **Papert, S.,** *Mindstorms: Children, Computers and Powerful Ideas,* Basic Books, New York, 1980.
19. **Park, O., Perez, R. S., and Seidel, R. J.,** Intelligent CAI: old wine in new bottles, or a new vintage?, in *Artificial Intelligence and Instruction: Applications and methods,* Addison-Wesley, G. Kearsley, Ed., Reading, MA, 1987.
20. **Peachey, D. R. and McCalla, G. I.,** Using planning techniques in intelligent tutoring systems, *Int. J. Man-Mach. Stud.,* 24, 77, 1986.
21. **Raeder, G.,** A survey of current graphical programming techniques, *IEEE Comput.,* 11, August 1985.
22. **Rundensteiner, E. A.,** The Development of a Fuzzy Temporal Relational Database (FTRDB): An Artificial Intelligence Application, Master's thesis, Florida State University, Tallahassee, 1987.
23. **Sleeman, D. and Brown, J. S., Eds.,** *Intelligent Tutoring Systems,* Academic Press, London, 1982.
24. **Snodgrass, R. and Ahn, I.,** Temporal databases, *IEEE Comput.,* 35, September 1986.
25. **Sweller, J. and Cooper, G. A.,** The use of worked examples as a substitute for problem solving in learning algebra, *Cognit. Instruct.,* 2, 59, 1985.
26. **Tanimoto, S. L.,** *The Elements of Artificial Intelligence: An Introduction Using LISP,* Computer Science Press, Rockville, MD, 1987.
27. **Wenger, E.,** *Artificial Intelligence and Tutoring Systems: Computational and Cognitive Approaches to the Communication of Knowledge,* Morgan Kaufmann, Los Altos, CA, 1987.
28. **Zemankova-Leech, M. and Kandel, A.,** *Fuzzy Relational Data Bases — A Key to Expert Systems,* Verlag TÜV Rheinland, Köln, Germany, 1984.

Chapter 18

EXPERT SYSTEM ON A CHIP: AN ENGINE FOR APPROXIMATE REASONING*

Masaki Togai and Horoyuki Watanabe

TABLE OF CONTENTS

* The earlier version of this paper was presented at the second Conference on Artificial Intelligence Applications, Miami Beach, FL, December 11—13, 1985. This paper is based on "Expert System On A Chip: An Engine For Approximate Reasoning" by M. Togai and H. Watanabe which appeared in *IEEE Expert,* Vol. 1, No. 3, pp. 55—62, Fall 1986.© 1986 IEEE.

I. INTRODUCTION

The information used in decision-making or reasoning processes may be uncertain, imprecise, or even vague and incomplete. Inference procedures with uncertainty are becoming more important in rule-based expertlike systems since the knowledge given by a human expert is often uncertain or imprecise.

In this paper, we report on the design of an inference architecture and a chip using a VLSI technology to cope with uncertainty and perform approximate reasoning. To deal with uncertainty, we employ fuzzy logic based on fuzzy set theory.[14] An inference structure suitable for hardware implementation is proposed and realized as a custom VLSI chip by using two simple units: circuits to calculate maximum and minimum elements. The design is also extensible to handle a large number of rules, and the speed of inference is almost independent of the number of rules. We will describe the simplicity, extensibility, and efficiency of the proposed design.

Fuzzy logic has been successfully used in several expert systems.[5,15] One example is CATS, a diesel locomotive diagnosis system.[2] This system currently contains approximately 530 rules and will soon be increased to approximately 1200 rules. Fuzzy inference is also proposed in real-time decision making in the area of command and control[7] to select the most suitable guidance algorithm for intercepting missiles. Selection is done by considering a constantly changing environment, that is, the relative angular positions, accelerations, and distances of an evader and a missle. These examples show the need for an efficient inference engine to handle large rule sets and for real-time use.

II. EXPERT SYSTEMS

A. REPRESENTATION OF IMPRECISE KNOWLEDGE

Expert systems consist of a body of knowledge (knowledge base) and a mechanism (inference engine) for interpreting the knowledge. A knowledge base consists of facts and rules. The information in the knowledge base is a propositional statement of the form:

The (attribute) of (object) is (value)

or alternatively expressed as an ordered triple of the form:

(attribute, object, value)

Examples of such a triple are

        AGE, JOHN, 50)...The age of John is 50
        (DISTANCE, (L.A., S.F.), 300)...The distance between L.A. and S.F. is 300
        (TEMP, PATIENT_____10, 102)...The temperature of the Patient_____10 is 102

The attribute value may not be precisely known. In such a case the mutually exclusive possible values of the attribute of the object can be represented by a weighted distribution on the attribute domain. Practically, two kinds of characteristic functions can be used: probability and grade of membership.[15] The grade of membership is especially appropriate for encoding vague concepts expressed in a natural language. The possible value of an attribute will be well illustrated in the following examples:

(AGE, JOHN, >50) . . . The age of John is over 50

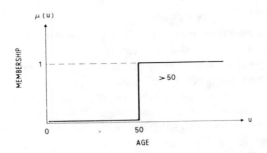

FIGURE 1. Knowledge representation with crisp membership over 50.

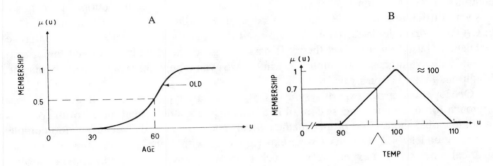

FIGURE 2. Knowledge representation with fuzzy membership. (A) Old; (B) approximately 100.

where ">50" is represented by a crisp membership function defined on the universe of discourse, i.e., U = [0,120] by:

$$\mu(u) = \begin{cases} 1 \text{ if } u > 50 \\ 0 \text{ } otherwise \end{cases}$$

The membership function for this attribute value is illustrated in Figure 1.

The attribute value could be more vague and imprecise:

(AGE, JOHN, OLD) . . . (The age of) John is old
(TEMP, JOHN, $\approx$100) . . . The temperature of John is approximately 100

where "old" and "$\approx$100" are represented by membership functions as shown in Figure 2A and B, respectively.

Notice that at this point we diverge from the conventional representational approach to expert systems. In the conventional systems, such as MYCIN, the value of the attributes are left as symbols, words, or values with no meaning. That is, the data:

The temperature of John is high

is left in this form; no attempt is made to give any meaning to the attribute value "high". The approach we suggest is based upon the idea of fuzzy subsets introduced by Zadeh, where the value of attribute, i.e., high is characterized by the grade of membership or membership function.

A precise value, on the other hand, can be represented as a membership function which is equal to one at the point where precise value is known and is zero elsewhere.

B. RULE-BASED EXPERT SYSTEMS

The rules supplied by a human expert are conditional statements normally expressed as *if-then* rules. A rule is also regarded as an ordered pair of symbol strings, with a left-hand side (LHS) and a right-hand side (RHS), respectively. LHS and RHS are regarded as a situation recognition part (antecedent) and an action part (conclusion), respectively. The situation part expresses some condition on the state of the data base and whether it is satisfied at any given point. The action part specifies changes to be made to the data base whenever the rule is satisfied. The inference engine is an interpreter of the rule base. Its task is to monitor the facts in the data base and execute the action part (RHS) of those rules that have their situation part (LHS) satisfied. The interpreter operates by scanning the LHS of each rule until one is found that can be successfully matched against the data base. If no LHS rules are true, then no action is taken. Therefore, a set of rules has to cover all possible combinations of situations, otherwise, the interpreter cannot take any action. In most cases it is very difficult or infeasible to form an exhaustive search for all possible situations. Thus, in CATS, a fuzzy logic technique is employed to avoid such deadlock. Since an inference mechanism based on fuzzy set theory is basically generalized multiple-valued logic, some action is always suggested through fuzzy inference even though there are no exactly matched conditions in a LHS of a rule.

One application of expert systems is the rule-based process controller.[6,8,9] A usual approach for automatic process control is to establish a mathematical model of the process. In some cases, there is no proper mathematical model because the process is too complex or poorly understood. In other cases, experimenting with plants for construction of mathematical models is too expensive. For such processes, however, skilled human controllers may be able to operate the plant satisfactorily. The operators are quite often able to express their operating practice in the form of rules which may be used in a rule-based controller. The rule-based controllers model the behavior of the expert human operator instead of the process. The following is a rule from a cement kiln controller.[9]

If
1. The temperature in burning zone is O.K.
2. The oxygen percentage in exhaust gas is *low*
3. The temperature at the back-end of the kiln is *O.K.*

then:
Decrease fuel rate *slightly*

As indicated by terms *O.K.*, *low,* and *slightly,* rules in expert systems may contain propositions with imprecise expressions. They can be represented by using fuzzy membership functions.

This is the reason why we employ fuzzy logic which can deal with the expert knowledge as a distribution of belief. In the following sections, we discuss how to process vague information and how to make inference.

III. INTRODUCTION TO FUZZY LOGIC

For the convenience of the reader, a brief summary of some of the relevant aspects of the theory of fuzzy sets and the linguistic approach is presented in this section. For more detail, see Reference 14.

A. NOTATION AND TERMINOLOGY

The symbol U denotes a universe of discourse, which may be an arbitrary collection of subjects or mathematical constructs. If A is a finite subset of U whose elements are $u_1, \ldots, u_n$, A is expressed as

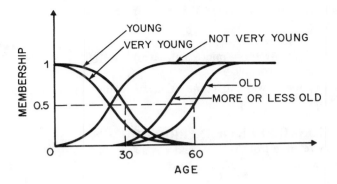

FIGURE 3. Various linguistic values are expressed by fuzzy sets.

$$A = \{u_1,...,u_n\} \tag{1}$$

A finite fuzzy subset A of U is a set of ordered pairs:

$$A = \{(u_i, \mu_A(u_i))\}; \quad u_i \in U \tag{2}$$

where the $\mu_A(u_i)$ represent *grades of membership (or membership functions)*, which indicate the degree of membership. If all $\mu_A(u_i) \in \{0,1\}$, the "fuzzy subset" will be understood as a "nonfuzzy subset" or "ordinary subset". The functions $\mu_A(u_i)$ are then binary Boolean functions. However, unless stated to the contrary, the $\mu_A(u)$ are assumed to lie in the interval [0,1], with 0 and 1 denoting *no* membership and *full* membership, respectively.

B. LINGUISTIC VARIABLES AND FUZZY SUBSETS
We will use fuzzy subsets to represent linguistic variables. Informally, a linguistic variable, L, is a variable whose values are words or sentences in a natural language or in a subset of it. If *age* is interpreted as a linguistic variable, then its *term-set* T(age) might be

$$T(age) = \{young,\ old,\ very\ young,\ not\ young,\ very\ old,$$
$$very\ very\ young,\ more\ or\ less\ young,...\} \tag{3}$$

where each of the terms in T(age) is expressed by a fuzzy subset of a universe of discourse, i.e., U = [0,100]. For example, given the sets *young* and *old* in T(age), one can construct other sets, such as *very young, more less young,* and *not very young* as shown in Figure 3.

C. FUZZY RELATION
In classical propositional calculus, the expression *if A then B*, where A and B are propositional variables, is written as $A \rightarrow B$, with $\rightarrow$ regarded as a connective which is defined by:

$$A \rightarrow B \equiv \bar{A} \cup B \tag{4}$$

where $\bar{A}$ means the complement of A.

A more general concept, which plays an important role in our approach, is the *fuzzy conditional statement: if A then B*, or i short, $A \rightarrow B$, in which A (the antecedent) and B (the conclusion) are fuzzy subsets rather than propositional variables. A relation R from A to B is a fuzzy subset of the Cartesian product $U \times V$, where $A \in U$ and $B \in V$. The conditional statement "if X is A then Y is B" is represented by the fuzzy relation R and defined as follows:

$$\mu_R(u,v) = \min(\mu_A(u),\mu_B(v)); \quad u \in U, \quad v \in V \tag{5}$$

Note that there are many ways to define a fuzzy relation. Here we introduce the most commonly used one.

D. COMPOSITIONAL RULE OF INFERENCE

If R is a fuzzy relation from U to V, and x is a fuzzy subset of U, then the fuzzy subset y of V which is induced by x is denoted by:

$$y \equiv xoR \tag{6}$$

and defined as follows:

$$\mu_y(v) = \max_{u \in U} \min(\mu_x(u), \mu_R(u,v)) \tag{7}$$

Note that when $R = A \rightarrow B$ and $x = A$ we obtain:

$$y = xo\,(A \rightarrow B) = B \tag{8}$$

as an exact identity. Thus, Equation 6 may be viewed as an extension of *modus ponens*.

IV. LOGIC ARCHITECTURE OF FUZZY INFERENCE ENGINE

In the current section, we propose an inference mechanism suitable for hardware implementation. It is based on the concepts of fuzzy implication, or fuzzy rules, and the compositional rule of inference. A fuzzy rule is defined by the relation between observation (or antecedent) and action (or conclusion). For a given set of fuzzy rules, the action is inferred from both a certain observation and the fuzzy relation which is composed from the rules.

If A and B are fuzzy subsets representing linguistic labels, over the universe of discourse U and V, respectively, then a decision rule in the form of "if A then B" is defined by binary membership functions of A and B so that:

$$\mu_{A \rightarrow B}(u,v) = f_{\rightarrow}(\mu_A(u), \mu_B(v)); \quad u \in U, v \in V \tag{9}$$

More specifically, suppose $A_1, A_2, \ldots, A_N$ are fuzzy subsets of U and $B_1, B_2, \ldots, B_N$ are fuzzy subsets of V, a fuzzy relation is defined by rules so that:

Rule 1.　　　　If A_1 then B_1
Rule 2.　ELSE　If A_2 then B_2
.　　　　　　　.
.　　　　　　　.
.　　　　　　　.
Rule N.　ELSE　If A_N then B_N

Then each rule is combined by the ELSE connective to yield an overall fuzzy relation R.

R_i is a fuzzy relation constructed from rule i, and linguistic values A_i and B_i. The connective ELSE is denoted by the function f_{ELSE}, then the overall relation R is defined by:

$$\mu_R(u,v) = f_{ELSE}\{\mu_{R_i}(u,v)\}$$

$$= f_{ELSE}[f_{\rightarrow}(\mu_{A_i}(u), \mu_{B_i}(v))] \text{ for } i = 1,...,N \qquad (10)$$

The connective f_{ELSE} is interpreted as an **or** connective in order to derive an overall fuzzy relation R because the relation R should consist of rule 1 **or** rule 2 **or** . . . **or** rule N. Hence the overall relation R is denoted and defined as:

$$R = \underset{i}{\cup} R_i$$

$$= \underset{i}{\max} f_{\rightarrow}(\mu_{A_i}(u), \mu_{B_i}(v)) \text{ for } i = 1,...,N \qquad (11)$$

Suppose we have a fuzzy observation A' and the overall relation R; then the resultant action B' is inferred by the compositional rule of inference, that is:

$$B' = A' o R \qquad (12)$$

The membership value of B' is calculated by so called the "max-min operation" defined by (7):

$$\mu_{B'} = \underset{u \in U}{\max} \min(\mu_{A'}(u), \mu_R(u,v)) \qquad (13)$$

To follow up the preceding discussion, we propose an architecture of a fuzzy inference engine suitable for a hardware implementation.

Let us consider the i-th rule of a set of N rules. Given an observation A' and a rule R_i, the action B'_i is inferred and defined as follows:

$$B'_i = A' o R_i; A' \in U, B_i \in V, \text{ and } R_i \subset U \times V \qquad (14)$$

$$\mu_{B'_i}(v) = \underset{u \in U}{\max} \min(\mu_{A'}(u), \mu_{R_i}(u,v))$$

$$= \underset{u \in U}{\min\max} [\min(\mu_{A'}(u), \mu_{A_i}(u)), \mu_{B_i}(v)]$$

$$= \min(\alpha_i, \mu_{B_i}(v)) \qquad (15)$$

where

$$\alpha_i = \underset{u \in U}{\max} \min(\mu_{A'}(u), \mu_{A_i}(u)) \qquad (16)$$

The operation given by Equation 14 and Equation 15 is illustrated in Figure 4. Then the maximum of $B_1, B_2, . . . , B_N$, determine the overall resulting decision (or action) B', that is

$$B' = \underset{i}{\cup} B'_i \qquad (17)$$

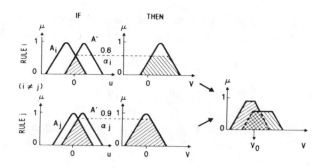

FIGURE 4.   A mechanism of "max-min operation" for $B' = A'$ o $(A \rightarrow B)$.

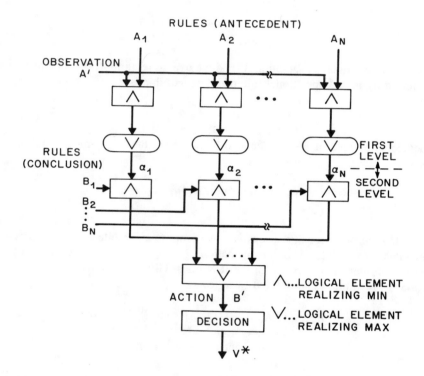

FIGURE 5.   A functional architecture of fuzzy inference mechanism.

The inference mechanism given before can be realized by the logical architecture with two-level hierarchy as shown in Figure 5. The proper choice of α_is is the first-level decision, while the proper choice of B' is the second-level decision. The operation serially done in each column of the structure is equivalent to the operation which is illustrated in Figure 4. It has been shown that the logical architecture obtained here is basically the same even for multiple observation and multiple action cases.[12] The advantage of the proposed structure is that max-min operations are realized by the ordinary *or* and *and* gates, respectively.

V. VLSI DESIGN

The VLSI inference engine consists of three major parts: a rule set memory, an inference processing unit, and a controller. The inference mechanism described in the previous section executes all the rules in parallel. This high degree of parallelism requires a wide bus between the rule set storage unit and the inference processing unit of the engine. Because of the very high rate of communication between these two elements, we decided to store the rules on a chip. Otherwise, the restriction of the pin counts prevents the effective use of the parallelism.

Since this is the first version of the design, we emphasize simplicity. One important decision is to process individual rules serially. This simplifies the design and increases its extensibility. The logic structure of the fuzzy inference mechanism is mapped nicely to the VLSI structure. There is an one-to-one correspondence between the basic operations of fuzzy logic and the basic units of the VLSI inference processing unit. The three major parts of the inference engine are described in the following section.

A. STORAGE AND FORMAT OF RULE SET

The rule set can be stored using either a random access memory (RAM) or a read only memory (ROM). The advantage of using RAM is its flexibility. Depending on the application, the rule set can be loaded from off-chip. On the other hand, ROM takes much less area for the same amount of data and operates faster. The control unit of the inference engine can be very simple since we do not need to load a rule set from off-chip. Also, we have a well-engineered ROM generator in our design system, so that we used ROM for the storage of the inference rules.

We considered the size of fuzzy subset and the grade of fuzziness for a practical use. In most cases, the size of fuzzy subset has been chosen to be from 3 to 16; the grade of fuzziness, 3 to 12 [e.g., 9, 11]. For this implementation, we limit the universe of discourse of a fuzzy subset to be a finite set with 31 elements. The membership function is made discrete in 16 levels (i.e., four bits). That is, 0 represents no membership, and 15 represents a full membership and other numbers represent points in the unit interval [0,1]. We used 124 bits for digitization of the membership function.* The format of the rule representation is as follows:

$$\text{Rule i: } A_i \rightarrow B_i$$

	u_1	u_2	u_3	$\cdots$	u_i	$\cdots$	u_{31}
A_i:	0010	0100	1111	$\cdots$	$\mu_{Al}(u_i)$	$\cdots$	0000

	v_1	v_2	v_3	$\cdots$	v_i	$\cdots$	v_{31}
B_i:	0000	0001	0011	$\cdots$	$\mu_{Bl}(v_i)$	$\cdots$	1100

* Actually, we used 128 bits of ROM for each fuzzy subset in the rule set, since the generator requires the column length of ROM to be a power of two. First location of 128 bits always contains zero and is used to inactivate the **if** part or the **then**-part of the inference processor when they are not computed. The 3 other extra bits are dummies, and we have only 124 bits (31 elements) of active data.

Here, each four bits represents degree of membership for each element of the universe. For example, u_1 has a degree of membership $^2/_{15}$ and u_3 has a full degree of membership in a subset A_i. Each four bits integer is stored, most significant bit first. Each word of ROM stores one bit from all the rules stored. All the rules are, therefore, accessed in parallel. An individual rule is, however, accessed in the serial manner. Two ROM modules are used for storing the antecedent As and the conclusion Bs of the rule set.

B. DATA PATH FOR INFERENCE

The inference processing unit consists of multiple data paths. One data path is allocated for each rule. The data path consists of two basic units: a minimum unit and a maximum unit. The minimum unit takes two integers and produces the smaller number; the maximum unit produces the larger number. This units process integers serially. The operation of the minimum unit is shown in Figure 6 as a finite state machine of three states. The minimum units and maximum units are basic units used to implement the fuzzy intersection and fuzzy union operations.

The basic data path of the inference engine for processing a single rule is shown in Figure 7. This directly corresponds to a single data path of the inference engine described in Figure 5. The shift register is used for keeping the maximum after the fuzzy intersection operation in the first level. This is necessary since within an individual rule operations are performed serially. The register records the value of the maximum point, a value α_i, when the first level has finished its operation. The last operation of the second level requires taking the maximum membership function over all the data paths (i.e., all the rules). This operation is accomplished by connecting the maximum units in the binary tree structure as shown in Figure 8.

C. CONTROL UNIT

Because of the simplicity of the architecture, the controller of the inference engine is straightforward. It consists of two counters for accessing two ROM modules serially. The controllers generates a reset signal for the minimum and maximum elements for every four cycles. The controller starts to access the conclusion parts of the rules as soon as the processing of the antecedent parts are finished. It also notifies a user of the beginning of the valid output.

D. CIRCUIT DETAIL

The layout of the first implementation stores and processes 16 rules. Each rule consists of 124 bits of antecedent and 124 bits of conclusion. An observation and an action consist of 124 bits each and they are loaded and produced serially. A nonoverlapping two-phase clocking scheme, supplied from off-chip, is used. The operation is initiated by a reset signal that must last one clock cycle and resets the entire circuit. Input of an observation should be started two clock cycles after the reset signal, that is, on the third clock cycle. The inference engine begins to produce the result on the 133rd cycle after the reset signal. The beginning of the valid output is signaled by the controller.

The active area size is 2.99 × 3.48 mm. A 68-pin package is used. Only eight pins are used for the operation of the chip. They are VDD, VSS, ϕ-1, ϕ-2 clock signal, serially loaded observation, action, reset signal, and a signal indicating the beginning of the valid output. To output the value of the critical nodes in the processor for debugging purposes 30 pins are used. The mask layout is shown in Figure 9.

E. CHIP PERFORMANCE

Timing tests on fabricated chips indicate a 20.8 MHz (48 nsec cycle time) operating rate. With the current data format of 124 bits per rule, a single inference process takes 256

FIGURE 6. Serial minimum unit.

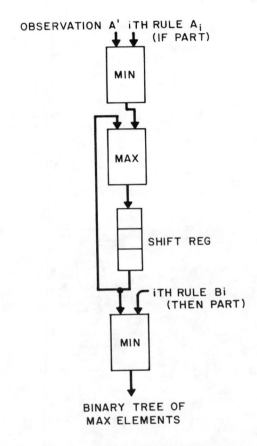

FIGURE 7. Single data path of inference processor.

clock cycles. Thus, the engine can perform approximately 80,000 fuzzy logical inference per second (FLIPS), which is more than 40,000 times faster than the simulation conducted on VAX-11/785. The chip has a drastic speed advantage over the conventional approach. If the resolution of rules is doubled, the speed of the inference engine is halved. However, by assigning two data paths per rule, we can reduce the slowdown to a few clock cycles per inference.

VI. EXTENSION AND INTEGRATION

A. RULE FORMAT

The advantage of the architecture is its simplicity which in turn makes this architecture easily extendable. Most important extensibility is to cope with the deferent format of rules. In the simplest case, we would like to increase or decrease the resolution of membership function. In the current format of the rule representation, we used 31 integers of four bits

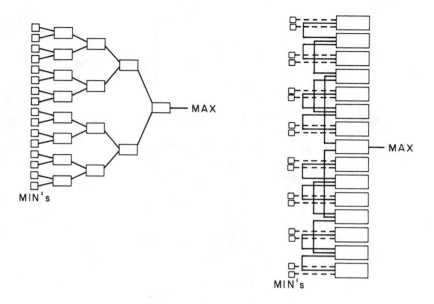

FIGURE 8.   Binary tree of maximum units.

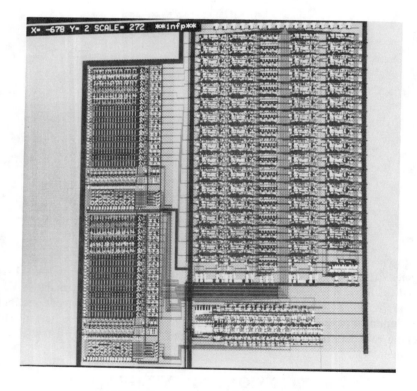

FIGURE 9.   Mask layout of inference engine.

each. This can be modified with minor changes in the layout. For example, the number of elements in the universe of discourse (number of data points) can be increased to more than 31. It can be achieved with only a minor modification of the controller without any change in the inferencing part of the architecture. We can use higher resolution for digitization of membership function using more than four bits. We can achieve this by increasing the length of the shift register and modifying the controller. The number of rules can be increased by laying out more data paths and modifying the binary tree layout accordingly.

We can also have more than one clause in the *if* part of the rule. The prototype chip has only one clause in the if-part of the rule and one clause (i.e., action) in the *then* part of the rule. As we have seen in an example of the cement kiln controller (Section II.B), rules often have more than one clause in the *if* part. Extension of the data path to handle multiple clauses is fairly straightforward. We can expect that multiple clauses are connected by the logical connective and. If clauses are connected by the logical connective **or**, we can divide a single rule to independent multiple rules. For each clause in a rule, we compute an α value independently. A minimum of them are used for truncation of the *then* part of a rule. Each data path still consists of the same basic units. Figure 10 shows the architecture of a single data path for the four clause rule. We can compute the minimum of the α values by a binary tree of serial minimum units.

B. SPEED UP

We can speed up the throughput of the inference processor by a minor modification of the architecture. By introducing another shift register, we can make a pipelining processor. The antecedent (first level) and the conclusion (second level) are performed simultaneously. The second shift register should be introduced right after the current shift register. The first shift register is used to keep the running maximum for α_i in the processing of the antecedent. The second shift register keeps the computed α_i for processing of the conclusion part. The throughput of the processor is doubled by this pipelining.

Another architectural possibility is the use of multiple data paths per rule. In this case, there is an area vs. speed trade off. For example, by assigning two data paths per rule, we can almost double the processing speed and a single clock cycle is absorbed by the overhead. We store each half of the rule independently. The corresponding half of the observation is loaded from two pins. In the first level and the second level of the processor, each half of the rule set and observations are processed almost independently. An extra maximum operation is necessary to combine the result of each half for α_i at the end of the first level. We now need two independent binary trees of the maximum units at the second level. The sketch of this architecture is shown in Figure 11.

Notice that the execution of each rule is done independently by a single data path except in the last step of the second stage where the selection of the maximum elements is done. In order to cope with the large number of rules, a chip set can be designed in a rule-sliced manner. Each chip executes a subset of rules (for example, 32 rules per chip). A very large number of rules can be executed in parallel by cascading them. The selection of the final result can be done by designing a maximum number selector as an independent chip. For example, if we have a 32 input maximum number selector (see Figure 8), we can combine 1024 data paths using two layers of such selectors.

C. SYSTEM INTEGRATION

We are designing an integrated system based on the VLSI fuzzy inference engine. In order to achieve general purpose interface, we use a single board computer as a host system. The fuzzy inference engine is integrated as a co-processor for approximate inference. The host processor communicates with a controller, a sensor, and a user. In order to make a rule set programmable, a ROM is replaced by a high speed static RAM. The overall configuration of the integrated system is shown in Figure 12.

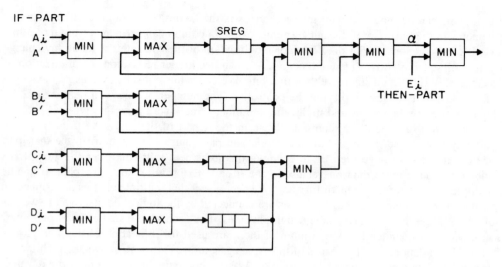

FIGURE 10.   Four clauses in condition.

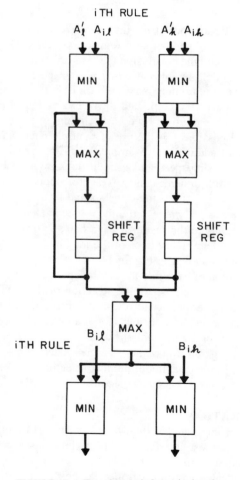

FIGURE 11.   Two data path for a single rule.

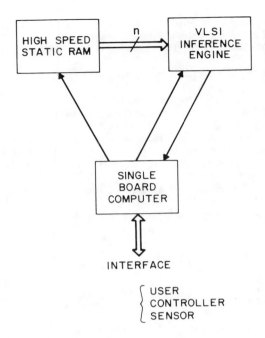

FIGURE 12.   Integrated system.

VII. CONCLUSION

We have designed and fabricated a chip for a real-time approximate reasoning based on the "max-min operation" of fuzzy set theory. The architecture and chip design of this inference engine have been presented. An actual inference engine has been fabricated on a chip using custom CMOS technology. The VLSI design emphasizes simplicity, extensibility, and efficiency. Preliminary timing tests have been performed and indicate that the chip can perform approximately 80,000 fuzzy logical inferences per second, which is more than 40,000 times faster than conventional software approach. The potential applications of this inference engine for real-time use include decision making in the area of command and control, intelligent controllers, and intelligent robot systems (e.g., References 2 and 4).

REFERENCES

1. **Ackland, B. and Weste, N.,** Functional Verification in an Interactive Symbolic IC Design Environment, Proc. 2nd Caltech Conf. VLSI, 1981, 285.
2. **Bonissone, P. P. and Johnson, H. E., Jr.,** Expert system for diesel electric locomotive repair, *Hum. Syst. Manag.,* 4, 255, 1984.
3. **Buchanan, B. G. and Shortliffe, E. H., Eds.,** *Rule-Based Expert Systems,* Addison-Wesley, Reading, MA, 1983.
4. **Dufay, B. and Latombe, J.-C.,** An approach to automatic robot programming based on inductive learning, *Intl. J. Robotics Res.,* 3(4), 3, Winter 1984.
5. **Gupta, M. M. and Sanchez, E., Eds.,** *Approximate Reasoning in Decision Analysis,* North-Holland, Amsterdam, 1982.
6. **Holmblad, L. P. and Ostergaard, J. J.,** Control of a cement kiln by fuzzy logic, *Fuzzy Information and Decision Processes,* Gupta, M. M. and Sanchez, E., Eds., 1982, 389.

7. **Kawano, K., Kosaka, M., and Miyamoto, S.,** An algorithm selection method using fuzzy decision-making approach, *Trans. Soc. Instrument Control Engineers,* 20(12), 42, 1984 (in Japanese).
8. **King, P. J. and Mamdani, E. H.,** The application of fuzzy control systems to industrial processes, *Automatica,* 13, 235, 1977.
9. **Larkin, L. I.,** A Fuzzy Logic Controller for Aircraft Flight Control, Proc. 23rd IEEE Conf. Decision and Control, December 12—14, Las Vegas, NV, 1984.
10. **Shortliffe, E. H.,** *Computer-Based Medical Consultations: MYCIN,* American Elsevier, New York, 1976.
11. **Sugeno, M. and Murakami, K.,** Fuzzy Parking Control of Model Car, Proc. 23rd IEEE Conf. Decision and Control, December 12—14, Las Vegas, NV, 1984.
12. **Togai, M.,** Principles and Applications of Fuzzy Inference: A New Approach to Decision-Making Process in Ill-Defined Systems, Ph.D. dissertation, Duke University, Durham, NC, 1982.
13. **Weiss, S. M. and Kulikowski, A.,** *A Practical Guide to Designing Expert Systems,* Rowman and Allanheld, Totowa, NJ, 1984.
14. **Zadeh, L. A.,** Outline of a new approach to the analysis of complex systems and decision processes, *IEEE Trans. Syst., Man, Cybern.,* SMC 3, 28, 1973.
15. **Zadeh, L. A.,** The role of fuzzy logic in the management of uncertainty in expert systems, *Fuzzy Sets Syst.,* 11, 199, 1983.

Chapter 19

A PROBABILISTIC LOGIC FOR EXPERT SYSTEMS

Arie Tzvieli

TABLE OF CONTENTS

I. INTRODUCTION

Modeling uncertainties is one of the most difficult and controversial issues in current expert systems research. This difficulty arises from the fact that, in the process of system development, some assumptions have to be made about "its world"; those are the axioms of the system. When a system is developed, it is not always known where and when it will be used, and whether the axioms will be valid in the environment of the application. The current literature on the subject of uncertainty management abounds with examples of incompatibility between the system premises and the application environment. It is, therefore, extremely important to specify each system's axioms explicitly and to check their validity, in a particular enviroment, *before* using it in an expert system. However, this is not always possible, because of lack of information about the applications. We believe that a productive approach to choosing a methodology for the management of uncertainty should attempt to compare the different approaches based on the range of applicable environments.

A formal logical system of reasoning with uncertainties can serve as a tool for studying different axiom sets, and later be used for tuning a shell to the application environment. This chapter describes a formal logical system (probability logic [PL]) which uses estimates based on probabilities (or beliefs). These estimates are assumed to be independent of each other. Logical systems model the world by assigning truth values to formulas. A logical formula is a very powerful tool for capturing objects and their relationship. It includes individual variables and constants to model objects, atomic formulas to express basic relationships, connectives (and/or/not) to connect subformulas, and quantifiers (exist/for-all) to capture existence and uniformity of a property over a set of objects. Logical systems use formal derivation rules to infer conclusions from a given description of the world.

In PL, formulas have a truth value, ranging between true and false. The truth value "true" is expressed by the constant one, "false" by zero, and intermediate truth values are expressed by numbers between zero and one. The higher the truth value of a formula ϕ, the more we believe that the fact expressed by ϕ is true. Tuning a PL system to an environment is achieved by two actions: choosing a definition of satisfaction, and choosing a corresponding set of axioms and inference rules. The definition of satisfaction involves a procedure for arriving at the truth value of a compound formula from the truth value of the constituent subformulas. The procedure also depends on the environment; by substituting one procedure for another, we change, in the logical model, the interpretation of the external world. So far, over 100 different satisfaction procedures have been proposed (e.g., Dubois and Prade[5] and Weber[16]). Yet, more research is required before an automatic system could be developed to decide which satisfaction procedure to choose, depending on the application (one of the difficulties is that different procedures correspond to the same axiom set). A practical approach to resolving this type of conflict may involve testing those candidate procedures, *sharing* a chosen set of axioms, against a set of test cases, and choosing the best fit.

The second component of tailoring a logical system to a specific environment is to choose the axioms and specific derivation rules. As in the satisfaction interpretation issue, more research is needed before automation of the process will become feasible.

The version of PL described in this chapter should be viewed as an example system. We have not elaborated on the *environment tailoring* process, but we believe that the case discussed here could correspond to a situation where we have only little knowledge of the application, and where it is reasonable to assume independence between truth values of formulas, regardless of their intention.

PL can be combined with other quantitative uncertainty management systems to enhance their reasoning capabilities. A probabilistic (or belief) approach can be used to assign probabilities to a core of formulas, and then PL can be used to reason about the probability of logical consequences of these formulas. This is demonstrated by using an example from Shafer[14] and adding derived consequences to his results.

The logical model advocated in this paper was motivated by the goal of applying the notion of probability in the context of expert systems. Another use of PL might be development of a theory of correctness of probabilistic systems. Lack of reliability is a very important criticism of expert and other software systems (e.g., Parnas[12]) and a crucial factor influencing their adoption and utilization. It is desirable that we be able to prove formally the correctness of an expert system. Logic-based expert systems seem to be most promising for this goal. Both the input and the output can be described by formulas, and the logical relationship between them is easier to express in a logical framework. In this chapter we outline a formal semantics for PL.

There have been previous attempts to combine probability and logic. Keisler[10] generalized the proportional calculus with probabilistic quantifiers. An important difference between his approach and ours lies in the adoption of the probability axioms (sum of probabilities is one, probability of a composite event is the sum of probabilities of components), which we do not consider to be in good correlation with the real world. A few other words (e.g., Hoover[9]) are based on Keisler.[10] Halpern and Rabin[8] discuss a logic of likelihood, which is the propositional calculus augmented with some modal operators (likelihood, necessity, and transitivity of likelihood). Their work is very interesting, but the focus is different from ours. Fuzzy logic, discussed by Zadeh[17] and others, is another generalization of logic in the presence of uncertainty. This methodology does not assume the probability axioms. In its initial stages it used to assume a continuum of numerical truth values (the interval [0,1]), but later versions resorted to symbolic truth values (true, somewhat true, very true), whose semantics are defined in terms of fuzzy numbers. Another deviation of fuzzy logic from the traditional spirit is the avoidance of syntactic derivations — only a semantic derivation rule is adopted. Nilsson[11] describes a very interesting probabilistic logic. The proposed semantics is a probability distribution of possible consistent worlds. The formalism is restricted to a finite number of formulas and finite number of possible worlds, and only entailment is discussed among the connectives and quantifiers. Nilsson also discusses a few other issues (e.g., approximation of the probability of a derived formula).

Section II of this chapter describes a formal logical system, based on the classical first order predicate calculus (FOPC), aiming at reasoning about the probability of correctness formulas in models. Each atomic fact is accompanied by a probability value that this fact is true. Facts can be combined using the connective "and" $\land$, "or" $\lor$, "not" $\lnot$, "implication" $\rightarrow$, "biconditional" $\leftrightarrow$, and the quantifiers "exists" $\exists$ and "for-all" $\forall$. The probability of satisfaction of a combined formula can usually be computed using the probabilities of its components.

Section III discusses the semantics, soundness, and completeness of PL. Our axiomatization of PL, together with the derivation rules, are compatible with FOPC. Consistency in PL is defined based on the notion of "not being able to derive both a formula and its complement." Logical implication, soundness, and completeness of PL are also studied.

This chapter is a summary of research which is expected to be continued in the future. Many possibilities for generalizations of the issues considered here exist, and the choices made here should be considered as representatives of the possibilities, rather than a best version.

Familiarity with basic FOPC is assumed in the following section. The interested reader may consult any of the many texts on FOPC, e.g., Enderton.[6]

II. A PROBABILITY LOGIC PL

In this section we describe a formal system of probability logic. *Webster's New World Dictionary* defines *probable* as: "likely to occur or to be so; that can reasonably be expected or believed on the basis of the available evidence, though not proved or certain." In what follows, the term "probability" will be used in this sense.

As in any logical system, we use variable and constant symbols to describe object and/or properties/values. Atomic formulas describe simple sets, and compound formulas are used to model more complex phenomena. The probability of a formula denotes the likelihood that the formula is satisfied in a model.

Two assumptions are characteristic of discrete probability systems:

1. $\sum_{a \in U} p(a) = 1$ (*U* is the universe)

2. $p(A) = \sum_{a \in A} p(a)$ for singleton components of A

Assumption 1 implies that complete knowledge of all possibilities is available. Assumption 2 assumes that basic probabilities are assigned only to singleton sets (i.e., the probability of a set of events is the sum of the probability of its constituents). In the Dempster/Shafer theory, only Assumption 1 but not 2 is adopted. An objection to the first axiom is based on possible ignorance of some of the domain values. Objection to the second axiom is related to ignorance in another way — we may know that a set of outcomes is possible with some probability without knowing the internal distribution of probabilities. We are reluctant to adopt either of these axioms as constraints on our model. The choice of axioms for a system seems to be a thorny point; we do not claim that our choice is *correct* in any sense. It should be left to the user who is responsible for an expert system to adopt a favorable axiom set. The choice of a different axiom set should be followed by an appropriate modification of derivation rules and of the definition of satisfaction.

For simplicity, assume that the probabilistic characteristics of a fact (from the statistical point of view) are described by a single numerical value — its probability (but consider Shafer[14] and others for an interval approach, and Cheeseman[3] for a description which consists of both a probability value and a standard deviation value).

Our interpretation of the logical connectives and quantifiers correspond to the probabilistic approach which assumes independence between the truth values of formulas. One deviation that we make is not to assume that implication is functionally dependent on the truth of the components. What follows is the formal description of the logic, starting with the syntax.

A. THE SYNTAX OF PL FORMULAS

A PL type is a relational type, i.e., a set of constant and relation symbols. Note that we are not using functions in PL.

Example 1: A PL Type

A PL type is {*red, Suppliers, Parts, SP, PS, Family*}. Here all relation symbols stand for binary relations. SP is the relationship between suppliers and parts and PS is a partnership relationship between suppliers. In what follows we shall use acronyms instead of full relation names. Reasoning about a very large data base of facts necessitates the use of many constant symbols, and a practical compromise is to accept any specially syntactically designated symbols as constants, without explicitly declaring them in the type. A relation symbol included in the type is called **a base relation**. The symbols in the language are those in the type, the equality relation symbol $=$, variable symbols denoting individual elements, the logical symbols ($\bigcap$, $\wedge$, $\vee$, $\rightarrow$, $\leftrightarrow$. $\forall$, $\exists$), and parentheses. We may also use new relation symbols (not in the type) to denote auxiliary relations. A term in PL is either a constant symbol or a variable. **Formulas in PL are only those of FOPC.**

Example 2: PL Formulas

Consider a data base with the type given in preceding Example 1:

1. $S(s1,N.Y.)$
2. $[x = y] \wedge P(y, red)$
3. $[\bigcap[y = red]]$
4. $[\bigcap[P(x,red)]]$
5. $[\forall xz \, \exists wy \, S(x,z) \wedge [[SP\,(x,w) \wedge P\,(w,y)] \vee [z = N.Y.]]]$
6. $\forall x \, [P\,(x,red) \leftrightarrow \exists y \, SP\,(y,x)]$
7. $\exists x \, [SP\,(x,p1) \leftrightarrow SP\,(x,p2)]$
8. $PS(x,y) \vee \exists z \, [PS(x,z) \wedge PS(z,y)]$

B. PROBABILISTIC MODELS

A structure in FOPC may be viewed as a model of the relevant aspects of the world. It is composed of two components: the extension, the model universe and its organization, and the intention, the correspondence between object names and objects in the model extension. An alternative modeling tool is the Kripke structures, often used in modal logic. The main difference of concern to us between Kripke and FOPC is that the former uses a single structure whereas the latter uses a set of "FOPC-like" structures to model each state of the world. The "extra" structures are used in interpreting a formula of probability values as the number of worlds in which it is satisfied divided by the total number of possible worlds. Kripke structures may be used in the future for PL modeling, but the present work is based on FOPC.

The main differences between FOPC structures and PL models are

1. Relations are probabilistic; i.e., there is a probability for tuples to belong to a relation.
2. The probability values of implication formulas are defined by the model (in other words — by an expert, possibly using some alternative uncertainty management paradigm), and not as a function of the truth values of their components.
3. Constraints (i.e., formulas with no free variables) are included in models as part of the world state description. The constraints are a tool for a concise description of some properties of the world which must be reflected by the model. An alternative approach separates the constraints from the model (maybe adds them as axioms) without any difference in the properties of PL. Since a model to us is a reflection of the world, we have chosen to make the constraints part of the model.

Let T be a type. Let I be a set of implication formulas (i.e., formulas whose most external connective is implication). A probabilistic model S of probabilistic type T is a function with domain $T \cup (I \cup (\{\emptyset\} \cup \{\{\emptyset\}\}))$ so that:

1. $S(\{\emptyset\})$ is a nonempty set — the universe, or domain. It is denoted by $|S|$, and although in practice it is many sorted, for simplicity of presentation we shall use a one-sorted notation.
2. $S(c_\alpha) \in |S|$ for each constant symbol $c_\alpha \in T$. Every constant symbol is mapped by the model to an element in the universe.

3. For each n-ary relation symbol $R \in T$, $S(R)$ is a probabilistic relation $<|S|^n, p_R>$, where p_R is an n-ary partial function from the n cartesian product of the domain to [0,1], assigning to tuples a probability of belonging to the relation.

4. $S(\{\{\emptyset\}\})$ is $<C, p_c>$ where C is a set of constraints (i.e., closed formulas) and p_c is a function from C to [0,1].

5. For $\phi \in I$, $S(\phi)$ (if specified) is in [0,1], and is the truth value of ϕ. We further impose a consistency condition on assignments of probability values to implications: if ϕ is $\Psi \rightarrow \theta$ and $S \models_1 \Psi$ (i.e., the probability of satisfaction of Ψ is 1) and $S \models_0 \theta$ then $S(\phi) = 0$.

Models will be denoted by bold capital letters. *Finite models* are those models whose universe is finite. The *support* of a probabilistic relation R is the set of tuples whose probability of membership in R is not zero. For safety, we are considering only finite models, or models where all the relations have a finite support.

Remarks

First, note that the probability assignment is a partial function; it is possible for our knowledge of the world to be partial and it is possible to express ignorance in PL simply by failing to specify unknown information (this is in contradiction to the closed world assumption,[13] which, paraphrased, states that anything not explicitly expressed is false). As a consequence, summation in PL of the probabilities over all known possibilities may be <1. An alternative formalism is to make the probability function assign a special value, $\perp$ (bottom, or unknown) in case a value is unknown. $\perp$ is considered to be smaller than any value in [0,1] (this order corresponds to "less known," for future "higher orders" corrections to be probabilities), and its multiplication with any number in [0,1] gives $\perp$. Second, as our real experts are human beings, we do not have complete knowledge of probabilities of all implications. In many cases it is impractical to specify, in the model definition, all these probabilities, even if known. Therefore, we under-specify model definitions, and assign probabilities only to implications seemed relevant to the problem at hand.

Example 3: A Probabilistic Model

This is the type that is given in Example 1. The universe is the set of elements given as values in Table 1. The constant symbol "red" is mapped to the element 'red'. The relation symbols are mapped to probabilistic relations, according to Table 1. The relevant constraints that are F and PS are symmetric relations. The probability of F(x,y) is the same as the probability of F(y,x), and likewise for PS. We do not specify in this example probabilities of implications. Note that in most cases, the sum of the probabilities over the attributes of an entity in each relation is not equal to one, meaning that the other values are unknown (e.g., in what other cities s2 may be located). Note also that although in Table 1 the relations PS and F do not appear symmetric, this is so just to save space, and the omitted tuples that would make these relations symmetric are implied by the constraints.

C. SATISFACTION IN PL

An assignment in FOPC is a mapping of variables to elements of the universe. We could have defined in PL a probabilistic assignment, which would have assigned each variable x to a set of elements so there is a probability p_c that c is the element assigned to x. This would correspond to uncertainty in the modeling process, rather than uncertainty in the external facts. However, the above choice would complicate the model without any evident benefits. Therefore, we define assignment in PL as it is defined in FOPC.

TABLE 1

P				SP			PS	
p#	Color	p	S#	P#	p	S_1#	S_2#	p
p1	Red	.5	s1	p1	.7	s1	s1	1
p2	Red	.3	s1	p2	1	s1	s3	.7
p2	Blue	.7	s1	p3	.2	s2	s2	1
p3	Green	.5	s1	p4	.6	s3	s3	1
p4	Red	1	s2	p1	1	s4	s4	1
			s2	p2	1			
	S		s2	p3	.5		F	
S#	City	p	s2	p4	.7	S_1#	S_2#	p
s1	N.Y.	.7	s3	p1	.4	s1	s1	1
s1	N.O.	.3	s3	p2	1	s1	s2	.8
s2	N.Y.	.4	s3	p3	.5	s2	s2	1
s3	N.Y.	.6	s4	p2	.7	s3	s3	1
s4	N.O.	1	s4	p4	.6	s4	s4	1

A formula ϕ is satisfied (in FOPC) in a model A for an assignment z, and we write $A \models \phi <z>$, if for the assignment z ϕ expresses a fact which is correct (true) in the world corresponding to the model A. FOPC is dichotomic: every formula is either satisfied in a structure or not; there are only two truth values: true or false. In PL, we have at least two possible directions of generalization of FOPC. The first one retains the traditional truth values true and false, and a formula ϕ is satisfied if the probability of the facts expressed by ϕ is at least a threshold p. The p may be 1, or in some cases less than one, depending on the application for which we tailor the logic. An alternative approach is to extend the range of truth values to the closed interval [0,1]. The 0 corresponds to false (probability of satisfaction is zero); the 1, to true (probability of satisfaction is one); and the intermediate values express intermediate probabilities of satisfaction of the formula in a given model (similar to fuzzy logic). In the following we formalize the second approach.

The probabilistic satisfaction of a formula ϕ in a probabilistic model A of type T with assignment z, written $A \models_\alpha \phi<z>$ (α being the probability of satisfaction) is defined as follows:

1. $\phi = [R (t_1, \ldots ,t_n)]$ and $\alpha = p_R <x_1, \ldots ,x_n>$ where $z(t_i) = x_i$, $1 \leq i \leq n$

2. $\phi = [t_1 = t_2]$, and $\alpha = \begin{cases} 1 & z(t_1) = z(t_2) \\ 0 & z(t_1) \neq z(t_2) \end{cases}$ where t_i, i = 1,2 is a term

3. $\phi = [\neg \Psi]$, $A \models_\beta \Psi < z >$, and $\alpha = 1 - \beta$

4. $\phi = [\Psi_1 \vee \Psi_2]$, $A \models_{\beta_1} \Psi_1$, $A \models_{\beta_2} \Psi_2$ and $\alpha = \beta_1 + \beta_2 - (\beta_1 \cdot \beta_2)$

5. $\phi = [\Psi_1 \wedge \Psi_2]$, $A \models_{\beta_1} \Psi_1$, $A \models_{\beta_2} \Psi_2$, and $\alpha = \beta_1 \cdot \beta_2$

6. $\phi = [\Psi_1 \leftrightarrow \Psi_2]$, $A \models_{\beta_1} \Psi_1$, $A \models_{\beta_2} \Psi_2$, and $\alpha = \begin{cases} 1 & \beta_1 = \beta_2 \\ 0 & \beta_1 \neq B_2 \end{cases}$

7. $\phi = [\forall v_n \; \Psi]$, $\alpha = \prod_{z'} \{\beta | \; A \models_\beta \Psi < z' >\}$, where z' ranges over all assignments which agree with z on every variable except possibly on v_n, and no two assignments z' agree on v_n

8. $\phi = [\exists v_n \Psi]$, $\alpha = \sup_{z'} \{\beta | \; A \models_\beta \Psi < z' >\}$ where z' ranges over all assignments which agree with z on every variable except possibly on v_n

Remarks

If any of the components in a compound formula has probability $\bot$ (unknown), the whole formula has probability $\bot$ except for existential formulas (which get $\bot$ only if no positive probability value exist for any assignment of the variable). This rule is a candidate for reexamination in the future.

When probabilities are restricted to be only 0 or 1, these rules are compatible with the FOPC interpretations.

The definition of implication is part of the model, not of satisfaction. There is a big controversy concerning the functionality of the conditional connective, for example, Anderson and Belnap.[1] To support our choice, note that it is rare in the context of expert systems reasoning that we compute the truth value of an implication based on the truth value of the components.

The probability of the disjunction of independent formulas could have been defined as the sum of the respective probabilities. This would require building into the model tools for the analysis of independence of formulas. Similarly, the probability of conjunction of formulas could have been defined as the product of the probability of the first formula and the conditional probability of the second formula. We adopted the definitions given in preceding Statements 4 and 5 for the sake of simplicity of implementation.

Satisfaction of $\phi \leftrightarrow \Psi$ may have a probability value different from that of $[\phi \rightarrow \Psi] \wedge [\Psi \leftrightarrow \phi]$. Tautologies of FOPC involving implication do not necessarily hold in PL.

The relationship between the probability α of $\exists v \phi$ and the probability β of $\forall v \; \neg \; \phi$ may satisfy $\beta \neq 1 - \alpha$.

Equality in our model is strict, not probabilistic. An alternative and more general approach is to allow elements of the universe to be equal with probability p (corresponding to a world in which facts are sometimes distinguishable, and sometimes indistinguishable). This is one of many possible generalizations of the proposed logic.

The proposed interpretation of the connectives is independent of the meaning of formulas. A sophisticated system may be able to determine the dependence/independence of the connected formulas (based on their meaning) and modify the rules accordingly (e.g., probability of "and" of complements may be set to zero).

Just as in FOPC,[6] one can show that if two assignments z and z_1 agree on all free variables of a formula ϕ, then $A \models_\alpha \phi<z>$ $<->$ $A \models_\alpha \phi <z_1>$. This implies that for assertions (formulas without free variables), the probability of satisfaction is independent of the specific assignment.

Example 4: Satisfaction of Formulas

We shall use the formulas of Example 2 and the model of Example 3:

1. $S(s1, N.Y.)$. Probability of satisfaction is .7.
2. $[x = y] \wedge P (y, red)$. If x and y are assigned to the same element, then the probability of satisfaction is the same as that of $P (y, red)$, which is .5 if y is assigned to p1, .3 if to p2, 1 if to p4. If x and y are assigned to different elements, the satisfaction probability is 0.

3. $[\square[y = red]]$. Probability of satisfaction is 1 if y = blue, 1 if y = gren, 0 if y = red.
4. $\forall x \, [P \, (x, \, red \,) \leftrightarrow \exists y \, SP(y, \, x]$. Probability of satisfaction is 0.
5. $\exists x \, [SP \, (x, \, p1) \leftrightarrow SP \, (x, \, p2)]$. Probability of satisfaction is 1.
6. $PS(x, \, y \lor \exists z \, [PS(x, \, z) \land PS(z, \, y]$. PS is identical to its transitive closure, and thus the satisfaction probability equals the probability that the tuple of values assigned to $<x, y>$ is in PS.

III. SEMANTICS, SOUNDNESS, AND COMPLETENESS OF PL

A. A DEFINITIONAL APPROACH TO ASSIGNING MEANING TO FORMULAS

Meaning is defined by a mapping of syntactical objects to semantic ones. An FOPC structure may be viewed as an assignment of meaning to constant and relation symbols. When assigning meaning to formulas, we first have to identify suitable semantic objects, i.e., the range of the meaning function. One way of creating a semantic domain is to view formulas as definitions of relations in the sense of definability theory. A formula without free variables defines in FOPC a singleton set containing either true or false, and in PL it defines the probability that it is satisfied. If a formula has free variables, it defines in FOPC a relation whose degree is the number of free variables, and whose components satisfy the formula. In PL, each formula defines a probabilistic relation. Note that since we do not assume the axioms of probability, relations in PL may be under-defined in the sense that we are unable to decide the belonging of some tuples to a relation. In these cases we adopt the cautious (minimal) approach, and do not include those tuples in the relation. Natural semantics in this context may be the mapping from formulas to the relations defined by them in a given model. This topic is discussed further in References 20 and 21.

Example 5: Meaning of Formulas

We use again the formulas presented in Example 2, and the model of Example 3.

1. $S(s1,N.Y.)$ (supplier s1 is in N.Y.). This formula has no free variables; it defines its probability of satisfaction, .7.
2. $[x = y] \land P \, (y, \, red)$. This defines a relation with two identical components in each tuple, the part numbers of the red parts (p1, p2, p4); the probability is the same as in the corresponding P tuple.
3. $[\square[y = red]]$. This formula defines the set of colors (probability 1 for each entry) which appear in the color domain and are not red.
4. $\forall x \, [P \, (x, red) \leftrightarrow \exists y \, SP \, (y, \, x)]$. This formula defines a probability that every part is red if supplied by some supplier.
5. $\exists x \, [\, SP \, (x, \, p1) \leftrightarrow SP \, (x, p2)]$. This defines the probability of satisfaction of the condition: there exists a supplier x who supplies part p1 with probability p, if x also supplies p2 with probability p.
6. $PS(x,y) \lor \exists z \, [PS(x,z) \land PS(z,y)]$. This defines $PS \cup PS^2$, which in the example model is equal to PS.

B. AXIOMS AND DERIVATION RULES IN PL

The new derivation rule in PL is based on the idea that if formulas ϕ and Ψ are satisfied in a model with probability p_ϕ and p_Ψ, respectively, then their conjunction $\phi \land \Psi$ is satisfied with probability $p_\phi \cdot p_\Psi$. Another rule which may be used in PL is substitution; since equivalent formulas have the same probability, they may be substituted for each other. In addition, we retain the derivation rules of FOPC (including both modus ponens and modus tollens, to accommodate forward and backward chaining).

Let Γ be a set $\{<\gamma_i,p_i>\}$ where γ_i has probability p_i of being true. Note that we may abuse notation and refer to $\{\gamma_i\}$ as Γ. Let C be a set of constraints, and AX a set of axioms (to be described).

Derivation (or **proof, deduction**) in PL of a formula ϕ with probability α $(\neq \perp)$ from Γ and C is a finite sequence of tuples $<\phi_i,p_i>$, i = 1, . . . n, where p_i is the probability associated with ϕ_i; $<\phi_i,p_i> \in \Gamma \cup C \cup AX$ or is obtained from prior tuples in the derivation using a derivation rule, and $<\phi_n,p_n> = <\phi,\alpha>$. In addition, each formula in the sequence, except ϕ, is used in the derivation of a later formula. Note that what is derived is both a formula and its probability.

Following are the axioms and derivation rules of PL. All the axioms have probability 1, which is omitted.

1. All tautologies of FOPC which involve only the implication, negation, and universal quantifier symbols
2. $\forall x\phi(x) \rightarrow \phi(y)$ where y is substitutable for x in ϕ
3. $\forall x (\phi \rightarrow \Psi) \rightarrow \forall x\phi \rightarrow \forall x\Psi)$
4. $\phi \rightarrow \forall x\phi$ where x is not free in ϕ
5. $x = y \rightarrow (\phi \rightarrow \Psi)$ where ϕ is atomic and Ψ is obtained from ϕ by replacing x in zero or more (but not necessarily all) places by y

Derivation Rules

1. **Conjunction:** $\dfrac{<\phi, \alpha>, <\Psi, \beta>}{<\phi \wedge \Psi, \alpha \cdot \beta>}$

2. **Substitution:** $\dfrac{<\phi, \alpha>, <\phi \leftrightarrow \Psi, 1>}{<\Psi, \alpha>}$

3. **Modus ponens:** $\dfrac{<\phi, \alpha>, <\phi \leftrightarrow \Psi, \beta>}{<\Psi, \alpha \cdot \beta>}$

4. **Modus tollens:** $\dfrac{<\neg\Psi, \alpha>, <\phi \rightarrow \Psi, \beta>}{<\neg\phi, \alpha \cdot \beta>}$

5. **Transitivity of implication:** $\dfrac{<\phi \rightarrow \Psi, \alpha>, <\Psi \rightarrow \theta, \beta>}{<\phi \rightarrow \theta, \alpha \cdot \beta>}$

6. **Resolution:** $\dfrac{<\phi\vee\Psi, \alpha>, <\neg\phi\vee\theta, \beta>}{<\Psi\vee\theta, \alpha \cdot \beta>}$

7. **Specification:** Let x be a free variable in ϕ, and $\phi(a)$ be any substitution of a constant symbol a to x in ϕ. then: $\dfrac{<\phi(x), \alpha>}{<\phi(a), \alpha>}$

Remarks

If all formulas in $\Gamma \cup C$ have probability values of either 0 or 1 corresponding to classical FOPC, the definition of derivation from Γ in PL is equivalent to the FOPC notion of derivation from $\Gamma \cup C$. We get the following:

Theorem: A formula ϕ is derivable from a set of formulas Γ in FOPC if ϕ is derivable with probability 1 in the corresponding PL in which all formulas in $\Gamma \cup C$ have probability 1. The proof is straightforward.

In some cases, more than one derivation of a formula may exist. Each derivation may assign the formula a distinct probability, thus creating a conflict. This may cause inconsistency in operations and reduce the usefulness of a system utilizing PL. Conflict resolution may be achieved in several ways:

- Additional information may be used to resolve the conflict. Among such additional information, probabilities expressing the probability estimates correctness (and possibly also higher levels) may be included. The more reliable estimate may be given priority.
- Compute the weighted average of the estimates over all possible derivations.
- We can also take the maximal value of the probability of derived formula, since the independence assumption tends to lower probabilities.
- Yet another approach may be to apply a relaxation process (similar to what is done in truth maintenance systems[4]).
- In the context of nonmonotonic logic and semantic nets, it was suggested to give preference to the shortest derivation path. Touretzky[15] further suggests to ''minimize'' a theory T from which the conclusions are drawn by eliminating from T formulas which may be deduced from the rest of T.
- When experience is gained with a specific theory, we may assign preference specific to the application and/or the theory.

We do not allow the inference of $< \neg \phi, 1 - p_\phi >$ from $<\phi, p_\phi>$, since when coupled with some of the conflict resolution policies, this deduction might lead to $p_\phi + p_{\neg\phi} > 1$.

Example 6: Consequences of Constraints

Let Γ include the set of atomic formulas satisfied in the model of Example 3 (the diagram). For each possible atomic formula not included in SP and F in Example 3, the probability of the formula is 0 (a closed-world-assumption for the relations SP and F).

Let the constraints set C include

C_1: $\forall xy\ F\ (x,y) \leftrightarrow F\ (y,x)$ (symmetry of F) — probability 1
C_2: $\forall xy\ PS\ (x,y) \leftrightarrow PS\ (y,x)$ (symmetry of PS) — probability 1
C_3: $[F\ (x,y) \wedge SP\ (x,z)] \rightarrow SP\ (y,z)$ (family members supply the same parts) — probability .9
C_4: $[SP\ (x,y) \wedge P\ (y,blue)] \rightarrow \neg S\ (x,N.Y.)$ (New Yorkers do not supply blue parts) — probability .8

First, let us discuss the consequences of C_3. The only two distinct suppliers which are family related are S1 and S2 (with probability .8). We shall check whether this constraint implies higher probability for some of the tuples in the relation SP. Using the conjunction derivation rule, we can compute the probability of ϕ, the left-hand side of C_3. Then we can use modus ponens to derive the probability of the right-hand side of C_3. Consequently, some of the probabilities are increased; in case the constraint implies a lower probability than was given, we retain the higher value. The following tuples probabilities of the relation SP are changed: $<s1, p1>$ to .72. $<s2, p3>$ to .36.

A similar analysis for the consequences of C_4 shows that in the relation S, the following tuples probability is reduced (since in this case the complement is the derived formula): $<s1, N.Y.>$ to .44, $<s3, N.Y.>$ to .44. Notice that although the probability of these tuples decreased, the probability of the other tuples (e.g., $<s1, N.O.>$) is not changed.

Example 7:

This example shows how a useful combination of PL and probability or belief theory may be achieved. We take Shafer's example, *The Burglary of the Sweetshop*, from Reference

14. The highlights of the case include some evidence that the thief is left-handed, and another evidence that the theft was an inside job. Using Shafer's theory we can compute the combined effect of the evidence against the single left-handed clerk LHC of the sweetshop. Assume now that the following additional information is available:

- If LHC is guilty then a new clerk must be hired (probability α).
- There is a shortage of clerks (probability 1).
- If there is a shortage of clerks and a new clerk is hired, then salary of new clerk > salary of previous clerk (probability β).
- The sweetshop, economically, just breaks even, and any additional expense will force its bankruptcy (probability 1).
- Now conclude that the sweetshop will go bankrupt with probability $\alpha \cdot \beta$.

C. CONSISTENCY, SOUNDNESS, AND COMPLETENESS OF PL

Whenever conclusions have to be derived using inferences (in the context of ES, for example), very important characteristics of the system are its soundness and completeness. In FOPC, a logical system is sound if, given any consistent set of formulas, only true formulas may be inferred. It is complete if, given a set of formulas Γ, every formula logically implied by Γ is derivable from Γ. We have to translate these FOPC notions to the context of PL, in which a formula, rather than being true in a model, has a probability of being true.

Let Γ be a nonempty set of formulas. We define the probability of Γ by $p_\Gamma = \Pi_i \, p_i$ where $i \in I$, I is an index set for Γ, and p_i is the probability of satisfaction of the ith formula in Γ. Obviously, for p_Γ to be positive, only a finite number of formulas in Γ may have probability <1. An alternative definition is $p_\Gamma = inf_i \, p_i$. Clearly, for each $\phi \in \Gamma$, $p_\phi \geq p_\Gamma$. In both cases FOPC satisfaction of Γ is equivalent to having $p_\Gamma = 1$.

Let ϕ be a consequent of Γ, and let Γ' be $\Gamma \cup \phi$. Whichever of these definitions we adopt, the probability of Γ' may be less than that of Γ. To see why, consider first the *inf* definition. The derived formula probability is a product of other probabilities and so may have a probability less than the *inf*. The same phenomenon might occur even if we adopt the first definition. Consider for example $\Gamma = \{\phi, \Psi, \phi \to [\phi \to [\Psi \to \theta]]\}$. If all formulas in Γ have probability p, than Γ has probability p^3 but θ has probability p^4, which is strictly less than p_Γ when $p < 1$. We conclude that both definitions are potentially valuable for the following discussion.

To study consistency, let us consider first logical (semantic) implication:

- **Definition:** A set of formulas Γ *logically implies* (or simply implies) a formula ϕ, denoted by $\Gamma \models \phi$, if in every model $p_\phi \geq p_\Gamma$.
- **Definition:** A set of formulas Γ *logically implies to a degree* α a formula ϕ, denoted by $\Gamma \models_\alpha \phi$, if in every model $p_\phi \geq \alpha \cdot p_\Gamma$.

A set of formulas Γ in FOPC in consistent if we cannot derive from Γ both a formula ϕ and *not* ϕ. This concept can be generalized to PL in more than one way. A starting point is to require that PL consistency be compatible with FOPC consistency whenever all the probabilities are 1. This number (1) may be interpreted as the difference between the probability of satisfaction of a derived formula and that of its complement.

Let Δ_Γ be the set of formulas derivable from Γ:

- **Definition:** Γ is *consistent to a degree* α if $\alpha = \overset{inf}{_{\phi \in \Delta_\Gamma}} |p_\phi - p_{-\phi}|$. If we cannot derive ϕs complement, we say that we derive at with probability $\perp$. For any α define: $\alpha - \perp = \alpha$.

Let us consider another approach to consistency. In FOPC, the consistency of Γ is equivalent to having a model for Γ. This can be translated in PL to having a positive probability for Γ. Note that changes in probability are possible as derived formulas are added to Γ, thus making the degree of consistency dependent on whether or not the consequences are added to the set (the first definition of consistency does not have this drawback).

For a second motivation for adopting the first definition of consistency, consider the following case: Γ has probability 1 and there are no constraints, and we may derive from Γ both a formula and its negation (this is possible in PL since implications are assigned probability by the model). In this case, all formulas derived from Γ will have probability 1. The first definition will assign in this case a consistency 0 while, according to the second definition, the consistency is 1. Therefore, in the following, "consistency" will be used in the meaning assigned by the first definition.

It is easy to see that consistency of Γ to degree 1 implies that $p_\Gamma = 1$ (and the same goes for the constraints). The converse is not true, as the formulas ϕ, $\phi \rightarrow \Psi$, and $\phi \rightarrow \neg \Psi$ may all have probability 1, but the consistency of the set is 0.

Consistency check may be used to assess the quality of the probability estimates of the experts (probability assignments of the model) to rules (implication formulas). Note that an expert system may be inconsistent in the everyday usage of the word by assigning different probability values to the same formula when different derivations are used. This is typical of expert systems execution, where many outputs may correspond to a single input, since the choice of rules to be fired is not completely deterministic.

The notion of *extended consistency* is also relevant to expert systems. This is the consistency of a set of implication formulas (rules). If Γ is a set of implication formulas, then only few conclusions can be drawn: the original formulas and their conjunctions, transitive closure, and substitution of constant symbols. Let Σ_1 be a set of formulas consisting of all the left-hand-side parts of implications in Γ, and let Σ_2 be a set of formulas consisting of all the negated right-hand-side parts of implications in Γ. When Σ_1 or Σ_2 is joined with Γ, additional conclusions may be drawn. It is the minimal consistency of $\Sigma_1 \cup \Gamma$, $\Sigma_2 \cup \Gamma$ which corresponds to our notion of the extended consistency of the rules. Obviously, extended consistency to degree α implies consistency to degree at least α, but not vice versa.

We are only interested in derivation from sets of formulas which are consistent to some positive degree. Given such a set of formulas Γ, define:

- Relative soundness: a derivation of a formula ϕ from a set of formulas Γ is relatively sound if $p_\phi \geq p_\Gamma$.
- Uniform soundness: a derivation of a formula ϕ is uniformly sound to degree k $(k \geq .5)$ if $p_\phi \geq k$.

It is easy to see that only consistency to degree 1 ensures both relative and uniform soundness. For a counter example consider the following Γ: $\{<\phi, p>, <\phi \rightarrow (\phi \rightarrow \theta), 1>\}$, and let $p < 1$; $p_\Gamma = p$ but $p_\theta = p^2 < p$, so we do not have relative soundness. This example can be modified (by substituting for the second formula $\phi \rightarrow \ldots \phi \rightarrow (\phi \rightarrow \theta)$) to show that absolute soundness is not ensured (for any positive value of k). Consistency to degree k implies uniform soundness to degree k. Derivations from sets of formulas consistent to some positive degree less than one (the common case for expert systems) may be sound, but each one must be carefully checked.

Completeness: PL is complete if for each Γ and formula ϕ, if $\Gamma \models_\alpha \phi$ then $\Gamma \vdash_\alpha \phi$. In the case of consistency to degree 1 and all constraints probability 1, corresponding to FOPC consistency, one can show that PL is complete for $\alpha = 1$, i.e., if $\Gamma \models_1 \phi$ then $\Gamma \vdash_1 \phi$. The proof idea is to use FOPC completeness, and we outline the technical details. Let Σ denote the union of the constraints C, Γ, and ϕ. Given a PL model S satisfying Σ with probability 1, construct a first order structure S' satisfying Σ in the following way: each atomic formula in S' will be true provided it has a positive probability in S. By structural induction on formulas in Σ it can be shown that S' satisfies them (use that every formula in Σ is satisfied to degree 1; the interesting part is implications, and the consistency condition on implications is the required key). This mapping from PL models to FOPC structures is onto. By completeness of FOPC, there exists a derivation in FOPC of ϕ from $\Gamma \cup C$, which can be easily translated into a proof of ϕ from Γ in PL (since the axioms of FOPC were made the axioms of PL). Therefore we have outlined the following:

Theorem — For PL theories where all probabilities are 1, derivation is sound and complete. In other words, PL is a generalization of FOPC. When the degree of consistency is less than 1, PL is not complete. For a proof consider $\Gamma = \{<\phi \lor \neg \phi, 1>, <\phi \to \Psi, .9>, <(\neg \phi) \to \Psi, .9>\}$; $\Gamma F_{.9} \Psi$ (since either ϕ or $\neg \phi$ has probability 1, and by modus ponens Ψ's probability is .9). Yet, there is no derivation of Ψ from Γ.

IV. SUMMARY AND PROPOSED RESEARCH

No expert system shell can be considered reliable without taking proper care of the uncertainty issue. In this work we suggest a formal logical system, PL, based on probability (or belief) estimates, to monitor uncertainties. The language and notions of the first order predicate calculus are used as a starting point on which we generalize to accommodate probabilities. We have shown that PL is compatible with FOPC.

The system described in this paper represents one out of many possible alternatives. In making an appropriate choice of a set of axioms, and a corresponding definition of satisfaction and derivation rules, the developer of the expert system could taylor a shell to his specific application. Further research is necessary to assist the developer in this task by improving the logical system modeling power, developing criteria for comparing between methodologies, and providing a set of test cases to validate the tuning of the uncertainty management system. In particular, we consider the following points to be important:

1. Study other interpretations of the logical symbols in PL; present the user with a comprehensive set of alternatives.
2. Study other axiomatizations of PL (especially the intuitionistic approach).
3. Use an interval approach to probability values to achieve greater accuracy.
4. Add probability of probability estimates (can be used as a second order correction to probability) for conflict resolution, in case different reasoning paths give conflicting conclusions.
5. Generalize PL formulas to reason about formulas (auto-epistemic capabilities) by including the probability of formulas in the model.
6. Improve the expressive power of the language by use of the modifiers of fuzzy logic.[20] Compare PL and fuzzy logic.[18]
7. Use modal logic for interpreting PL, especially the use of Kripke structures to assign semantics to modifiers.

ACKNOWLEDGMENT

This research was conducted while the author was with the Computer Science Department at Louisiana State University.

REFERENCES

1. **Anderson, A. R. and Belnap, N. D., Jr.**, *Entailment: The Logic of Relevance and Necessity,* Vol. 1, Princeton University Press, New Jersey, 1975.
2. **Buchanan, B. G. and Shortliffe, E. H.**, *Rule-Based Expert Systems,* Addison-Wesley, Reading, MA, 1984.
3. **Cheeseman, P.**, In Defense of Probability, Proc. 9th IJCAI, 1985, 1002.
4. **Doyle, J.**, A truth maintenance system, *Artif. Intell.,* 12, 231, 1979.
5. **Dubois, D. and Prade, H.**, Fuzzy logics and the generalized modus ponens revisted, *Cyber. Syst.,* 15, 293, 1984.
6. **Enderton, H. B.**, *A Mathematical Introduction to Logic,* Academic Press, New York, 1972.
7. **Grosof, B. N.**, An inequality paradigm for probabilistic knowledge, in *Uncertainty in Artificial Intelligence,* Kanal, L. N. and Lemmer, J. F., Eds., Elsevier, New York, 1986, 259.
8. **Halpern, J. Y. and Rabin, M. O.**, A Logic to Reason About Likelihood, Proc. 15th STOC, 1983, 310.
9. **Hoover, D. N.**, Probability logic, *Ann. Math. Logic,* 14, 287, 1978.
10. **Keisler, H.**, Hyperfinite model theory, in *Logic Colloquium 76,* Gandy, R. O. and Hyland, J. M. E., Eds., North-Holland, Amsterdam, 1977, 5.
11. **Nilsson, N. J.**, Probabilistic logic, *Artif. Intell.,* 28, 71, 1986.
12. **Parnas, D. L.**, Software aspects of strategic defense systems, *Am. Scient.,* 73, 1985.
13. **Reiter, R.**, On closed world databases, in Gallaire, H. and Minker, J., Eds., Logic and Databases, Plenum Press, New York, 1978.
14. **Shafer, G.**, *A Mathematical Theory of Evidence,* Princeton University Press, New Jersey, 1976, 80.
15. **Touretzky, D. S.**, *The Mathematics of Inheritance Systems,* Morgan Kaufmann, Los Altos, CA, 1986.
16. **Weber, S.**, A general concept of fuzzy connectives, negotiations and implications based on t-norms and t-conorms, *Fuzzy Sets and Syst.,* 11, 115, 1983.
17. **Zadeh, L. A.**, Fuzzy sets, *Inf. Control,* 8, 338, 1965.
18. **Zadeh, L. A.**, The role of fuzzy logic in the management of uncertainty in expert systems, *Fuzzy Sets Syst.,* 11, 199, 1983.
19. **Zvieli, A.**, Queries and Definability, D.Sc. thesis, Technion-IIT, Haifa, Israel, 1984.
20. **Zvieli, A.**, On Fuzzy Complete Relational Query Languages, Proc. N. Am. Fuzzy Inf. Proc. Soc. Int. Meet., New Orleans, June 1986.

Chapter 20

COMEX—AN AUTONOMOUS FUZZY EXPERT SYSTEM FOR TACTICAL COMMUNICATIONS NETWORKS

Mordechay Schneider, Joseph M. Perl, and Abraham Kandel

TABLE OF CONTENTS

I. INTRODUCTION

Reliable and efficient communication is crucial in many situations where the success or failure of an endeavor depends on the ability to transmit and receive important information. Such situations abound in military applications as well as in civilian ones such as in the merchant navy, oil exploration, aviation, etc. At present, although a variety of communication means and equipment is available, the success or failure of the communication depends largely on the operator's expertise. This is particularly evident in ionospheric HF communication,[1] although it is true on other communication channels as well.

COMEX is an intelligent expert system designed to perform the tasks presently performed by an expert operator. Thus, it selects the proper communication channel (HF, VHF, UHF, satellite, telephone) for the desired communication (voice, data); it selects the proper path (direct, relay) and equipment to be used (modem, radio, antenna); it selects the equipment operating parameters as well as the frequency and protocol to be used; and finally, it performs the communication. COMEX employs an initially loaded data base, which is automatically updated by the expert system learning capability. In addition, it collects and updates statistical data concerning channel and equipment availability and communication performance. COMEX communicates with utility programs running on the same processor (IONCAP, MINIMUF, satellite visibility prediction and BER performance evaluation) as well as with external units such as frequency management systems.[1-7]

These frequency management systems measure channel occupancy, signal-to-noise levels, (pseudo-) bit error rates, fading rates on time varying channels, multipath propagation conditions, and Doppler shifts. They can operate autonomously or under the control of the expert system, depending on the operation mode of COMEX.

COMEX runs on an IBM compatible PC, which also acts as the communication station controller. When the station equipment (radios, modems, terminals) has remote control capability, it will remotely control the equipment, set the selected operation modes, and perform the communication.

COMEX is intended for the use of novice operators and is not limited to communication station or equipment constraints. During the initial setup, COMEX prompts the operator to provide all the information necessary for an intelligent and expert decision on solving the communication problem at hand. Its intelligence enables it to interface with a variety of channel monitoring sensors, and accept and use the information provided by these autonomous sensors to achieve an optimal solution to the communication problem at hand.

II. DESCRIPTION OF COMEX

Figure 1 presents the architecture of COMEX, which is composed of a number of modules. The man-machine-interface (MMI) module handles the interactions between the user and COMEX. Through the MMI module, the user loads and updates the "user data base". This data base accepts data from external sensors as well. Such external sensors are frequency management and communication monitoring devices,[1] which communicate with COMEX via a standard RS-232 serial I/O port. For example, in the HF band, COMEX can operate with a variety of sensors, including MESA,[1] AUTOCAL,[7] AUTOLINK,[6] SELSCAN,[5] etc.

The main part is the CONTROLLER, which contains the overall operating system that controls the operation of the various modules. The CONTROLLER operates the EXECUTE module, which carries out the communication with the communication station equipment via a serial I/O port. Thus, when the expert system is ready to commence communication (transmit or receive), the EXECUTE module performs this task. In addition, the CONTROLLER supervises the operation of a "statistical data base", which accumulates com-

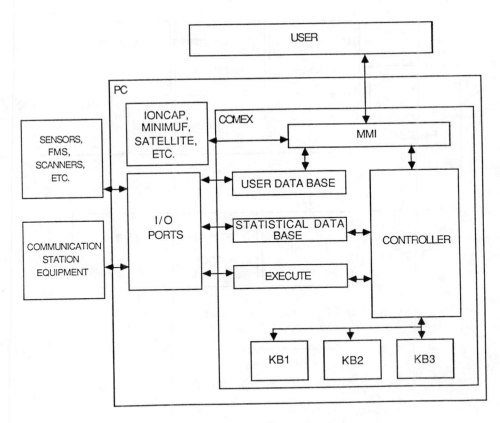

FIGURE 1.   COMEX architecture.

munication statistics in real time. Certain information, such as the sample distributions of various channel parameters, are also accumulated in this data base by the external sensors (see Figure 2). It also supervises the activities of MMI, especially the handling of the data base provided by the user and by the external sensors. Finally, the CONTROLLER employs the three knowledge bases (KBs) during the inference procedure. The CONTROLLER is in fact an expert system which also manages the various functions performed by COMEX.

A. THE MMI

The MMI is a self-prompted, human-friendly interface, which leads the user through the various steps of operating COMEX. Since COMEX requires information concerning various aspects of the communication network, net members, available communication equipment, traffic type, frequency assignments, topological conditions, system configuration, etc., the MMI interrogates the operator with respect to these and additional aspects. The MMI classifies and stores this information in its data base and transfers pertinent parameters to the CONTROLLER expert system.

In addition to the information loaded by the operator, the MMI accepts relevant information from external sensors, such as frequency management systems,[1-7] and from resident utility programs running on the same processor. Such utility programs include Fresnel zone propagation, IONCAP, MINIMUF, satellite visibility, and BER performance evaluation programs.

The following examples (see Figures 3 through 6) present a sample of the numerous screens presented to the user by the MMI. Through these interactive, self-prompted screens the communication station operator loads the various data that are relevant to the operation of COMEX or instruct COMEX to perform the communication.

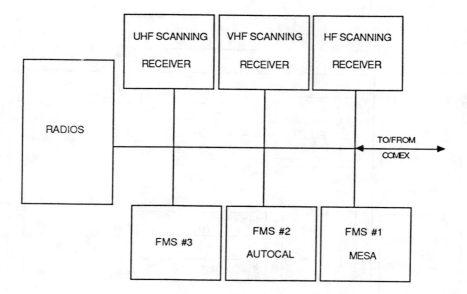

FIGURE 2.   The external sensors attachable to COMEX.

```
      t to transmit

      p to change/update parameters




ESC to quit
```

FIGURE 3.   Starting screen of COMEX.

```
    selcall
    broadcast
    roll-call




ESC to return to main menu;   Up-Down arrows to move on screen;
RETURN to confirm selection.
```

FIGURE 4.   Screen following the selection of transmit option.

```
┌─────────────────────────────────────────────────────────┐
│                                                           │
│     Updating Station: Own                                 │
│                                                           │
│     Topological Cond.................sea/planes/hills/mountains │
│                                                           │
│     Comm. Channel.................. undef.                │
│                                                           │
│     Mobile/Fix............................ undef.         │
│                                                           │
│     Comm. Path......................... undef.           │
│                                                           │
│     Comm. Protocol.................. undef.               │
│                                                           │
│     Comm. Means..................... undef.              │
│                                                           │
│                                                           │
│  ESC to return to menu;        Up-Down arrows to move on screen; │
│                                                           │
│  RETURN to confirm selection;  PgUp/PgDn to select screens. │
│                                                           │
└─────────────────────────────────────────────────────────┘
```

FIGURE 5.   Screen following selection of parameter change option. In this case the user's own parameters are being updated. The sea option was selected out of four possibilities, the rest of parameters still being undefined.

```
┌─────────────────────────────────────────────────────────┐
│                                                           │
│          Updating Station: Own                            │
│                                                           │
│          Topological Cond...............sea               │
│                                                           │
│          Comm. Channel...............automatic            │
│                                                           │
│          Mobile/Fix.......................mobile          │
│                                                           │
│          Comm. Path.....................relay via Joe     │
│                                                           │
│          Comm. Protocol................ARQ               │
│                                                           │
│          Comm. Means.................manual              │
│                                                           │
│              Trans. Power............low                 │
│                                                           │
│              Receiver...................auxiliary         │
│                                                           │
│              Trans. Ant...............LPA                │
│                                                           │
│              Receive Ant..............active             │
│                                                           │
│                                                           │
│  ESC to return to menu;        Up/Down arrows to move on screen; │
│                                                           │
│  RETURN to confirm selection;  PgUp/PgDn to select screens. │
│                                                           │
└─────────────────────────────────────────────────────────┘
```

FIGURE 6.   Full first screen of parameters for user Own.

In Figure 7 the operator has selected a Manual FMS operation mode. Thus, the communication channels recommended by the frequency management (FMS) must have better communication conditions than the conditions set in Figure 7. These parameters are being measured by external sensors (see Figure 8) and updated periodically.

B. THE CONTROLLER

The CONTROLLER is the intelligent subsystem which controls the operation of the various other COMEX components. It accepts action commands from the operator via the MMI. Such commands include "Call Joe"; "Call 4231"; "Broadcast"; "Call Group 12"; "Monitor Channels 1, 16, international emergency channels (EMERG)"; etc. The decision

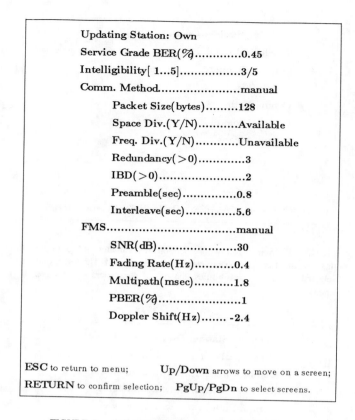

```
          Updating Station: Own
          Service Grade BER(%)............0.45
          Intelligibility[ 1...5]................3/5
          Comm. Method.....................manual
              Packet Size(bytes).........128
              Space Div.(Y/N)...........Available
              Freq. Div.(Y/N)............Unavailable
              Redundancy(>0).............3
              IBD(>0).......................2
              Preamble(sec)..............0.8
              Interleave(sec)..............5.6
          FMS...................................manual
              SNR(dB)....................:...30
              Fading Rate(Hz)...........0.4
              Multipath(msec)...........1.8
              PBER(%).......................1
              Doppler Shift(Hz)....... -2.4


  ESC to return to menu;      Up/Down arrows to move on a screen;
  RETURN to confirm selection;   PgUp/PgDn to select screens.
```

FIGURE 7.   Second screen of parameters for user Own.

as to how best establish the desired communication is reached by the expert system, with the execution by the EXECUTE module.

The CONTROLLER also continuously controls the statistical data base. This data base collects and updates station statistics concerning the success/failure of communication attempts, channel qualities, correlation of predicted and achieved performance, channel parameters analysis and scaling, channel model and time constants, connectivity matrices, etc. This data base is used by the CONTROLLER subsystem to reach the solution for the communication problem. The various statistics accumulated in the data base are also used to update the ''certainty factors/membership grades'' of the fuzzy parameters used by the expert system for uncertainty management.

COMEX contains three knowledge bases (KBs), on which the expert system operates in solving the communication problem. KB1 is used in the CHANNEL TYPE SELECTION — where according to the desired communication task and constraints of the available equipment, range, available channels, etc. — the expert system selects and recommends the use of telephone, MF, HF, VHF, UHF, or satellite communication channels. The CONTROLLER then utilizes KB2 in the EQUIPMENT and PATH SELECTION. Thus, according to the size and type of message to be transferred (the range, the antenna coverages, etc.), the expert system selects and recommends the specific antenna, transmitter, receiver(s), modem, and terminal to be used. It also selects and recommends the path and the relay station (if needed) to be used.

Finally, the expert system employs KB3 to perform the OPERATING PARAMETERS SELECTION by selecting the modem operating parameters (modulation, data rate, code, interleaving depth, preamble length, IBD, OBD, etc.); the radio operating parameters (power,

```
┌─────────────────────────────────────────────────────────────┐
│                                                               │
│        Updating Station: Own                                  │
│        Phone(Y/N).........................Unavailable         │
│        HF/MF(Updating)................Available               │
│        VHF/UHF(Y/N)..........................Undef            │
│        Satellite(Y/N)...............................Undef     │
│                  # XMTRs(<5).............1                    │
│                  # XMTR ANT(<5).......2                       │
│                  # RCVRs(<5)..............2                   │
│                  # RCVR ANT(<5)........2                      │
│                  # Modems(<5).............1                   │
│                  FMS Available(<5)........2                   │
│                        FMS # 1..............MESA              │
│                        FMS # 2...................Undef        │
│                                                               │
│                                                               │
│      ESC to return to menu;      Up/Down arrows to move on screen; │
│      RETURN to confirm selection;   PgUp/PgDn to select screens.  │
└─────────────────────────────────────────────────────────────┘
```

FIGURE 8.   Equipment updating screen for user Own. The external HF frequency management sensors are in the process of being defined, the first one being MESA with the other still undefined.

bandwidth, diversity, etc.); operating frequencies and communication protocol (FEC, ARQ, packet size, etc.). The CONTROLLER has now all the necessary knowledge to perform the communication in an "optimal" and "expert" manner, just as an expert operator would. This described cycle of decisions based on the three separate knowledge bases may be performed a number of times before the CONTROLLER reaches the optimal solution and decides to operate the EXECUTE module (if in AUTOMATIC operating mode) or notify the human operator of its conclusions (if in MANUAL operating mode).

III. THE INTERNAL STRUCTURE OF THE CONTROLLER

This section describes the internal structure of the CONTROLLER. The expert system employs three distinct knowledge bases and is divided into four main logical parts,[8-27] as shown in Figure 9:

1.   Blackboard
2.   The inference engine
3.   The question-answer program (QAP)
4.   Communication channel for communicating with other modules

The blackboard is a global data structure (i.e., the contents of the blackboard is reachable from any part of the software) that stores intermediate results from the inference procedure and contains important information about the data. Before we describe the components of the blackboard, it is important to understand conceptually its function. Suppose we have a knowledge base containing two rules:

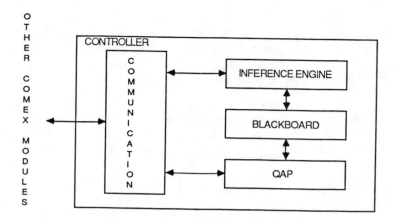

FIGURE 9.   The general structure of the CONTROLLER expert system.

if a then b

if b then c

and suppose that the user provides the system with the data a. First, since the user provides the data a, we put it on the blackboard. The blackboard now contains the truth sentence a. The inference engine starts to evaluate the knowledge base by matching the premise of each rule to the data placed on the blackboard. Since the premise of the first rule is matched against the data on the blackboard, the conclusion of that rule becomes true and is placed on the blackboard. Thus, the blackboard now will contain the data a and b. The inference engine matches the premise of the second rule against the blackboard (since the premise of the second rule is b and the blackboard contains b) so that the conclusion of the second rule, c, becomes true and is placed on the blackboard.

The blackboard plays a very important role in the inference process, since without storing the intermediate results on the blackboard the inference engine would not be able to perform its task. Each line on the blackboard contains:

1.   Key words (to be described later)
2.   Certainty factor (CF), which is used for the evaluation of the conclusion(s)[23]
3.   Rule number (RN), which is the same as the number of the rule that was fired and its conclusion placed on the blackboard

The expert system described here uses production rules as a knowledge representation method.

The general structure of a production rule is

$$IF \; P \; THEN \; C \qquad (1)$$

where P is the premise of the rule and C is the conclusion. The general structure of P is (using the BNF notation):

$$P ::= S \mid (P) \mid NOT \; P \mid S \; OR \; P \mid S \; AND \; P \mid$$

S ::= An English phrase in the form: the A of [the] B is [not] [adjective] C. Where A, B, and C are key words. An example to S can be "the range of VHF is 25 miles". In this case A = "range", B = "VHF", and C = "25".
The general structure of the knowledge base is

$$R \quad 1$$
$$\text{IF } P_1 \text{ then } C_1$$

$$R \quad 2$$
$$\text{IF } P_2 \text{ then } C_2$$

.

$$\text{IFP}_2$$

.

.

.

$$R \quad n$$

$$\text{IF } P_n \text{ then } C_n$$

For example, some of the rules in KB1 are

120. - - - - - - -
121. If range of target is long and satellite is available then type of ltransmit is satellite.
122. If range of target is long and not satellite is available and HF is available then type of ltransmit is HF.
123. If range of target is short and satellite is available then type of stransmit is satellite.
124. If range of target is short and not satellite is available and VHF is available then type of stransmit is VHF.
125. If range of target is short and not satellite is available and not VHF is available and HF is available then type of stransmit is HF.
126. - - - - - - -

The knowledge base is associated with a special matrix called the bit matrix (BM). BM is an N by N matrix where N represents the number of rules in the knowledge base, so that:

$$BM\ [i,j] = \begin{cases} 1 \text{ if the conclusion of rule } i \text{ is} \\ \quad \text{involved in the evaluation of rule } j \\ 0 \text{ otherwise} \end{cases} \tag{2}$$

and

$$if \sum_{j=1}^{N} BM\ [i,j] = 0 \text{ then the conclusion of rule } i \tag{3}$$
$$is\ a\ terminal\ in\ the\ decision\ tree$$

The bit matrix has three functions:

1. It is used by the question-answer program (QAP) to trace the inference path when describing how a conclusion was reached.
2. It helps the inference engine to guide the user to reach a conclusion when the data which was provided by the user is insufficient.
3. As described in Equation (3), the BM can determine which rule is a concluding rule (i.e., a terminal) and which rule is an intermediate rule.

The bit matrix is created during the initial start-up procedure and updated when a new rule is inserted into the knowledge base or a rule is deleted from the knowledge base. As shown in Equation (2), an entry $BM[i,j]$ accepts the value 1 if the conclusion of rule i can contribute to the firing of rule j and 0 otherwise. In other words, if the conclusion of rule i is similar to a clause in rule j then 1 is inserted in $BM[i,j]$. Thus, a matching process takes place to determine the similarity between the conclusion of rule i and a clause in rule j. As was stated earlier, each clause in a production rule is in the form:

THE A of [the] B is [not] [adjective] C

where A and B are key words and C is either a key word or a number or a range of numbers describing B.

The matching process is performed as follows. If the key word A in the conclusion of rule i or its synonym is the same as the key word A in the clause of the premise in rule j or its synonym, then we say that the A part is matched successfully. The same matching process is conducted for the key words B and C. If C is a number or a range of numbers, then we perform the evaluation of C in the conclusion of rule i and in the clause of the premise in rule j and perform the matching as will be described later. If all three key words do match, then we assume that conclusion i may contribute to the firing of rule j. When the bit matrix is created, a final check is performed:

$$If\ BM[i,j]\ =\ BM[j,i]\ =\ 1\ then\ flag\ an\ error$$

If the conclusion of rule i may fire rule j and the conclusion of rule j may fire rule i, then we have identity and we do not need this rule. Consider the following two rules:

IF a THEN b

IF b THEN a

In this case the reasoning path will be

b because of a (rule 2)

a because of b (rule 1)

b because of a (rule 2)

a because of b (rule 1)

.

.

As can be seen, we have an infinite loop.

```
PROCEDURE conclusion (k);  (* k is the rule number to be retrieved *)
FOR I = 1 to n do          (* n is the number of rules in the knowledge base *)
IF BM[I,K]='1' THEN
BEGIN
     find out if rule I is in LR;   (* if rule I was used in firing rule K *)
     if so then
     BEGIN
         print the conclusion of rule I;
         activate procedure ===> conclusion(I);
     END;
END;
```

FIGURE 10.   The QAP algorithm.

Another problem that can be created is that since each line will contain the value 1 ($BM[1,2] = BM[2,1] = 1$), the expert system will not be able to reach a conclusion. It will fire both rules but will terminate the inference process unsuccessfully because there is not at least one concluding rule in the knowledge base.

A question-answer program (QAP) is used to retrieve the chain of firings in the inference procedure and display all rules and data that were involved in the firing of that rule. The user may invoke the program by typing:

explain rule s

where s is the desired rule. First the QAP checks to ensure that rule s is in the list of fired rules (LR). If the rule was not found in the LR then the program is terminated (after an error message is displayed to the user). If the rule number is found in LR, the process continues.

The QAP uses the bit matrix (BM) to retrieve rules and data. As was shown in Equation 2:

$$BM[i,j] = 1$$

means that rule i may contribute to the firing of rule j. But it also means that rule j can be fired from rule i. Using this logic, we can derive a general recursive algorithm for retrieving the rules and data which were involved in the firing of the rule in question. The algorithm is presented in a Pascal-like language in Figure 10. If column k in BM consists of only zeros then, necessarily, conclusion k came from user-supplied data.

Since this procedure is activated recursively, it is guaranteed to find all the rules and data which were involved in the firing of rule k.

IV. COMMUNICATION IN COMEX

One of the major issues in constructing cooperative expert systems is the communication between the expert systems and within the expert system. Requests and replies must arrive to the right expert system, they must be understood, and it is desirable for them to be simple and short. In COMEX the communication is performed via a shared memory area.

Since COMEX uses a communication protocol to communicate between MMI and the CONTROLLER to pass data and knowledge, it can be treated as a local network of processes. To reduce the complexity in the design, most networks are organized as a series of *layers*, each one built upon its predecessor. The number of layers and the functions of each layer may differ from network to network. However, in all networks, the purpose of each layer is to offer certain services to the higher layers, shielding those layers from the details of how the offered services are actually implemented.

The communication process in COMEX belongs to the "data link" layer of the ISO-OSI model, with three sublayers:

First sublayer — This consists of three key words:

1. A key character: "R" for request or "A" for answer
2. The address of the sender
3. The designated address

Thus the first sublayer looks like:

A MMI CONTROLLER

which means a reply from MMI to CONTROLLER.

Second sublayer — This contains the code number for the request. In other words, there are several code numbers which are understood by both the sender and the receiver. This code number must be unique for both expert systems. For example, if the user wants to see a rule, then he asks MMI to print a rule. This request is made in a natural language. MMI interprets the request and sends the request to CONTROLLER in the following way:

1. R MMI CONTROLLER {request from MMI to CONTROLLER}
2. 3   {the code for printing a rule}

Third sublayer — This consists of the appropriate data, which are associated with the code number just described. Each code number must be followed by the appropriate data. If, for example, a user wants to print a rule, then he will enter the request in the following way:

print rule 6

MMI interprets the request and sends it to CONTROLLER in the following way:

R MMI CONTROLLER

3

6

which is interpreted by CONTROLLER as "**Print rule #6**".

In order to avoid deadlock and interference problems, it was necessary to build a semaphore. A semaphore is a data structure which is designed to block a common memory from being accessed by more than one processor simultaneously. Before each request or reply, the communication file is blocked and the request is processed. After the completion of the communication process, the communication file is unblocked. Let LOCKFN be the procedure that locks the communication file (called COMFILE) and let UNLOCKFN be the procedure that frees COMFILE. The communication will be performed as presented in Figure 11.

```
LOCKFN;                        {lock COMFILE}
rewrite(COMFILE);              {open COMFILE to write}
write(COMFILE,'R MMI CONTROLLER');    {request from MMI to CON-
TROLLER}
write(COMFILE,'1');                {request to initialize the system}
UNLOCKFN;                  {free COMFILE for CONTROLLER to read request}
done:=false;                   {control variable}
while not done do
begin
    LOCKFN;
    reset(COMFILE);            {open COMFILE for reading}
    read(COMFILE,st1);         {read the code name}
    read(COMFILE,st2);         {read the sending address}
    read(COMFILE,st3);         {read the receiving address}
    if(st1='A')and(st2='CONTROLLER')and(st3='MMI') then
    done:=true;                {if CONTROLLER acknowledged then quit}
    UNLOCKFN;
end;
```

FIGURE 11. Example of communication between two expert systems.

V. CONCLUSIONS

In this chapter, a fuzzy intelligent autonomous expert system, named COMEX, is presented, which utilizes a variety of autonomous external sensors to collect and update channel and network communication parameters. COMEX runs on an IBM compatible PC and can be used by any operator to achieve expert results in communication. Thus, the expert system will "study" the network and the available equipment in order to use that knowledge to establish the desired communication in an optimal manner. This learning process will initially use the information supplied by the operator concerning the communication station capabilities and will continue with on-line knowledge accumulated by the expert system itself and by the external sensors. COMEX uses three separate knowledge bases for three distinct phases of the decision process. This architecture has enabled a simpler and more efficient implementation of COMEX.

ACKNOWLEDGMENTS

Research supported in part by NSF grants IST 8405953 and by the Florida State University through use of its Cyber 205; by the Florida State University Supercomputer Computations Research Institute which is partially funded by the U.S. Department of Energy through Contract No. DE-FC05-85ER250000; by the Florida High Technology and Industrial Council grant UPN 85100316; and by Tadiran Ltd., Holon, Israel, where part of this work was performed.

REFERENCES

1. **Perl, J. M.,** MESA — A Real Time Adaptive Communication System, AFCEA West Conference, February 1987.
2. **Kagan, D. and Perl, J. M.,** Real time HF channel parameters estimation, *IEEE Trans. Commun.*, COM 34(1), 54, 1986.
3. **Perl, J. M. and Kagan, D.,** HF Modem Performance Prediction Based on Channel Parameters Estimation, MONTECH '86 Conf., October 1986.
4. **Perl, J. M. and Rainish, D.,** Optimal Signaling Rates for HF Communication, MIL COM '86 Conf., October 1986.
5. **Harmon, J. W.,** Automatic Connectivity for Low Speed Data Communication Systems, IEE Conf. HF, London, February 1979.

6. AUTOLINK, Harris Publication, 1984.
7. AUTOCAL, HF — 2000 Radio, Tadiran Publication, 1987.
8. **Apte, C. and Weiss, S. M.,** An approach to expert control of interactive software systems, *IEEE Trans. Pattern Anal. Mach. Intell.,* PAMI 7(5), September 1985.
9. **Baldwin, J. F.,** Fuzzy sets and expert systems, *Inf. Sci. (N.Y.),* 36, 123, 1985.
10. **Bandler, W.,** Presentation and Manipulation of Knowledge in Fuzzy Expert Systems, Workshop on Fuzzy Sets and Knowledge-Based Systems, Queen Mary College, University of London, 1983.
11. **Barr, A. and Feigenbaum, E. A.,** *The Handbook of Artificial Intelligence,* Vol. 1, Morgan Kaufmann, San Mateo, CA, 1981.
12. **Carroll, J. M. and McKendree, J.,** Interface design issues for advice-giving expert systems, *Commun. ACM,* 30(1), 1987.
13. **Davis, R., Buchanan, B., and Shortliffe, E.,** Production rules as presentation for a knowledge-based consulting program, *Artif. Intell.,* 8, 15, 1977.
14. **Gaines, B. R. and Shaw, M. L. G.,** From fuzzy logic to expert systems, *Inf. Sci. (N.Y.),* 36, 5, 1985.
15. **Hayes-Roth, F., Waterman, D. A., and Lenat, D. B.,** *Building Expert Systems,* Addison-Wesley, Reading, MA, 1983.
16. **Harmon, P. and King, D.,** *Expert Systems,* John Wiley & Sons, New York, 1985.
17. **Kandel, A. and Byatt, W. J.,** Fuzzy sets, fuzzy algebra, and fuzzy statistics, *Proc. IEEE,* 66(12), 1619, 1978.
18. **Kandel, A.,** *Fuzzy Techniques in Pattern Recognition,* Wiley Interscience, New York, 1982.
19. **Kandel, A.,** *Fuzzy Mathematical Techniques with Applications,* Addison-Wesley, Reading, MA, 1986.
20. **Negoita, C. V.,** *Expert System and Fuzzy Systems,* Benjamin/Cummings, Menlo Park, CA, 1985.
21. **Schneider, M. and Kandel, A.,** Properties of the fuzzy expected value and the fuzzy expected interval, *Int. J. Fuzzy Sets Syst.,* 26, 1988.
22. **Perl, J. M., Schneider, M., and Kandel, A.,** COMEX — A Communication Expert System, submitted to the 1988 Conference on Expert System Applications, Avignon, France.
23. **Schneider, M. and Kandel, A.,** On the Theory of Fuzzy Expected Intervals and their Applicability to Fuzzy Expert Systems, Proc. Int. Symp. Fuzzy Syst. Knowledge Eng., Guangzhou Guiyang, China, 1987.
24. **Waterman, D. A.,** *A Guide to Expert Systems,* Addison-Wesley, Reading, MA, 1985.
25. **Weiss, S. M. and Kulikowski, C. A.,** *A Practical Guide to Designing Expert Systems,* Rowman and Allanheld, Totowa, NJ, 1984.
26. **Zadeh, L. A.,** The rule of fuzzy logic in the management of uncertainty in expert systems, *Fuzzy Set Syst.,* 11, 199, 1983.
27. **Zimmermann, H.,** *Fuzzy Set Theory and Its Applications,* Kluwer-Nijhoff Publ., Dordrecht, Germany, 1985.

Index

INDEX